AF566668

COAL EXPLORATION 2

Volume Two

Proceedings of the second International Coal Exploration Symposium Denver, Colorado, USA, October 1978

Edited by George O. Argall, Jr.

Also published by Miller Freeman Publications for the mining and minerals processing industries

Evaluating and Testing the Coking Properties of Coal
by Raymond E. Zimmerman

Stability in Coal Mining: Proceedings of the first International Symposium on Stability in Coal Mining, Vancouver, British Columbia, Canada, 1978
Edited by C. O. Brawner and Ian P. F. Dorling

Waste Production and Disposal in the Mining, Milling, and Metallurgical Industries
by Roy E. Williams

Copper Porphyries by Alexander Sutulov

Coal Exploration, Volume 1: Proceedings of the first International Coal Exploration Symposium, London, England, 1976

Tailing Disposal Today
Proceedings of the International Tailing Symposiums
Volume 1: Proceedings of the 1972 symposium, Tucson, Arizona, USA
Volume 2: Proceedings of the 1978 symposium, Denver, Colorado, USA

Minerals Transportation, Volume 2: Proceedings of the 1973 International Symposium on Transport and Handling of Minerals, Rotterdam, Netherlands

World Mining Glossary of Mining, Processing, and Geological Terms
Over 10,000 alphabetized, cross-referenced terms and phrases in English, Swedish, German, French, and Spanish
Edited by R. J. M. Wyllie and George O. Argall, Jr.

World Coal Industry Report and Directory
An in-depth report on the major coal producing nations including an international directory of over 1,000 key coal mines

World Mines Register
International directory of active metal and mineral mining companies, operations, and key personnel

World Mining Copper Map
Shows every known important porphyry copper and molybdenum mining operation in the world

Map of World Coal Resources and Major Trade Routes

Library of Congress Catalog Card Number 76-41151
International Standard Book Number 0-87930-107-4

TABLE OF CONTENTS

Coal Exploration: Education and Organization

Exploration Drilling: Onshore and Offshore

Review of Exploration Practices

United States: Concepts and Practices

Seismics and Geochemistry

Australia: Methods and Results

Exploration in Western Canada

USSR: Exploration Drilling

Coal Logging Techniques

Computer Technology

Exploration in India

LIST OF ILLUSTRATIONS

Chapter 11

Chapter 12

Chapter 13

Chapter 14

Chapter 15

Chapter 16

Chapter 17

Chapter 25

LIST OF TABLES

FOREWORD
BY THE EDITOR

The Second International Coal Exploration Symposium, held in Denver, Colorado, USA, in October 1978, gave the international coal exploration group the opportunity to report on and learn about techniques and equipment perfected after the First Coal Exploration Symposium. The first Symposium, held in London, England, in May 1976, reviewed the search for and development of coal deposits, often by newcomers to the coal industry, and took special notice of the growing importance of oil companies in coal exploration.

Denver was chosen as the locale of the second Symposium because of its growing importance as a center of coal exploration in the Western Hemisphere. Coal is making a very strong comeback in the United States, and western coal is widely sought because of its low sulphur content. Oil companies in both the United States and Canada have aggressive exploration programs, and details of these programs and results were outlined in several papers. In fact, oil company geologists presented two papers.

Exploration, particularly using geophysics in the United Kingdom and West Germany, was reported in several papers at the first Symposium. At the second, the improved techniques developed recently for this exploration tool were reviewed in detail.

While the first Symposium provided a brief overview of world coal deposits, the second was directed more to the growing use of electronics, primarily for evaluating and measuring coal deposits rather than for discovering coal deposits. The second Symposium, however, reported specific and successful explorations in Canada, Australia, the United States, and India, using case histories as examples.

The rapid development of seismic exploration systems in the short interval between the two Symposiums was reported in detail for both underground in-seam and surface prospecting. The presentations discussed exploration for abnormalities in the coal beds—faults, sand channels, discontinuities, bad roof, and others. Similar methods are being used to locate man-caused abnormalities, mine workings, and oil and gas well casing, ahead of mining.

The second Symposium was heavily oriented to the measuring, evaluating, and checking of the ash, sulphur, moisture, and carbon contents and rank of coal deposits. Methods to determine the strength of coal and its enclosing rocks were also reported.

New developments in coal logging were detailed with new techniques permitting faster logging of drill holes with four measurements being made in one run. Two measurements developed by the oil industry have been scaled down for coal logging. They are the sonic log and the dip meter log.

The computer-based logging truck, which was introduced about the time of the first Symposium, has proved to be very valuable. On-site logging is more than the mere plotting of logs to speed analysis of data. The new system uses digital methods.

It is not surprising that work to develop new and better exploration equipment is an important part of government research in many countries. The British National Coal Board, the United States Geological Survey, the United States Bureau of Mines, the Soviet Institute Sojuzlegeologia, and the Central Mine Planning and Design Institute of India sent delegates to the Symposium to report on new developments in their respective countries.

The organizing committee is grateful for the advice and assistance of many geologists, geophysicists, and exploration and mining companies. It was their interest and cooperation which made the Symposium such an outstanding success. There is no question that the Coal Exploration Symposiums are a well-established facet of the coal industry, and planning has already started for a third international meeting.

Volume 2 of Coal Exploration will complement the first volume of these Coal Exploration proceedings and, it is hoped, will be helpful in the search for and use of the oldest energy mineral.

George O. Argall, Jr.
Symposium Chairman
February, 1979

Denver, Colorado October, 1978

Symposium Committee

George O. Argall, Jr.
Symposium Chairman
Senior Editor, World Mining
Consulting Editor, World Coal
San Francisco, California, USA

George H. Roman
North American Chairman
Publisher, World Mining/World Coal
Editor, World Coal
San Francisco, California, USA

W. L. G. Muir
European Chairman
Mining Consultant
High Wycombe, Bucks, England

Bernard W. Lansdowne
Managing Director
Miller Freeman Publications
London, England

R. J. M. Wyllie
Editor
World Mining
Brussels, Belgium

Keith Whitworth
International Editor
World Coal
London, England

I. P. F. Dorling
North American Editor
World Coal
San Francisco, California, USA

AUTHORS & ABSTRACTS

G. O. ARGALL

SYMPOSIUM CHAIRMAN

George O. Argall, Jr.,
Miller Freeman Publications,
San Francisco, California, USA

George O. Argall, Jr., editor of this volume, is an internationally known mining engineer. He has been associated with *World Mining, World Coal,* and affiliated publications since 1950. During his 28 years as editor of Miller Freeman Publications mining magazines, he has visited mining, milling, and smelting operations in 63 countries, publishing reports about remote and unusual developments in these fields.

He joined Miller Freeman Publications as editor of *Mining World* and, when that publication was sold, continued as editor of *World Mining.* From 1974 through 1977 he was publisher of *World Mining* and *World Coal* in addition to his editorial duties. At the present time he is senior editor of *World Mining* and consulting editor for *World Coal.*

George Argall is the author of two books—*Occurrence and Production of Vanadium* and *Industrial Minerals of Colorado*—and the editor of *Minerals Transportation,* Volumes 1 and 2, *Tailing Disposal Today,* Volumes 1 and 2, and the *World Mining Glossary of Mining, Processing, and Geological Terms.* In addition he has written numerous articles on all phases of the minerals industry. In 1969 his special issue of World Mining, "Japan and the World of Mining," won the Associated Business Publications' Jesse H. Neal award for editorial excellence.

He is a professional engineer, a member of the Mining and Metallurgical Society of America, the Colorado Mining Association, the Northwest Mining Association, the Nevada Mining Association, and the American Institute of Mining, Metallurgical, and Petroleum Engineers.

T. E. BOETTGER

1 KEYNOTE ADDRESS: EXPLORING THE KINGDOM OF COAL

By Thomas E. Boettger, Executive Vice President, Western Associated Coal Corporation, Denver, Colorado, USA

Born in Philadelphia, Pennsylvania, Thomas Boettger received an A.B. degree in economics from St. Vincent College in 1960 and was graduated from Dickinson School of Law in 1964. From 1964 to 1970 he was a partner in the Pittsburgh law firm of Rose, Schmidt, Dixon, Hasley & Whyte. He then joined Eastern Associated Coal Corp. in Pittsburgh as vice president, general counsel and secretary, becoming senior vice president-administration before leaving the organization in 1978. Now executive vice president of Western Associated Coal Corp., a newly formed subsidiary of Eastern Gas & Fuel Associates, he is responsible for developing and operating coal operations in the midwestern and western United States.

C. C. MATHEWSON

2 HOW AND WHERE TO EDUCATE COAL EXPLORATION GEOLOGISTS

By Christopher C. Mathewson, Associate Professor of Geology, Texas A & M University, College Station, Texas, USA

Christopher Mathewson is a Registered Professional Engineer in Texas and Arizona, a Registered Professional Geologist, Certified Professional Engineering Geologist in Oregon and a member of over 10 professional societies. His educational background includes a B.S. degree in civil engineering from Case-Western Reserve University and M.S. and Ph.D. degrees in geological engineering from the University of Arizona.

He has carried out research into the subsidence of reclaimed surface mine land, the geology and engineering geology of lignite exploration and mining, impacts of surface mining on the groundwater hydraulic system, groundwater resource evaluation, and the analysis of mining risk due to groundwater.

Dr. Mathewson's teaching responsibilities include a graduate level course in lignite geology. He has directed or is presently directing eight graduate student theses or dissertations related to lignite development in Texas. He has over 40 publications and presentations in the field of engineering geology and has directed more than $300,000 in research grants at Texas A&M University.

ABSTRACT: The coal exploration geologist is a geologist who is aware of the social, economic, environmental, and engineering aspects of the exploration and extraction of coal as well as the reclamation of mined lands. This geologist is a vital part of any coal development program and must act as the industry's "man" in the field. Responsibilities, ranging from classical exploration geology to ecology, must be directed toward maximizing each exploration dollar.

Industry and/or university educational and research associations must be developed in order to overcome the effects of the "coal depression," which left few geology departments with trained and experienced coal geologists who can provide the necessary expertise to teach and carry out research. Through industrial support these necessary faculty members can be encouraged and developed. Research support breeds knowledge which in turn produces coal-educated geologists.

J. E. McNULTY

3 ORGANIZING AND MANAGING COAL EXPLORATION PROJECTS

By James E. McNulty, Vice President, Paul Weir Company, Chicago, Illinois, USA

Jim McNulty, a Certified Professional Geologist, joined the Paul Weir Company as a junior geologist in 1961, and now heads the geologic group within that company.

His career has been devoted to the exploration for and development of reserves of coal and industrial minerals, throughout the United States, western Canada, and abroad. He is regarded as an authority on the methods and management of coal exploration.

Mr. McNulty holds a bachelor's degree from the University of Iowa and is a member of the Society of Mining Engineers of the American Institute of Mining, Metallurgical and Petroleum Engineers, the Illinois Mining Institute, the Rocky Mountain Coal Mining Institute, the Geological Society of America, the Association of Professional Geological Scientists, and the Society of Economic Geologists.

ABSTRACT: The creation and operation of a coal exploration program is discussed. Cost-limited, time-constrained, and specific-objective exploration projects can influence program planning. Problems encountered in preparing a time budget include setting completion dates, data interpretation, and activities in the planning period. The effect of the Surface Mining Control and Reclamation Act on premining overburden sampling costs and planning time is projected as expensive and lengthy. Some aspects of a typical budget include staff supervision, communication and documentation. On-site supervision by field management can provide the desired control. Simple forms supply the necessary documentation and eliminate all but important communications.

R. W. PIDSKALNY

4 UTILIZATION OF OILFIELD RIG EQUIPMENT AND TECHNOLOGY IN COAL EXPLORATION

By Ron W. Pidskalny, Managing Director, Kenting Drilling Services Limited, Wilford, Nottingham, United Kingdom

R. W. Pidskalny received his B.Sc. degree in mechanical engineering from the University of Manitoba in 1969. Upon graduation he joined Mobil Oil Canada Limited and, during the following three years, gained substantial experience in western Canadian drilling and completion operations. He then became senior drilling engineer for Canadian Reserve Oil and Gas Limited where, for two years, he was responsible for all drilling operations and related engineering.

In late 1974 he joined Kenting Drilling Company Limited as a drilling engineer based in Calgary, Canada. His prime responsibilities were the preparation of bids and drilling programs for the company's operations in the United Kingdom.

In November 1976, Mr. Pidskalny was transferred to the United Kingdom and appointed managing director of Kenting Drilling Services Limited.

ABSTRACT: In 1975 the National Coal Board, in a bid to accelerate its exploration program, mobilized oilfield drilling equipment from Canada. Deep boreholes, which up to that time required three months to complete, are now being routinely completed in three weeks or less. Along with this achievement was the recovery of very acceptable coal seam core.

This chapter describes rig design, site preparation, open-hole drilling, coring, diversion and abandonment techniques that are successful in coal exploration in the United Kingdom.

A. R. BECKERING

5 OFFSHORE NOVA SCOTIA COAL EXPLORATION DRILLING

By Allan R. Beckering, Director, Contracts/Engineering Liaison, Global Marine Drilling Company, Houston, Texas, USA

Allan Beckering, who holds a master's degree in petroleum engineering from the University of Southern California, is an engineer with Global Marine Inc., Houston, Texas, an offshore drilling contractor operating worldwide.

He spent four years as staff engineer for a deep-sea coring vessel and engineer in the subsea equipment section, working on equipment installation and repair projects. Now assigned to liaison duty between the engineering and contracts departments, he is responsible for financial and profitability analysis of new projects and operating cost estimates for drilling contract proposals. No abstract of Mr. Beckering's paper was provided.

G. R. WALLIS

6 A CRITICAL VIEW OF SOME COAL EXPLORATION PROGRAMS

By Graham R. Wallis,
Chief Coal Geologist,
Robertson Research (North America) Limited,
Calgary, Alberta, Canada

G. R. Wallis graduated from the University of New South Wales, Sydney, Australia, in 1958 with a B.E. in applied geology. He worked for the Geological Survey of New South Wales for 11 years before resigning from the position of assistant director in 1970. As a geologist in the Survey, he worked in the fields of sedimentary and coal geology, both in the field and from the managerial and evaluation aspects. He served with the Australian National Antarctic Research Expedition in the summer of 1964-1965 as part of a four-man geological team working in Antarctica.

Between 1970 and 1976, when he was senior associate with the consulting geological firm of Clifford McElroy and Associates Pty. Ltd. in Sydney, Australia, he was involved in all phases of the coal industry from the management of exploration projects through consultant advice in producing coal mines. During this period his time was almost evenly divided between Australia and Canada.

In 1977 he joined Robertson Research (North America) Limited as chief coal geologist for North America in Calgary, Canada, with the technical responsibility for the company's consultant activities on the North American continent. Since that time he has worked in western Canada and eastern and western United States on coal evaluations and exploration projects.

ABSTRACT: Weaknesses observed in coal exploration programs conducted in various parts of the world are commonly related to three elements: management control, experience of personnel, and financial constraints. The effect that these three functions have on exploration programs is illustrated by use of examples from the countries studied.

When weaknesses occur in the planning stage of a program, they result from a lack of appreciation of the task, the consequent failure to clearly define the objectives, and the lack of technical capabilities of the persons involved. The lack of geological mapping resulting in an unnecessary use of funds is one example of this.

In conducting a drilling program, management at all levels is operative from the calling of tenders or bids to final move-out of a drill rig. Experience is needed in gathering the necessary data in the required detail and finance, in ensuring that there are sufficient funds.

The ramifications of inadequate core recovery are discussed; and the value of the use of triple inner tubes, recovery clauses in drilling contracts, and core diameter size are reviewed. The ability to determine the location of core losses by the use of geophysical logs has been a practicality for some two years, as has the verification that part of the core recovered from the coal seam was reversed in the core box by the driller.

The collecting of miscellaneous samples of coal, and regarding them as representative, produces analytical results which are of questionable value, at best, and misleading at worst; yet such inconsistencies still are to be found in technical reports. The final phase of the exploration program requires complete synthesis of results, accurate reporting and, of importance, the safe and recoverable storage of the raw data.

G. E. VANINETTI

7 COAL EXPLORATION CONCEPTS AND PRACTICES IN THE WESTERN UNITED STATES

By Gerald E. Vaninetti, Director of Exploration, Department of Mining and Exploration, Utah Power and Light Company, Salt Lake City, Utah, USA

Gerald Vaninetti has participated in the exploration and development geologic analysis of energy minerals since 1971. His experience includes exploration drilling programs in the Kaiparowits Plateau and Wasatch Plateau coalfields of Utah as well as the evaluation of drill hole data concerning coal properties located within most of the major coalfields of the western United States. He has also been responsible for the organization and implementation of a successful uranium exploration program that has resulted in the delineation of several uranium ore bodies. In addition to his geologic experiences, he has been involved in the generation and interpretation of down-hole geophysical logging data and Mr. Vaninetti organized a geophysical logging company in 1978.

ABSTRACT: Prior to World War II, steam coal was used extensively as an energy source for electric-generating power plants, homes, and steam locomotives. The coalfields of the eastern United States supplied the bulk of the steam coal, although minor contributions were made from the coalfields of the western states. The postwar expansion in the United States, however, was primarily fueled by oil and gas derivatives because of greater ease in handling, cleaner burning, better availability, and lower cost advantages over steam coal. Accordingly, after the war, U.S. steam coal production steadily decreased.

However, the energy crisis of the 1970s (when energy users became aware that oil and gas supplies were limited and available at radically escalated prices) has resulted in the revitalization of the U.S. coal industry. The effect has been expansion of coal production in the eastern states to prewar levels and unprecedented growth in the western coalfields.

The rapid expansion in exploration activities in the coalfields of the states of Montana, Wyoming, Utah, Colorado, Arizona, and New Mexico has had a profound impact upon the exploration concepts and practices utilized in the western coalfields. The intent of this chapter is to define the evolution of exploration concepts in terms of accumulation and interpretation of data and to present case examples of the implementation of these concepts.

J. A. LUPPENS

8 EXPLORATION FOR GULF COAST UNITED STATES LIGNITE DEPOSITS: THEIR DISTRIBUTION, QUALITY AND RESERVES

By James A. Luppens, Area Geologist, Phillips Coal Company, Tyler, Texas, USA

James A. Luppens attended the University of Toledo, Ohio, receiving a B.S. degree in geology in 1968 and an M.S. degree in geology in 1970, then spent the following two years at New Mexico Institute of Mining and Technology doing additional graduate work. In 1973 he joined Phillips Petroleum Company in Casper, Wyoming, as supervisor of coal exploration drilling programs in North Dakota and Montana. In 1974 he was transferred to Tyler, Texas, to help in Gulf Coast lignite exploration. Until recently development supervisor, responsible for detailed property evaluations, he is now in charge of all exploration and development drilling and evaluation programs.

ABSTRACT: With the rapidly dwindling supplies of petroleum and natural gas, the demand for alternative energy fuels has stimulated the exploration for lignite in the Gulf Coast region of the United States.

Lignites in the Gulf Coast primarily occur in three Tertiary stratigraphic units: the Wilcox, the Claiborne, and the Jackson groups. Ancient depositional systems represented by these three groups, which have significant lignite deposits, are fluvial, deltaic and strandplain-lagoonal.

Analogues for all three depositional environments can be found in the modern Mississippi delta system and associated Louisiana and Texas coastal deposits. An understanding of the environments of deposition is not only an aid in developing a basic exploration model, but also serves to explain and predict variability within an individual deposit of such parameters as areal extent, thickness, and quality. Lignites that accumulated in deltaic environments are the thickest and most extensive. They are generally characterized by a nonwoody composition, low ash content, moderate sulfur content, and high Btu values. Lignites that accumulated in the fluvial environment are generally characterized by a high percentage of coalified woody material, moderate ash content, low sulfur content, and low to moderate Btu values. Lignites that accumulated in the lagoonal environment are typically high in ash and sulfur, relatively low in Btu values, and are lenticular.

The rank of lignites with similar depositional environments generally decreases with progressively younger formations and also decreases eastward and northeastward from central Texas and Louisiana toward Arkansas, Tennessee, Mississippi, and Alabama.

The following state-by-state reserve figures are preliminary estimates. They include lignites 3 feet or greater and less than 200 feet deep. Total in-place short tons are as follows: Texas, 11.5 billion; Louisiana, 1.1 billion; Arkansas, 2.5 billion; Tennessee, 1.0 billion; Missisippi. 5.0 billion; and Alabama, 1.4 billion. The total estimated 22.5 billion tons for the Gulf Coast lignite trend represent an approximate increase of 18 billion tons to the estimated 141 billion tons of coal reserves in the United States that are potentially minable by surface methods, based on recent figures published by the National Coal Association.

W. P. DIAMOND

9 EVALUATION OF THE METHANE GAS CONTENT OF COALBEDS: PART OF A COMPLETE COAL EXPLORATION PROGRAM FOR HEALTH AND SAFETY AND RESOURCE EVALUATION

**By William P. Diamond, Supervisory Geologist,
Methane Control and Ventilation,
Pittsburgh Mining and Safety Research Center,
U.S. Bureau of Mines,
Pittsburgh, Pennsylvania, USA**

William Diamond received his bachelor's degree in geology from the University of Kentucky in 1968. Prior to entering graduate school, he worked for the Kentucky Geological Survey, Water Resource Section. A master's degree in geology was earned at the University of Kentucky in 1972, and he went to work for the Chevron Oil Company, New Orleans, Louisiana, as a production geologist.

In 1973 he joined the U.S. Bureau of Mines as a geologist in the Pittsburgh Mining and Safety Research Center, Pittsburgh, Pennsylvania. His present position is that of supervisory geologist in the Methane Control and Ventilation group. Areas of research responsibility include the investigation of geologic factors that influence the migration and storage of gas in coal measures, and management of directional and vertical drilling programs to degasify coalbeds in advance of mining.

ABSTRACT: The explosion hazard of methane-air mixtures has become an increasingly serious problem in mine planning. As mining progresses to greater depths or develops in new, previously unmined areas, an advance assessment of methane gas potential can be essential for a safe and economic mine development program.

The U.S. Bureau of Mines, as part of its Coal Mine Health and Safety Program, has developed a simple, inexpensive test to accurately measure the methane content of coal samples obtained from exploration cores. The gas content of the coal per unit weight determined by this "direct method" test can be used as a basis for a preliminary estimate of mine ventilation requirements and to determine if degasification of the coalbed in advance of mining should be considered.

The test results can also be used to estimate the methane resources of coalbeds in specific geographic areas. With the eventual decline in conventional domestic gas production and increased price of natural gas, commercial production of coalbed gas will become a reality.

D. G. PEACE

10 SURFACE REFLECTION SEISMIC—LOOKING UNDERGROUND FROM THE SURFACE

By David G. Peace, Western Geophysical Company of America, Isleworth, Middlesex, United Kingdom

David Peace started his geophysical exploration career in 1967, as a geophysical assistant with ESSO Exploration. During the past 11 years he has worked as a field seismologist, data processor, geophysical analyst, interpreter, and consultant geophysicist.

Since 1975 he has been closely associated with the use of high-resolution seismic methods, especially to help coal exploration. He has been a consultant to the United Kingdom's National Coal Board on seismic data processing and has been the author or coauthor of papers on using seismic for coal work. He is now working with Western Geophysical which has conducted several high-resolution seismic surveys in the United Kingdom and in America, both for coal and other projects.

ABSTRACT: Over the past few years the coal geologist has seen the gradual introduction of surface seismic techniques into the coal exploration field. These techniques are gaining rapid acceptance as one of the most important coal exploration tools for both new and known coal deposits.

However, there are many questions about seismic surveys that the coal geologist would like answered. They include: Where should he use seismic? Will it work in his area? Are there any problems he might encounter? What sort of logistics are there? How accurate is it possible to be?

This chapter discusses these and other questions in an attempt to give the coal geologist a seismic "benchmark" on which to base his own coal exploration requirements.

H. RÜTER

11 IN-SEAM SEISMIC METHODS FOR THE DETECTION OF DISCONTINUITIES APPLIED TO WEST GERMAN COAL DEPOSITS

By Horst Rüter and Reinhard Schepers, Institut für Geophysik der Westfälischen Berggewerkschaftskasse, Bochum, Federal Republic of Germany

Horst Rüter obtained a Diploma in geophysics from the University of Münster and a doctorate in natural sciences at the Ruhr-University in Bochum.

Since 1968 he has been research geophysicist in the German coal industry and, since 1977, director of the Geophysical Institute, Westfälische Berggewerkschaftskasse, Bochum. Dr. Rüter is part-time lecturer at the Ruhr-University, Bochum. His professional activities include geophysical coal exploration, rockburst, and environmental sciences.

R. SCHEPERS

Reinhard Schepers received his Diploma (M.S.) in geophysics from the University of Münster in 1969 and his Dr. rer. nat. (Ph.D.) from the University of Bochum in 1972.

From 1972 to 1978 Dr. Schepers was research assistant at the University of Bochum. During that time he taught applied and theoretical geophysics and conducted research in the application of the seismic method to engineering problems.

In 1978 he joined the Geophysical Institute, the Westfälische Berggewerkschaftskasse, Bochum, as geophysicist. At present he is responsible for the development of the geophysical method to be applied in coal mines. No abstract for Dr. Schepers' and Dr. Rüter's paper was provided.

R. E. ELLIOTT

12 EXPLORATION IN THE EAST MIDLANDS, UNITED KINGDOM: SOME PROCEDURAL AND INTERPRETATION PRINCIPLES

By R. E. Elliott, Regional Geologist, National Coal Board, East Midlands Region, Eastwood, Nottingham, United Kingdom

R. E. Elliott has been concerned with colliery geology and exploration for new mines for more than 28 years. After obtaining a Home Office Mine Surveyors certificate and serving as an assistant colliery surveyor for a few years, he became a Chartered Surveyor by examination. He continued his education as an external student in geology at the University of London, graduating in 1955.

Mr. Elliott joined the National Coal Board's geological organization in 1959 and became its East Midlands regional geologist in 1966. He was president of the East Midlands Geological Society for three years and has published articles in a number of geological and mining journals.

ABSTRACT: An outline of the context of exploration in the East Midlands, United Kingdom, is followed by a description of a five-stage exploration procedure. This description concerns a prospect from its initial conception, through phases of combined drilling and seismic reflection activity, to the development of a deep coal mine and the eventual exhaustion of its reserves.

The limitations and problems associated with high-resolution seismic surveys are briefly described, and the requirements of basic sedimentologic exploration by drilling are illustrated by discussing two features that are the key to many mining problems: seam-splits and sand-silt ribbons. These are introduced by statements of their origin, and details of two mapping techniques are given in appendices.

The author concludes by discussing the conviction that, although exploration starts on the surface, it continues underground during the life of each colliery. During this fifth stage, mine geologists play a part in the planning cycle, periodically feeding data interpretations in as a result of ongoing exploration and thereby helping to reduce mining risks.

J. L. CONDON

13 SEISMIC AND ELECTROMAGNETIC TECHNIQUES: TECHNIQUES FOR PREMINE PLANNING

By Joseph L. Condon, Research Supervisor, Premining Conditions Group, Bureau of Mines Denver Research Center, U.S. Department of the Interior, Denver, Colorado, USA

A native of the Minnesota Iron Range, Joseph Condon worked as a laborer in the iron mines during summer vacations from college. He graduated with a degree in geophysics from the University of Minnesota in 1963, and was employed in seismic petroleum exploration until 1965. He then joined the U.S. Bureau of Mines in Minneapolis, Minnesota, doing research on blast vibrations and detonation rates of explosives. In 1977 he became a research supervisor at the Bureau of Mines' Denver Mining Research Center, where he is responsible for research projects utilizing geophysical techniques for premine investigations.

ABSTRACT: Premining investigations are the activities necessary to acquire mining information for economic evaluation and engineering design of a mine, subsequent to reconnaissance exploration and prior to mining. The U.S. Bureau of Mines is doing research on methods to obtain premining information on problems such as detection of old mine workings, abandoned oil wells, channel sands, faults, and other geologic features that are neither productive nor safe.

Techniques being developed to acquire the necessary information fall into two primary categories: seismic and electromagnetic. The seismic methods include high-resolution seismic and acoustic methods applied from the surface and inseam channel waves applied underground. The electromagnetic, or radar, methods are applied from the surface, from boreholes, and from the working face underground. All of these methods are useful under certain site-specific conditions, but none are universally applicable for all coal mining conditions.

A. ORHEIM

14 GEOCHEMICAL AFFINITIES AS EXPLORATION AND CLASSIFICATION TOOLS

By Alv Orheim, Geologist, Store Norske Spitzbergen Kulkompani A/S, Longyearbyen, Norway

Born in 1947, Alv Orheim received his education at the Norwegian Institute of Technology, where he earned a degree in geology in 1971. He joined SNSK A/S as geologist in 1972 and, in 1974, was appointed head of the company's exploration services.

His work has been mainly in the fields of geology in mines and exploration areas, prospecting logistics, and run-of-mine quality. At present he is engaged in a total assessment of coal resources with regard to exploration efforts and future quality requirements and spent time in Essen, West Germany, to study coal petrology and market prospects.

ABSTRACT: The coal industry has a promising future in the opinion of most experts. It is expected that coal will take care of a larger share of the global energy demand and, in addition, will serve as a substitute for oil in the petrochemical industry.

This growth will undoubtedly influence market prices, and the market will become more selective. Emphasis will be placed on many criteria of quality that have little significance today.

This chapter deals with the chemical composition of coals and tells how it may play a part in this development related to petrographic and physical characteristics. The close relation between exploration effort and market prospects is evident, and is not discussed.

15 HUNTER VALLEY COAL: A SUCCESSFUL AUSTRALIAN EXPLORATION PROJECT

By George E. Edwards, Manager Marketing (Technical), and Brian W. Vitnell, Chief Geologist, Coal and Allied Industries Limited, Sydney, New South Wales, Australia

G. E. EDWARDS

Before joining Coal & Allied Industries Limited, George Edwards was chief of marketing and fuel technology at the Joint Coal Board in Sydney, N.S.W., Australia.

Graduated with a B. Sc. (technology) degree in Metallurgy from the University of New South Wales, Mr. Edwards is a Fellow of the Australian Institute of Energy and a member of the Institute of Fuels and of the Australian Institute of Mining and Metallurgy.

His activities also include chairmanship of the Standards Association of Australia's Committee for Coal and Coke and membership in the Executive of that association's Chemical Standards Board. He served as international chairman for the Seventh International Coal Preparation Congress and is a past chairman of the Coal Preparation Society of New South Wales. He is also a member of the Australian Committee of the World Energy Conference and a member of the Coal Technology Utilisation Committee of the National Energy Research Development and Demonstration Council.

B. W. VITNELL

Brian Vitnell, chief geologist for Coal & Allied Industries Limited, is based at Hexham, Newcastle, N.S.W., in the Lower Hunter Valley. Educated at Sydney grammar school and the University of Sydney, he graduated with a B. Sc. degree in 1950. He is a member of the Australasian Institute of Mining and Metallurgy and of the Geological Society of Australia.

In 1951 he joined the Joint Coal Board of New South Wales as geologist, spending the first year of his employment in the Western District coalfield, based at Lithgow. In 1952 he was transferred to the Hunter Valley and was based at Cessnock and Singleton before leaving in 1963 for a private study tour of coal exploration and mining in the United Kingdom and Europe. He joined Coal & Allied as foundation chief geologist in 1964. Mr. Vitnell is responsible for directing exploration operations for the company and providing geological advice and assistance to its operating mines. He made a further world study tour for the company in 1974 and has been a member of technical sales missions to the Far East.

ABSTRACT: This paper outlines the exploration and testing of an area of some 900 million metric tons of measured plus indicated coal reserves in Hunter Valley, New South Wales, Australia, where coal production will commence in mid-1979.

More than 350 slim cored holes (51 millimeters in diameter) have been drilled in the exploration program, together with almost 500 rotary open holes (114 millimeters in diameter) to delineate oxidation by analysis of chip samples and 23 large-diameter (203-millimeter) cored holes for bulk sample testing. A 2-meter by 1-meter shaft was sunk for further testing and to produce bulk samples for assessment by potential consumers.

Comprehensive chemical, physical, and float/sink analyses of coal bore cores, large and small, have been undertaken throughout the program, and all data obtained have been encoded for computer playback of plans, sections, and selected tabular and mining information.

Geophysical surface exploration, comprising ground magnetometer and resistivity surveys, has also been undertaken to provide greater detail of faults and igneous instrusion in the initial operations area at Hunter Valley No. 1 mine, to the north of the Hunter River. Operations will also take place at Hunter Valley No. 2 mine, south of the Hunter River, in the future.

The 13 seams occurring in the two areas contain high volatile coals of both coking and steaming potential, and marketing-related studies in both areas of utilization have been undertaken on the various samples produced during the exploration and testing programs.

D. SVENSON

16 EFFICIENT AUSTRALIAN PRACTICE IN EXPLORATION AND EVALUATION OF COAL PROSPECTS

By David Svenson, Projects Manager, R. W. Miller and Company Pty. Limited, Brisbane, Queensland, Australia

David Svenson is a geologist who is well known in both the coal mining and the civil engineering construction industries. Following war service, he received a degree in geology from Sydney University, then did some petroleum exploration work before he joined the Snowy Mountains Hydro-Electric Authority in 1952. He spent eight years as resident or project geologist for some of the authority's major undertakings prior to 1960, when he became engineering geologist in charge of Construction Materials Field Investigation Section.

Leaving the authority in 1969, he joined Thiess Holdings Limited as exploration manager and was involved in the evaluation of projects such as the South Blackwater and Callide Coal mines, exploration for coal and other minerals, and providing geotechnical services for Thiess in Australia and overseas.

In 1977 Mr. Svenson joined his former general manager, Howard Jones, in a mining consulting practice. Since August 1977 he has been responsible for overseeing R. W. Miller & Company Pty. Limited's increasing Queensland coal interests, including the mining consultancy for the Queensland State Electricity Commission's Curragh project in Central Queensland.

ABSTRACT: Over the past 10 years, the Australian coal industry has enjoyed an unprecedented growth resulting from the remarkable advance of one of Japan's major industrial bases, the steel industry. The consequent exploration effort of Australia's coal industry as a whole has been sustained with a high standard of professionalism in field and laboratory and in the presentation of the coordinated results to the world markets for both coking and fuel coal.

The industry's exploration effort has had solid technical support from both statutory authorities and the industry's own research unit, ACIRL. Advances in field exploration and laboratory evaluation techniques and equipment have been and will continue to be essential in sustaining the effort.

In the next decade, steam coal will become increasingly important in terms of exploration and evaluation, but as surface minable reserves of quality coking coal become fully developed, attention will again be directed to reducing the costs of exploring the underground potential.

E. W. BERESFORD

17 PROBLEMS AND SOLUTIONS FOR ROCKY MOUNTAIN COAL EXPLORATION IN CANADA

By Eric W. Beresford, Manager, Coal Division, Union Oil Company of Canada, Limited, Calgary, Alberta, Canada

Eric Beresford graduated in 1956 from Nottingham Technical College, England, where he specialized in coal mining, geology and mine surveying. Joining the United Kingdom's National Coal Board, he worked in the surveying and planning departments, holding various positions related to underground mines. He also was involved in major reconstruction programs in both the Nottinghamshire coalfield and Derbyshire coalfield.

Since emigrating to western Canada in 1969, Mr. Beresford has worked in coal operations and has held senior management positions in mine planning, engineering, and exploration for both surface and underground mines.

He is a qualified Surveyor and Surface Mine Manager in the Province of Alberta and a member of the Alberta Society of Engineering Technologists and the Canadian Institute of Mining and Metallurgy.

In his recent appointment as manager of the Coal Division, Union Oil Company of Canada Limited, he is responsible for the operations and planning of the Company's coal development in western Canada.

ABSTRACT: Problems encountered during coal exploration in the Rocky Mountains of western Canada are not uncommon in other parts of the world. In the Canadian Rockies, however, these problems are compounded, not only by the physical restrictions imposed by terrain, climate, and highly disturbed seam structure, but also by environmental and government restraints.

Large areas in the mountains have been reclassified as environmentally sensitive by recent government policies and this has resulted in virtual elimination of these areas for coal exploration and development.

Increased environmental legislation and reclamation regulations have resulted in a change of exploration techniques to adopt the new rules.

To avoid long and disruptive access roads in relatively unexplored territory, the use of helicopters for drill programs and for surface mapping parties is being more widely advocated.

Reclamation cost in the order of $4,900 per mile is involved in restoring roads and drill sites to a state equal to or better than their original condition. This cost is another factor in limiting the disturbance especially in first-phase exploration of an area.

Heavy snowfall, sub-zero temperatures, and high winds make winter exploration extremely expensive and normally limit the exploration season to about five months.

Often poor and inadequate core recoveries are experienced due to the soft and pliable nature of the coal, and heavy reliance is placed on adit sampling for representative washability tests.

Reclamation costs can now add some 40 percent to the exploration program when using conventional methods.

In spite of these problems western Canadian operators successfully carry out coal exploration programs and are using the most advanced techniques available in the industry to achieve worthwhile results.

I. P. DYSON

18 EXPLORING THE METALLURGICAL COAL DEPOSITS OF THE CANADIAN ROCKY MOUNTAINS

By I. P. Dyson,
Geological Consultant,
Paul Dyson Consultants and Holdings Limited, Calgary, Alberta, Canada

Paul Dyson was graduated from Cambridge University, England, receiving a B.A. in geology. From 1957 to 1963 he was geologist for a major oil company, covering western Canada and working mainly in the Foothills (disturbed) belt. In 1963 he joined a small independent firm with multiple mineral interests and, until 1967, was responsible for coal interests. Since 1967 he has been an independent geological consultant, working in western Canada and also in the United States, Colombia, and Australia. No abstract of Mr. Dyson's paper was provided.

V. S. BORISOV

E. V. TERENTYEV

19 COAL EXTRACTION IN DEEP MINES: IMPROVING GEOLOGICAL ESTIMATES OF RELIABILITY

By V. S. Borisov, Deputy Chief of Sojouzuglegeologia and E. V. Terentyev, Ministry of Coal Mining, Moscow, USSR

V. S. Borisov is deputy chief of Sojouzuglegeologia, the coal geology association of the USSR Coal Ministry. He was a member of the delegation of three USSR coal specialists who visited the San Francisco headquarters of World Coal magazine and participated in the symposium in Denver.

E. V. Terentyev has had extensive field experience during his service with the Ministry for the Coal Industry in the USSR, particularly in the Pechora Basin, which is within the Arctic Circle, and the Donets Basin, which is situated in the Ukraine and is the largest coal-producing basin in the Soviet Union. As chief engineer of the ministry's geological department, however, he travels to all Soviet coal basins, including those in Siberia, to provide geological investigative services on behalf of the Coal Ministry.

Mr. Terentyev has published a number of technical articles and was a member of the editorial board for the "Map of the Coal Fields of the World." No abstract of Dr. Borisov's and Mr. Terentyev's paper was provided.

D. R. REEVES

20 SOME IMPROVEMENTS AND DEVELOPMENTS IN COAL WIRELINE LOGGING TECHNIQUES

By D. R. Reeves, Managing Director, BPB Instruments Limited, East Leake, Leicester, United Kingdom

Following graduation from London University in 1956, Don Reeves spent a year in research at Cambridge University. Until 1960 he worked for Shell Oil Company and the National Coal Board (United Kingdom), then joined BPB Industries Limited, where he was involved in the development of borehole logging techniques. Mr. Reeves was made managing director of BPB Instruments Limited in 1974 and, in 1978, was appointed chairman of the company.

ABSTRACT: Continuing interest in coal logging over the past few years has resulted in further improvements in equipment and techniques. This chapter reviews some of the recent trends and is divided into four sections: (1) Consideration of improvements in techniques that are now relatively well established. (2) Discussion of the results and experience with some tools that are relatively new in the field of coal logging. (3) Application of new analytical techniques in handling the large volume of data now provided by coal logs. (4) Description of some of the items now being researched.

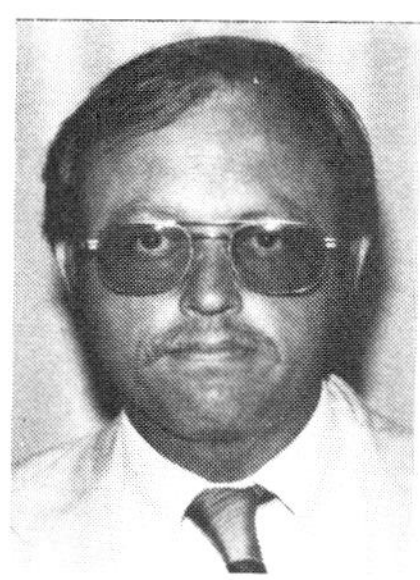

K. W. FISHEL

21 EXTREMELY HIGH RESOLUTION DENSITY COAL LOGGING TECHNIQUES

By Ken W. Fishel, Chief Geologist, Island Creek Coal Company, Lexington, Kentucky and Robert Mayer, Jr., Director, Well Reconnaissance Division, Gearhart-Owen Industries, Inc., Dallas, Texas, USA

Ken Fishel received his B.S. degree at California State University, Northbridge, in 1968, then earned an M.S. degree in geology at the University of Southern California in 1970. He then joined Consolidation Coal Company and, until 1977, was successively geologist for Consol of Canada, Western Division; resident manager for Eastern Exploration; and regional manager for Central Exploration.

In 1977 he joined Island Creek Coal Company and was appointed chief geologist for Island Creek Coal Pty. Limited of Australia, becoming manager of exploration in 1978.

R. MAYER

Robert Mayer, Jr. graduated from Harvard University in 1943. After service in World War II and experience in field geophysical operations, he worked on the development and manufacture of geophysical borehole logging systems and tools, becoming president of Well Reconnaissance, Inc., in Dallas, Texas, in 1952. This company provided logging systems for oil and gas, mineral, and groundwater exploration and production for use in many areas of the United States and in other countries; and, in the process, it developed a number of new techniques. In 1978, Well Reconnaissance, Inc. became part of Gearhart-Owen Industries, Inc. Mr. Mayer continues as director of the Well Reconnaissance Division.

ABSTRACT: A new approach based on the EHR, or extremely high-resolution, density probe provides significantly greater delineation of coal seams and partings. Its use has been primarily in the eastern United States. Extensive field experience has shown the technique to give much improved accuracy in thickness determinations, and improved definition of impure benches, providing the basis for improved ash content determinations and seam correlations.

W. G. MILLER

22 MICROCOMPUTER TECHNOLOGY FOR COAL EXPLORATION

By William G. Miller, Technical Director, and William H. Smith, President, GeoGraphics Corporation, Champaign, Illinois, USA

William G. Miller received his B.S. and M.S. degrees in 1971 and 1974, respectively, from the University of Illinois at Urbana-Champaign. While studying for his master's degree, he was a research assistant with the Illinois State Geological Survey, which is located on the U of I campus.

From 1971 to 1977 he was an assistant geologist with the Survey, working in trace element geochemistry, mineralogy of coal, and development of data processing software. His interest and expertise in computers led to his association with William H. Smith, consulting geologist, and the formation of GeoGraphics Corp., of which Miller has served as vice president and technical director since 1977. The firm is an associate of William H. Smith & Associates, Inc.

Mr. Miller has designed and implemented several hardware-software microcomputer systems for the analysis of coal-related information including reserves evaluation, base mapping, and geochemical statistics. He also has organized and implemented a computerized coal resource evaluation team in support of several U.S.G.S. CRO/CDP studies involving automated reserves evaluation, computerized base mapping, data base generation and management, and statistical analysis.

W. H. SMITH

William H. Smith received his B.S. and M.S. degrees in geology from Ohio State University with a minor in cartography. He served as an officer in the U.S. Navy during World War II, then joined the Ohio Geological Survey where he was in charge of the Coal Geology Division.

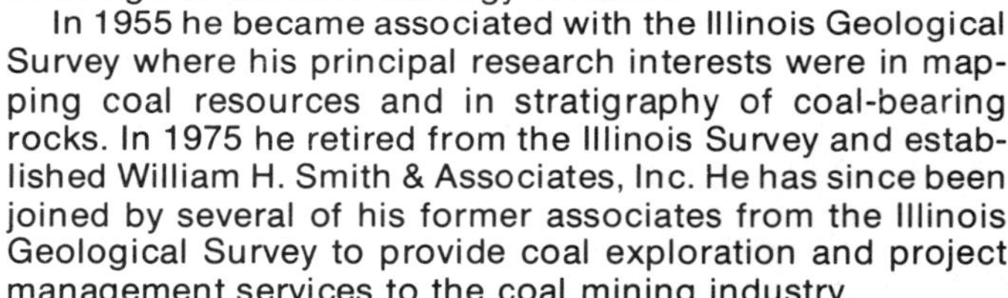

In 1955 he became associated with the Illinois Geological Survey where his principal research interests were in mapping coal resources and in stratigraphy of coal-bearing rocks. In 1975 he retired from the Illinois Survey and established William H. Smith & Associates, Inc. He has since been joined by several of his former associates from the Illinois Geological Survey to provide coal exploration and project management services to the coal mining industry.

Mr. Smith is a registered Professional Geologist and a fellow of the Geological Society of America. He is a member of the American Institute of Mining and Metallurgical Engineers, Society of Economic Paleontologists and Mineralogists, Society of Sigma Xi, Association of Engineering Geologists, American Congress on Surveying and Mapping, and the Association of Professional Geological Scientists in which he is now serving as vice president of the Illinois-Indiana Section. He is the author of numerous publications on coal geology and related fields. No abstract for Mr. Miller's and Mr. Smith's paper was provided.

23 COMPUTERIZED COAL LABORATORY MANAGEMENT

By Troy F. Stallard, President, and
Alan White, Vice President, Instrumentation,
Standard Instrumentation, Inc.,
division of Standard Laboratories, Inc.,
Charleston, West Virginia, USA

T. F. STALLARD

Troy Stallard holds a B.A. degree in psychology from Wake Forest University, Winston-Salem, North Carolina, and an M.S. degree in industrial psychology from North Carolina

A. WHITE

State University. Before becoming vice president and chief operating officer for Standard Laboratories in 1975, he was organization analyst for the Aluminum Company of America and manager of employee relations for Xerox Corporation.

Alan White graduated from West Virginia Institute of Technology with a degree in physics in 1971. Before establishing Standard Instrumentation, Inc. for Standard Laboratories, he sold coal laboratory and process control instrumentation. He also worked on design and development of special coal laboratory equipment. As vice president in charge of operations for Standard Instrumentation, he is engaged in design and analytical instrumentation for metallurgical coal and in microprocessor design and interfacing.

ABSTRACT: This chapter examines some of the problems associated with obtaining analytical data on exploration projects. A completely computer-controlled laboratory is offered as the best solution. A system is described which manages work flow and automatically operates balances, calorimeters, furnaces, ovens, and sulfur analyzers, minimizing operator input requirements. Laboratory data are collected and calculated, and reports are generated entirely by the computer system.

J. K. HALLENBURG

24 ON-SITE COMPUTER ANALYSIS OF COAL LOGS

By J. K. Hallenburg, Manager, Applications Engineering, Century Geophysical Corporation, Tulsa, Oklahoma, USA

After serving three years as a bomber pilot in the United States Air Force, Jim Hallenburg graduated from Northwestern University, Evanston, Illinois, with a B.S. degree in physics. In 1947 he joined Schlumberger Well Surveying Corporation (Schlumberger Well Services) in the research department. Later he was transferred to engineering.

At Schlumberger, he worked on dipmeters, self-potential (S.P.) problems, modeling, multiplexing, explosives, photographic, induction log, and focused log problems. He was project engineer on the West Texas system, and was concerned with the selective S.P., the micro log, the formation density log, the 1-1/16-inch gamma ray neutron, the high-temperature gamma ray neutron, and the calibration pits.

He left Schlumberger to go to the Mohole Project where he was in charge of logging operations.

At the close of the Mohole Project he became Chief Engineer of the Wire Line Division of the Western Company and later formed Data Line Logging Company in Casper, Wyoming. Data Line was bought by Teton Exploration (United Nuclear) and Mr. Hallenburg stayed as geophysical manager until 1975. In 1975 he joined Century Geophysical Corporation as manager of the Casper Region. In 1976 he was transferred to Tulsa, Oklahoma, as applications engineering manager.

ABSTRACT: On-site computer analysis of coal logs is reviewed in this chapter. The advantages of capturing raw data on tape and of on-site analysis are examined. It is shown that use of the on-site computer results in a significant saving in turn-around time for necessary data and processing. In addition, acquisition time is shortened and mistakes are fewer.

T. N. BASU

25 STRATEGY OF EXPLORATION IN THE CONCEALED COALFIELD OF KAMPTEE IN CENTRAL INDIA

**By T. N. Basu, Chief of Geology and Drilling,
T. K. Chandra, Deputy Chief of Geology, and
B. B. P. Shrivastava, Superintending Geologist,
Central Mine Planning and Design Institute Limited,
Ranchi, India**

T. N. Basu has had more than 33 years' experience in planning, organization, research, and utilization of India's coal resources. He obtained a master's degree in geology from Calcutta University in 1944, then worked for a time as a scientist in the Central Fuel Research Institute of India. While with the institute, he implemented a physical and chemical survey of the country's coal resources and engaged in research on coal geology, petrography, and other fundamental aspects of coal science.

Subsequently he joined the then National Coal Development Corporation Limited, a public sector coal producing company, as a superintending geologist. He became head of the company's Division of Geology in 1965 and was promoted to chief of geology and drilling in 1971. Later, with the formation of Coal India Limited, NCDC's Division of Geology (renamed Exploration Division), he was transferred to the Central Mine Planning and Design Institute, a subsidiary of Coal India Limited, in 1974. Mr. Basu directs all work related to coal exploration.

He visited the United Kingdom and France in 1959 to make a first-hand study of the techniques applied to the survey and utilization of coal resources. In 1972, as a leader of the team of experts from NCDC, he visited Poland to make a study of its Planning and Design Institute and, as a result, the Central Mine Planning and Design Institute was set up in India in 1974. He also visited the USSR in 1977 to present a paper at the Third IIASA Conference on Energy in Moscow.

Mr. Basu is the author of a number of papers on coal geology, petrography, fundamental research, and assessment of resources.

T. K. CHANDRA

T. K. Chandra obtained his M. Sc. degree in geology from the University of Calcutta in 1954. Before joining the Central Fuel Research Institute as geologist in early 1956, he served as geologist-cum-mines manager in a mica and beryl mine.

He joined the Coal Wing of the Geological Survey of India in 1958 and carried out a resurvey of the Pench-Kanhan

B. B. P. SHRIVASTAVA

Valley coalfield during the period 1958-1960. He was also associated with the detailed exploration for coal in the South Karanpura coalfield during 1960-1961.

In 1962 Mr. Chandra joined the Division of Geology of the then National Coal Development Corporation Ltd. (a government undertaking) as a senior geologist and has since been directly associated with detailed coal exploration activities in the country's coalfields. At present, as deputy chief of geology in the Exploration Divison of the Central Mine Planning and Design Institute, he is responsible for the planning, programming and technical guidance of all the departmental exploration activities as well as for other exploration agencies engaged on a contractual basis.

B. B. P. Shrivastava obtained his M. Tech. degree in applied geology from the University of Saugar in 1961 and, in 1962, joined the then National Coal Development Corporation Limited as geologist. Since then he has been actively associated with detailed coal exploration for mine planning and design in India's coalfields. At present he is working as superintending geologist in the Exploration Division, Central Mine Planning and Design Institute Limited, and is actively associated with detailed coal exploration.

ABSTRACT: The chance discovery of minable Permian (Lower Gondwana) coal deposits in a tubewell drilled in 1940 in the neighborhood of Kamptee township put this area on the country's coalfield map. The area is now known as the Kamptee Coalfield.

Since the first discovery of coal in this area, several mines have been opened, but, until recently, knowledge about the extent of the coalfield had been limited. This was mainly because of the fact that the area is blanketed by a thick (30 to 40-meter) alluvial cover with the exception of the detached exposures of the Kamthi Formation (younger than the coal-bearing formation) and the rare exposures of the Talchir Formation (older than the coal-bearing formation) of the Lower Gondwanas. Thus, the Kamptee Coalfield can truly be regarded as a "concealed coalfield."

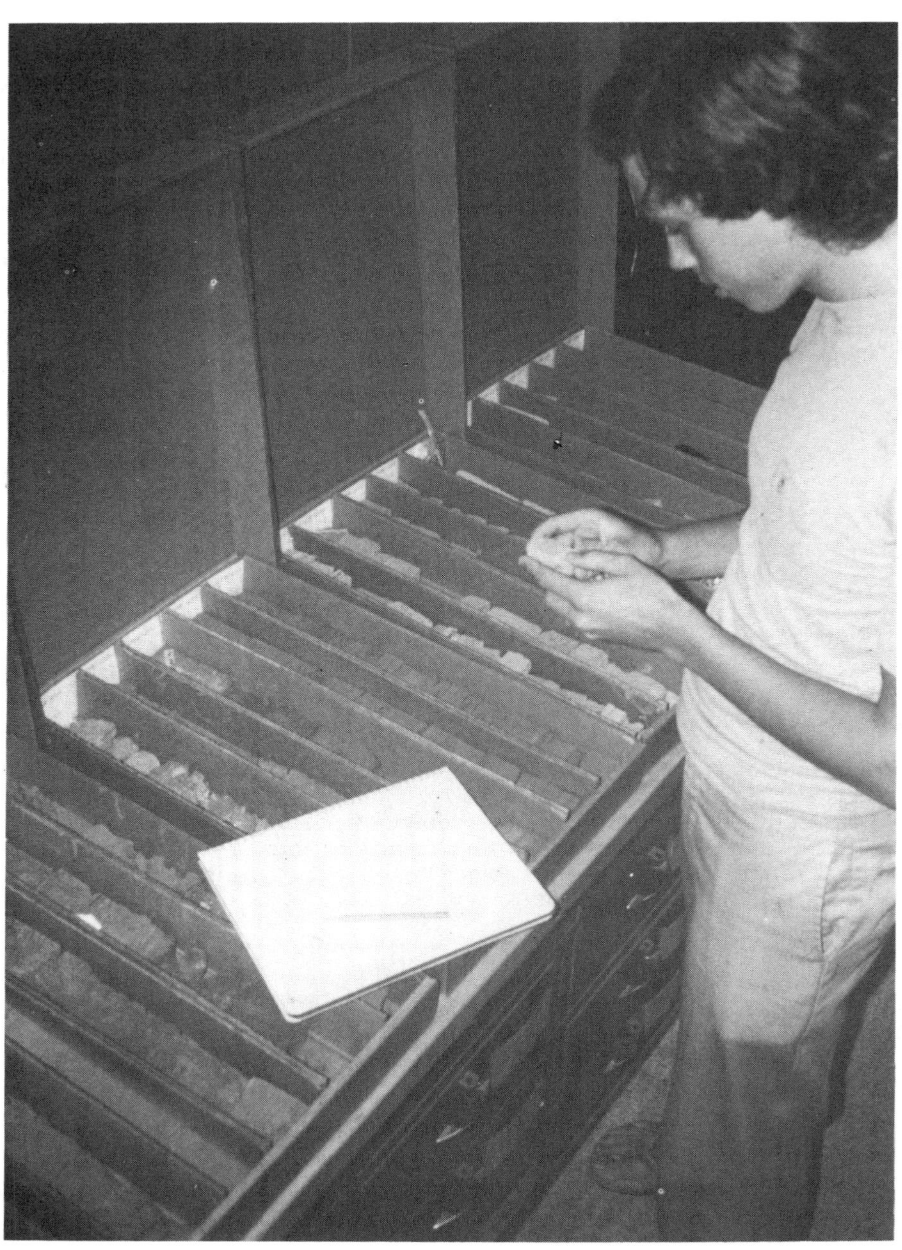

Detailed stratigraphic descriptions of overburden cores from an east Texas lignite exploration program provide the basic geologic data for Texas A&M graduate student Jim Kennedy's dissertation research. The cores were collected under the direction of the Paul Weir Company and made available to Texas A&M by Southwestern Electric Power Company. Photo courtesy Christopher C. Mathewson

Coal Exploration: Education and Organization

1

Keynote Address: Exploring the Kingdom of Coal

By Thomas E. Boettger
Executive Vice President
Western Associated Coal Corporation
Denver, Colorado, United States

I truly appreciate this opportunity to speak with the ladies and gentlemen present for this Symposium who represent the intellectual base for coal exploration activity in a large area of the world. We extend a particular welcome to our international guests. You have traveled long distances to share common goals and problems with your many colleagues. It is interesting and perhaps fitting that we are brought together by the bond of a natural resource which was formed long before the existence of nations and cultures as we know them now.

My remarks have been given what many of you will consider the corny title, "Exploring the Kingdom of Coal". I admit to the triteness of this caption, but in turn will agree, I'm sure, that it does reflect the fact that coal is now at the top of the energy picture.

In the United States, coal is the most abundant domestic fuel mineral. Coal resources here are larger in total heat value than the combined heat value of domestic petroleum, natural gas, and oil shale. Yet, most of the energy consumed in the United States is produced from oil and gas.

The potential shortage of energy related minerals has been a major national concern in recent years. Consequently, the coal industry is in the position of being among the most highly touted industrial groups in the nation. Newspapers and national magazines continuously advise us that coal is the answer to our energy problems.

To give you an idea of the growth potential the coal industry really has, consider the fact that the Federal Energy Administration and numerous other groups both in and outside of government have estimated that the demand for coal in the United States will roughly double from the present output of around 600,000,000 tons per year to over 1,000,000,000 tons per year in 1985. In order to reach these production goals, the National Coal Association tells us that 125,000 additional miners will need to be added to the coal mining payrolls in those same years, and this does not even include the many thousands of retirees and others leaving the industry during that period who must be replaced.

These and many other statistics with which you are familiar should present a very favorable climate for the future of coal. It is my own observation, however, that coal mining itself has a very problem-laden future, particularly in the United States. People are struggling within themselves to reconcile their desires for energy with their equally strong desires for clean air and water, safe and healthy working environments, and unspoiled wilderness. And all of this is taking place against a past history of inexpensive and readily abundant electricity and petroleum products which most people had learned to take for granted.

While the general population of the world is establishing its priorities in connection with energy, numerous other obstacles to coal production are continuing or developing.

The most significant portion of the forecasted new production I referred to will emanate from the western United States, which had its problems in dealing with Mineral Leasing Act moritoriums, strip mining laws, demands for Indian sovereignty, and local political impediments. But by the same token, a significant portion of the increased production will also be coming from surface and underground mines in states east of the Rocky Mountain area.

In the eastern and mid-western United States, particularly in the rugged mountain areas known as Appalachia, they have been mining coal in abundance since the early nineteenth century. Coal is the lifeblood of hundreds of cities and towns in states like West Virginia and Kentucky and is likely to remain so for many generations. What makes coal particularly important to the vitality of the community in that part of the country is the fact that coal mining there most often requires numerous workers, both at the mine site and in related capacities --- suppliers, contractors, and service men. So, when we talk about coal mining, whether it is in the east or the west, we are really talking about people --- coal people --- the miners, managers, support personnel, and their families.

Coal people have established a hallmark in the structure of American society bearing the identifying characteristics of hard work, independence, and plain straightforward honesty. Less known, perhaps, is the fact that the coal miners of today are sophisticated technicians who are required to operate and maintain complex machinery in an integrated materials movement system requiring a constant awareness of how each man's role fits into the coal mining effort. All of these factors taken together are both the strength and weakness of the coal industry, because no matter how sophisticated the machinery and no matter how well designed the flow of materials may be, the efficient application of human skills is an absolute ingredient for a successful mining operation.

Not surprisingly, providing properly for the needs of people in the coal community, particularly the coal miners themselves, is the foremost challenge facing the coal industry. That challenge is so monumental that it overshadows any other obstacle which could impede the orderly development of the industry over the next several years.

Also of importance to the coal industry is the absolute necessity of convincing coal people that safety, their own safety and that of their fellow workers, is the most important aspect of their job. No law, rule, or regulation, no punitive action, fine or civil penalty will have any meaning whatsoever until all coal people from top to bottom within every organization and throughout the community look upon safety as the fundamental requirement of their daily lives.

The puzzling phenomenon of wildcat strikes is perhaps the most well known coal industry problem. It is staggering to realize that in 1976 alone, there were a minimum of 3,500 wildcat strikes in the coal industry, including one major strike lasting 20 working days. Since 1974, there has been a major wildcat strike every year, one lengthy contract strike and thousands of shorter walkouts. Many of the protestations of the strikers have been over matters wholly unrelated to the operations of the coal mines. Without debating the question of who is the precipitating cause of these wildcat strikes, I can say one thing with certainty, and that is that court rooms and congressional chambers are not the forums for resolving the problem.

The impetus to end wildcat strikes will be the ultimate realization by all parties that their objectives and opportunities are the same. This will take a reasonable level of mutual trust and understanding and you present here must be willing to make a commitment toward that end.

Coupled with the matter of training coal miners in safety and effective working habits is the problem of imparting technical skills to the thousands of new miners and supervisors who must be assimilated into the mining system. Just as in the case with safety, technical skills must be taught and training programs established. Because of the fact that the coal mining business is not the pick-and-shovel operation it was years ago, the careful selection and training of young miners and the gradual absorption of those people into the system will be a true challenge to the industry over the next 10 years. It should be noted that supervisory people at all levels must be trained in management skills, too, so we are faced with a diverse need for training at several levels.

Many national leaders recognize that upgrading the total quality of life experienced by coal miners and their families is part of the solution to other industry problems. Although improvement in this area can be accomplished by known resources and available talent, the challenge of improving living conditions, recreational facilities, educational systems, and all the support facilities that make a community a satisfying place to live and work will take an enormous commitment.

Modern highways are opening new economic vistas in the mountain areas and improved wages have given the coal miner the economic strength to support commercial development. It is only a matter of time until these facts are realized by enterprising investors.

The next most significant impediment to the orderly mining of coal is the influence of government, both state and local. Most of you are aware of the fact that the Federal Coal Mine Health and Safety Act was passed in 1969 and revised in 1977 primarily as a result of political and public reaction to phenominal mine disasters. Also affecting public concern was the incidence of occupational disease in coal workers. The public was right. Stronger control was necessary in 1969. Unfortunately, however like many political constraints, the 1969 law has in some respects become unworkable, has been enforced inconsistently, and has unnecessarily impeded the orderly mining of coal. The 1977 amendments only worsened that situation. The incredible vacilation now developing in the promulgation of surface mining regulations is even more difficult to accept and understand.

There is hardly any similarity between the coal mine of the late 1960's and that of 1978. In underground mines dust levels have been reduced significantly, methane detection and control has been refined, every mine has a meticulously prepared and implemented roof control and ventilation system, detailed electrical standards are being met, water and sanitary facilities are made reasonably accessible to everyone, fire suppression and control has been improved, and without continuing on endlessly, it is sufficient to say that the underground coal mine is a considerably safer place to work today than it has ever been in the history of our nation or any other nation. Similarly, backfilling, revegetation, and hydrologic concerns are more important to the surface miner than most are willing to give credit for. There is no denying that these developments are responsive to the demands of a society which will not tolerate a lack of concern for the health, safety, and living environment of its citizens. But I believe the goals could have been reached with less disruptive impact and I submit that the present enforcement of existing laws is unnecessarily driving up costs without a proportionate improvement in safety or our living environment.

What concerns me the most about enforcement of these laws is the tendency on the part of many to believe that any law, rule, or regulation which "socks it to the coal industry" is good just for that reason alone. There are times when there are more federal and state mine inspectors in a mine that there are qualified supervisors, and you still will hear demands for more inspectors and more regulations.

You, as a group, are in a unique position to help change the direction of these distressing short-terms trends. First, the government itself admits that an accurate and reliable recoverable reserve estimate for federal coal does not exist. This gives them an excellent excuse to prolong indefinitely the leasing of new federal coal. There must be a strong effort toward establishing standard methodology and parameters for computing coal reserves and these guidelines should be applied promptly to the evaluation of federal coal. Hopefully, these would become a generally accepted industry standard. Second, because you are at the forefront of coal development, you can make a substantial contribution toward reversing public attitudes toward our industry.

Concern for the environment and local social needs should be paramount in all of your exploration work. As extra effort to explain what you can about the non-proprietary portion of your on-site activity can relieve a lot of the anxiety associated with coal exploration in undisturbed areas. Finally, you have as much opportunity as anyone to interact with those government officials who have developed hostile attitudes toward coal development. As frustrating as it may be, patience and meticulous professionalism will go a long way toward recasting some of these antagonistic viewpoints.

If I were standing here delivering this same talk five or six years ago, I would have ranked the solution of environmental problems along those posing boundless difficulty. I no longer can. From what I can see, coal operators have made tremendous strides in treating mine water pollution, reclaiming surface mined areas, eliminating impoundments, correcting burning refuse conditions, and effecting many other environmentally sound practices. These programs will continue and improve as time goes on.

The trend toward environmental awareness came about during a period when coal mining personnel were reeling from the constant flow of new laws and regulations which literally revamped everything they had learned. This is the kind of farsighted responsible view of our industry obligations which convinces me that the coal industry will be able to meet the challenges of improving personnel relationships and creating a safe working environment to the satisfaction and respect of its detractors. We trust we will accomplish our task with the encouragement and good will of everyone and that we all will find suitable rewards in the Kingdom of Coal.

2
How and Where to Educate Coal Exploration Geologists

By Christopher C. Mathewson
Associate Professor of Geology
Texas A & M University
College Station, Texas, United States

ABSTRACT

The coal exploration geologist is a geologist who is also aware of the social, economic, environmental, and engineering aspects of the exploration and extraction of coal as well as the reclamation of mined lands. This geologist is a vital part of any coal development program and must act as your "man" in the field. Responsibilities, ranging from classical exploration geology to ecology, must be directed toward maximizing each exploration dollar.

Industry/university educational and research associations must be developed in order to overcome the effects of the "coal depression," which left few geology departments with trained and experienced coal geologists who can provide the necessary expertise to teach and carry out research. Through industrial support these necessary faculty members can be encouraged and developed. Research support breeds knowledge which in turn produces coal-educated geologists.

INTRODUCTION

"Coal Geology? What's that?--no one uses coal anymore!--it's all oil and gas!" This was the general opinion of many university faculty and students since Spindletop and the great oil boom. The idea that coal was anything more than a solid fuel of limited value from restricted areas had become the bylaw of geologic training except at the few universities near the coal producing areas of the U.S.

The East Texas oil fields and the development of the petroleum industry marked the start of the "Coal Depression." Environmental sentiments for a clean fuel that was also cheap and limitless further contributed to the "Coal Depression." A new Ph.D. in the 1930's with a background in coal became a petroleum geologist simply because coal was on the way out.

A society based on petroleum and the technology which it allows (automobiles, aircraft, plastics) became the norm for 30 years. Even though there were voices in the background warning that oil was NOT a limitless fuel, the general public continued to believe in oil and the White Knight of Technology who would always find more. Then it occurred; the oil embargo was here and petroleum became a tool of political pressure. Long lines of people waiting for gasoline appeared and conservation became the new word of the hour. Some say all that is now past, that we have nothing to worry about, even the National Energy Plan is still just a plan and no action.

The increased cost of oil and gas and governmental limitations on the use of oil and gas for power production has turned many energy and utility companies toward lignite and coal. With increasing exploration efforts, the need for coal exploration geologists has increased. Many exploration and coal investment firms have recently formed, further increasing the demand for exploration geologists. Unfortunately, in some cases, the lack of adequately trained personnel has lead to the employment of "hole drillers" and "black box readers" who do not understand what they are looking for. Without a geologic background the exploration personnel are either blindly stabbing holes into the ground or following some pre-arranged exploration plan, without the knowledge necessary to maximize the benefits of each exploration dollar. These conditions can lead to serious problems and expensive re-exploration programs.

THE COAL EXPLORATION GEOLOGIST

If one is to educate a coal exploration geologist it is

first necessary to define the responsibilities of this person. The cost of exploration is a minor or minimal cost when compared to the costs of mine development and production and the construction costs of a power plant; however, it is the significant cost in an unminable deposit. Therefore the exploration geologist must be able to approach each exploration assignment with an awareness of what conditions can make a deposit non-recoverable. The largest unminable deposit in the area may be a great geologic find but it is not a mine. Without a recoverable reserve there is no return on the exploration dollar; this is simple economics of the minerals industry.

The coal exploration geologist is first and foremost a geologist with a complete academic background in geology. Primary job responsibilities require geologic knowledge; environment of coal deposition, stratigraphy, structural deformation, and regional geologic history, for example. This knowledge is vital to the exploration for coal deposits. The geologist must be able to recognize unusual or subtle changes in the geologic record as it is collected. Electric logs should be given a field evaluation as soon as they are run to insure that the best quality data is obtained. Preliminary interpretations, beyond simply determining the seam thickness, should be made and recorded in the geologist's field notes to provide data for planning modifications to the exploration program should they be necessary. In addition, the professional must also be aware of the full spectrum of factors that could close down an operation. Some of these are far from the field of geology and the geologist's training but are significant in the whole mining operation.

The coal exploration geologist, is therfore, more than a general geologist. Of primary importance is a basic understanding of mining and mine equipment capabilities and limitations. The geologist must be able to determine if the coal or lignite seam measured at the exploration site meets the minimum recoverable conditions. The geologist that reports eight seams of Texas lignite as 15 feet in 150 feet of hole, is misleading the client. If the seam is at an uneconomic depth, the geologist must be qualified to predict where to look for the seam or able to estimate structural conditions that may have displaced it. If thin seams or multiple seams exist, the geologist must understand the feasibility of a multiple seam operation.

Just as a series of very thin seams may be uneconomic,

so too is a very thick deposit of "black shale" that will not burn. The coal exploration geologist must therefore understand the basics of coal utilization and coal quality, because the chemical characteristics of the coal affect its utilization and the economics of its extraction. High quality metallurgical coal, for example, has a higher value than high sulphur steam coal, and therefore can be economically extracted from more complex deposits having greater geologic constraints. The coal exploration geologist must know the significance of these chemical differences and how their values can vary in response to changes in the conditions of the sample. This knowledge is critical during the sampling phase of the exploration program. Improperly collected, handled, packed, or stored samples may bias the results, because the coal quality values may be subject to serious error.

The coal exploration geologist must be aware of the geologic conditions that can adversely affect the mining operation, which requires a basic understanding of geological engineering or engineering geology. Overburden characteristics, first observed during drilling, should be recorded in the field notes, with particular attention given to the engineering characteristics, hardness, strength, roof rock conditions, and density. Field notes should provide sufficient geologic information for the geologist to advise the mining engineers about the expected mine conditions, slope stability, roof conditions, blasting, etc.

Hydrogeologic conditions have an important influence on mine operations and are a significant environmental consideration. The coal exploration geologist can provide valuable data for both mine design and the environmental analysis if the hydrogeology at each exploration site is recorded. Such information as the presence of flowing wells with estimates of pressure and rate, perched water tables, and groundwater levels should be recorded when possible. The stratigraphic relationship between the coal seams, aquifer sands, and aquitards should be considered and given a field evaluation in order to identify potential water resource pollution problems. As the geologist is moving from exploration site to exploration site features such as surface drainage, springs and water resources (wells) should be noted. The exploration geologist should be able to discuss the hydrogeology and hydrology of the area with the hydraulics personnel.

Beyond an understanding of mining and engineering geology

are the environmental considerations. During the drilling operations the geologist should be aware of the surrounding biological environment. Consideration should be given to unique organisms or to an endangered species that may inhabit the area. The geologist is not expected to be the environmental assessment team, but to be aware of the environment and therefore able to assist the environmentalist. In some cases it may be advisable for the geologist to call in an environmental specialist early in the program to investigate particular observations. This early biological or environmental reconnaissance may cost more in the beginning, but the savings may be extensive if an environmental limitation or constraint is confirmed early in the exploration program.

Finally, the geologist has the first feel for the characteristics of a site with respect to reclamation. Cuttings and cores should be viewed with an eye toward their potential as a medium for plant growth. Consideration should be given to units containing concentrations of acid forming minerals. Existing top soil characteristics, the characteristics of the overburden and how they differ from the top soils are significant observations of this phase. The geologist is not a soil scientist, but the soil forming processes are weathering processes and therefore related to the geologist's interests and training.

In order to carry out these responsibilities, the coal exploration geologist must be aware of environmental reclamation, and mining laws. An awareness that requires effort to keep up with the constantly changing system; what may be acceptable today may be illegal tomorrow.

The preceeding discussion of the coal geologist's responsibility may suggest that a coal exploration geologist is some sort of overeducated "supernut" who will cost too much to hire, particularly when viewed as an extra man or woman on the drill rig. It may appear to be less expensive to send out just a driller and logger and to make the decisions from an air conditioned office. However, the decision maker must remember that any decision is at the mercy of the data input to the system. Unusual or subtle conditions can easily be overlooked or ignored if their significance is unknown. Potentially more serious than a lack of first hand geologic observations is the loss of "your man in the field." The company loses any direct field experience relative to the entire project, thereby limiting the value of the exploration dollar. The exploration geologist probably visits more of any prospect than any other professional involved in the entire pre-mining activities. Because of these visits the geologist is "your man in the field" and is best qualified to act as your initial environmental, engineering, and economic observer. The coal exploration geologist is, therefore, one of the most important professionals involved in the exploration for and extraction of any coal deposit and the reclamation of the mine. It may cost more to start but far less to complete a resource analysis if the coal exploration geologist is on the team.

HOW AND WHERE TO EDUCATE THE COAL EXPLORATION GEOLOGIST

Where does a company go to find the personnel needed to carry out an exploration program? The few existing universities with recognized Coal Geology programs can only fill a few of the necessary positions. Many new geology graduates are petroleum oriented and trained and must be retrained as coal geologists at company expense, but on the job training programs are difficult to organize, often expensive and time consuming. Many universities would like to develop and introduce Coal Geology, but finding the faculty with the required degree (Ph.D.) and experience in coal is almost impossible. The academic community cannot afford to hire a good coal geologist away from industry. Herein lies the problem, how and where do you educate coal exploration geologists.

The coal depression has resulted in a number of problems in the academic community: 1) the number of coal geologists at universities is limited, 2) modern textbooks on coal are limited and expensive, 3) active coal research is only now beginning on a national scale, and 4) trained coal personnel are in high demand in the industry. The student interested in coal is hard pressed to find much support in his education program. As a result, the recent graduate with a degree in geology (except for those graduates from the presently existing coal programs) probably has little or no knowledge about coal. Hiring graduates from the existing schools will not solve the problem since the number of graduates available is apparently not sufficient to meet the need. Additional and expanded educational programs must be developed to provide the necessary professionals for the expanding market. This is particularly true if the coal exploration geologist is to adequately carry out the responsibilities previously discussed.

The significant problem is not where but how to establish the necessary educational programs. The "where" is any university that: 1) has an orientation toward applied research, 2) has a geology program, 3) has engineering and agricultural programs, and 4) a geology faculty member who is interested in applied research. The "how" is dependent upon the interested geology faculty member and the support provided by the industry in search of trained professionals. Unfortunately, much of the research funds and requests for research proposals are directed toward exotic energy sources and programs, and not the intermediate problems associated with coal. Geology faculty who may be interested in the field of coal may find it difficult to prepare proposals for coal research, because they lack the basic background in coal geology necessary to fully evaluate a problem and prepare the proposal. As a result, research that supports students in coal studies is not available, and a potential coal aware faculty member is not developed. In addition, faculty members who must find summer support through research are discouraged from coal research because of the risks associated with a new field.

Problems associated with coal exploration and extraction, and mine land reclamation exist in every company file cabinet. It is here that coal exploration geologists must start their educational training, through a closer association between the industry and the university (preferably your local university). Getting these problems out to the university community is a simple act of public relations. If the company geologist, mining engineer, etc. offered to present a talk at some student function or a department seminar at the university, you are filling the gap left by the coal depression. Remember, your speaker has been living with coal--the university has probably just discovered it. These seminars and slide presentations of your "Out Back Operation" may be the first exposure a prospective coal exploration geologist has to coal.

Interested faculty who become aware of your geologic, engineering, or environmental problems will probably approach the firm with a research proposal to study and solve the problems. Such research will provide answers for industry and experience for faculty and students. This industry/academic partnership is in my opinion the most efficient mechanism to reintroduce coal to the geologic curriculum. It is important for the interested faculty member to realize that the industry must receive a useful product from any research effort that they support. The research proposal must be for applied research, directed toward a solution to specific problems. A proposal that requests funding to support an in-depth stratigraphic analysis of an area having many thin seams (i.e., uneconomic) may make a great scientific constribution but it will not solve an applied problem and therefore will probably not be funded. In contrast, a proposal to study the stratigraphy of a proposed mine site to outline and define sand bodies (potential aquifers, slope stability problems, etc.) has direct value to the industry.

Active faculty research broadens horizons. For example, what started as an analysis of the stratigraphy of the sand bodies at a proposed mine, can lead to studies of the hydrogeology, geotechnical properties of the overburden, and reclamation potential. As the faculty gains practical experience in solving coal problems, they gain the basic knowledge necessary to better identify basic research needs. This knowledge allows them to submit research proposals to other funding sources to do the scientific research. Thus, industry seed money is harvested through the results of basic research supported by other agencies.

Research generates student interest, which in turn generates new courses and course programs. At the undergraduate level, the faculty member may initially provide a recommended list of electives for a student interested in coal exploration geology. The graduate from this program is first and foremost a geologist, and the degree given is in Geology. The student however, has taken courses that discuss coal geology, engineering, ecology, mining, and economics as electives to broaden the academic background. A selected list of elective courses for the undergraduate geology student is given in Table No. I.

The only new course to this list of electives is probably the coal geology course offered in Geosciences. If no mining program or mining courses exist at the university the coal geology course should include mining. Industrial support for this course is often greatly appreciated and very valuable. Mining personnel can present a series of lectures to bring mine experience into the classroom. With a tie between the industry and the classroom, you are making a valuable contribution to the educational experience and you are attracting new professionals into the industry.

Table No. I: Selected Undergraduate Electives For Coal Exploration Geology

Department	Course
Biological Sciences......	Ecology
Engineering..............	Geotechnical Engineering Mining Hydrology
Agronomy..................	Soil Sciences
Liberal Arts.............	Economics Environmental Law
Geosciences..............	Hydrogeology Coal Geology Engineering Geology

Summer employment programs for faculty and students can further enhance the benefits to the industry and encourage more geologists to consider coal.

Graduate level training in Geology, to the Master's level, is almost a prerequisite for a professional geologist, because the undergraduate program is by design very broad in scope. The graduate student is expected to gain a more specialized education and to learn how to identify, define, and solve a problem while writing a thesis.

An industry/university relationship works very well at the graduate level, because the industry has the problems that need to be solved, and the student needs a problem to solve. By working with the faculty to define and select suitable thesis research and then by providing some financial support, students may enter the job market with hands on knowledge of coal exploration and extraction, and reclamation of mined lands. In some cases the student's graduate committee could include a member from the industry, who would advise the student and review and assist in the research and writing of the thesis. This association further strengthens the industry/academic relationship, and enhances both the industry and the university programs.

A suggested Master of Science course program for a coal exploration geologist is given in Table No. II.

Table No. II: Suggested Master of Science Course Program For A Coal Exploration Geologist (Minimum)

Department	Course
Geology...........	Stratigraphy Structural Geology Coal Geology Hydrogeology Seminar Thesis Research
Geophysics........	Exploration Geophysics
Engineering.......	Fundamentals of Mining Soil Mechanics/Rock Mechanics
Agronomy..........	Reclamation of Drastically Altered Land (Soil Science)
Liberal Arts......	Environmental Law

Depending on the student's schedule and interests such additional courses as Ecology, Engineering Geology, Sedimentary Petrology, Sedimentology, Aqueous Geochemistry, and Hydrology may also be taken. The above course list is an arrangement of courses designed to first educate a geologist and then add the necessary supporting courses.

Once coal exploration geology is established at a university opportunities to further enhance the program still exist. Aside from supporting applied research and presenting occasional lectures, such programs as joint university/industry short courses, field trips, and student work-study programs help to maintain the relationship. If employees are being sent back to school for special graduate training in geology and the supporting fields, they could act as instructors for a semester and augment the existing faculty. In this manner both the university and the industry gains. The mutual benefits gained from any industry/university association are a function of the imagination of the parties and the willingness to strive for the same goal--educating coal exploration geologists.

THE TEXAS A&M UNIVERSITY EXPERIENCE

Texas has just recently recognized the potential energy resources of the Eocene lignites of the coastal plain. There are over 10 billion tons of surface minable lignite in these deposits. Although lignite was reported in geologic reports as far back as 1819, the first modern lignite fired power plant was not operational until 1954. During the late 1880's annual production from small underground mines was about 20,000 tons. Production reached a maximum of 1.4 million tons in 1914 and by the 1930's had declined to about 1/2 million tons. Texas Utilities burned lignite in their 40 megawatt Trinidad Station from 1925 to 1940 when it was converted to gas. Production practically ceased in the early 1950's. Only a few utilities have maintained an interest in lignite.

Extensive exploration and development programs were underway during the late 1960's and early 1970's, sparked by the decreasing oil and gas supplies. Texas, an oil state, was becoming a coal state. The state passed its first surface mining reclamation law in 1975. Mining in Texas and its impacts on the environment are unsolved problems, except in a few companies.

The Texas Utilities Generating Company, jointly owned by Dallas Power and Light Company, Texas Electric Service Company, and Texas Power and Light Company, began construction of the Big Brown plant in 1968. As part of this operation they established the Environmental Research Center and supported graduate research on environmental problems associated with the land, air, and water. "The purpose of the research program is learning those facts which will improve man's ability to have both a desirable environment and an adequate supply of electric energy." The Environmental Research Center formed an industry/university tie to solve environmental problems related to lignite utilization. The results of these and similar studies produced important data for the Texas surface mine regulations so that they were designed for Texas conditions.

In 1975, I was engaged by the Paul Weir Company to confirm a lignite deposit for the Texas Power Pool, Inc. (now the Texas Municipal Power Agency). This initial industrial contact kindled the fire--I developed an interest in the engineering geologic aspects of surface mining. This interest led to research proposals to support graduate research in hydrogeology, prediction of pit slope stability, and groundwater pollution potential, all of which were supported by the Power Agency. Further research into the settlement of reclaimed land was supported by the Texas Real Estate Research Center and the Aluminum Company of America. We are engaged in similar graduate research projects in another area of the State supported by the Southwestern Electric Power Company, and in a broad study of the impact of surface mining on ground and surface water funded by Texas A&M and the Texas Energy Advisory Coundil. These research projects brought in Dr. Kirk Brown from Soil and Crop Sciences to study reclamation, Ms. Vi Burke from Political Science to study the resource conflicts between land, groundwater and energy, and Dr. Wayne Dunlap of Geotechnical Engineering to investigate slope stability problems. As a result of an initial contact, lignite has become a recognized aspect of geology at Texas A&M University.

Industry let us know that they were interested in our Department because of our problem solving approach to research.

Interest created by research led to the formation of new coal geology courses. During the fall of 1977 a graduate seminar on Lignite, its geology, exploration, extraction, reclamation, and regulation was offered. Student interest resulted in a new course "Coal and Lignite Geology" which attracted 7 graduate students when offered in the spring of 1978. Dr. Earl Hoskins joined the faculty in Geosciences and Petroleum Engineering in 1977 to develop courses in mining. The Department of Soil and Crop Sciences has just introduced a graduate course in the reclamation of drastically altered lands. The University Administration responded to the State's research and educational needs and with the exception of a few new faculty members brought coal and lignite into the system by encouraging the existing faculty.

CONCLUSION

The education of coal exploration geologists is of vital interest to the coal industry. As the costs of exploration programs increase, the easy deposits are found and mined, and regulations and environmental protection requirements stiffen, the coal exploration geologist will become increasingly more valuable as your "man" in the field. The geologist must continue to expand and broaden his/her educational background as the responsibilities increase. The days of the hole driller and logger exploration team are ending in response to the increased responsibility of maximizing the exploration dollar.

How and where to educate the necessary personnel will require an industry/university association to bring the field experience into the classroom. The coal depression has created a general void in the national capability of the universities to meet the educational requirements of the coal industry. Industry support in the form of 1) public lectures, 2) seminars, 3) faculty and graduate research, and 4) summer employment programs will go a long way toward reintroducing coal to geology.

The coal exploration geologist must first be a geologist. Academic programs that dilute the geology are, in my opinion, doing both the industry and the student a serious disservice. Building on the strengths of our universities and encouraging industrial and faculty cooperation answers the question: "How and where do we educate the coal exploration geologist?"

ACKNOWLEDGEMENTS

I wish to express my appreciation to the administration of Texas A&M University, Dr. D. W. Stearns, Head of the Department of Geology, and Dr. Earl Cook, Dean of the College of Geosciences for their support of my interest in the engineering geologic aspects of lignite and lignite mining. Without the support of Mr. J. E. McNulty, Jr., Paul Weir Company, the Texas Municipal Power Agency, the Aluminum Company of America, and the Southwestern Electric Power Company, this interest would not have been fired and I would not have been a contributor to this volume.

DISCUSSION

QUESTION: Do you suggest that Coal Geology Ph. D. theses topics to be on a specific aspect (e. g. paleoenvironmental sedimentology, hydrogeology) or to be interdisciplinary in nature?

ANSWER: It is my opinion that a Ph.D. thesis or dissertation is a reflection of the particular interests of the student working towards his Doctor of Philosophy degree. The topic that a student may choose, under the advice of his faculty committee, is based very heavily on the student's own professional interests and previous experience in his undergraduate and Master's-level academic training. Therefore, a Ph.D. dissertation may be a specific problem-oriented thesis directed toward the understanding of some specific aspect of the science of coal geology. By the science of coal geology I mean an understanding of the environments of deposition, the geochemistry, the petrography, the sedimentology, and the like. On the other hand, a Ph.D. dissertation may be oriented toward the broader interdisciplinary aspects of coal geology in such problems as groundwater hydrogeology, pit slope stability, exploration programs, and the like. Their research, by necessity, would be oriented toward an interdisciplinary problem, because a Ph.D. student who does a dissertation study on the impacts of surface mining on the hydrogeology must be capable of communicating with 1) the civil engineer in hydrology, 2) the geochemist in aqueous geochemistry, 3) the agricultural engineer or soil scientist in the soil formation processes and the reaction of soil and water in soil physics.

In short, the thesis topic is a reflection of the interests of the student and the capabilities and interest of that student's academic department and institution. When we consider the general lack of active research in many aspects of coal, as a sedimentary rock, as an organic deposit, or as a geochemical system, we realize that specific research directed toward the solution to specific problems is as important as interdisciplinary research oriented to the solution of problems that are interdisciplinary in nature.

QUESTION: What information do you hope to get from your geophysical logging program?

ANSWER: The geophysical logs, from the mine exploration programs, are a major source of basic geologic information. I believe that it is impossible to adequately evaluate the problems of highwall stability or hydrogeology, for example, without a thorough understanding of the three-dimensional stratigraphy of the mine site. It is imperative that the hydrogeologist be able to predict the geologic characteristics of the mine site in three dimensions. A thorough understanding of the environments of deposition and the stratigraphy of the site, gained from a comprehensive analysis of the three to four hundred electric logs frequently collected by a Texas Gulf Coast lignite exploration program, is of primary importance in the prediction of the characteristics of the site. If the geophysical logging programs, carried out by companies in their exploration, strive not only to identify potentially minable lignite beds, but also to obtain adequate, quality information relative to sand bodies, overbank deposits, delta deposits, and other stratigraphic units that make up the underburden and overburden, the coal exploration geologist, hydrogeologist, and engineering geologist is much better prepared to make complete evaluations of impacts and potential mining problems. The geophysical logging program plus the overburden cores that are selected in association with the exploration program, provide the basic stratigraphic information for a stratigraphic model of the entire mine site. For example, to know that a lignite is a delta plain deposit and that the lignites are blanket peats, one immediately envisions the characteristics of the site in three dimensions. Channel sands should be expected in both the underburden and the overburden. Since this was a delta plain deposit, the possibilities of transgressive marine sequences, bringing in large strike trending sand bodies, has a significant implication in mine design, groundwater conditions, or highwall stability, and to a general understanding of the physical characteristics of the mine site.

ADDITIONAL COMMENT AND DISCUSSION

A third question related to this paper came up more in informal discussions than it did in the formal discussion period. The particular question dealt with a basic question of philosophy: What is the role of the university? Should the university be a training center or should the university be an educational center? In my opinion, it is impossible for the university to train a geologist to use every company's particular logging style, map coloring code, geophysical interpretation methods, equipment, and so on. Because this is an impossible task, I feel that the university's role is to provide the basic science, and engineering, behind the problem, the basic understanding of the theory of the problem and, most importantly, an appreciation and understanding of why certain basic professional practices (such simple practices as keeping good field notes and why good field notes are needed) are used. A recent graduate should be able to move into a company and with relatively simple training in the company policies and practices, become a contributing member of the organization. The student should not have to ask "Why do it that way?" because the student should have learned in his educational program 1) why that particular function is necessary and 2) why each company has a particular style. The university graduates should be capable of using the library, they should know how to write in a clear and complete form, and they should have a basic understanding of the theory and practice behind what they are doing. The university graduates should be capable of understanding the problems in the field and be capable of following the procedures established by the employer.

Jim McNulty of the Paul Weir Company summed up this discussion as:

The academician educates

with the promise of future prosperity

The chief geologist trains

with the threat of instant poverty.

3

Organizing and Managing Coal Exploration Projects

By James E. McNulty
Vice President
Paul Weir Company
Chicago, Illinois, United States

Exploration is defined as the work involved in looking for ore. Coal exploration is an activity that is largely different and distinct from the work of looking for most ores.

One doesn't have to look very far to find coal in the United States. In fact, coal bearing rocks are reasonably well mapped on a world-wide basis.

But merely finding coal is not enough. Because of coal's variability of chemical composition, its changing patterns of utilization, the capital intensive nature of its recovery systems and its precarious position in an energy market clouded by environmental concerns, the geologist involved in coal exploration must deal with concerns far more complex than simple discovery.

Coal exploration requires the development of data necessary to ascertain the commercial potentials of mining, the geologic factors which will affect mine design and the analytical data required for utilization and marketing concerns.

Add to these, the requirements imposed by regulatory agencies in the interest of satisfying environmental concerns.

More than a quarter of a century has passed since U.S. coal production reached its post war peak and began a long decline. Today, the industry can boast of a strong present and a hopeful future. Exploration activity will continue to be motivated by increased demand for fuel. We need to guide that activity with efficient and cost effective programs.

Exploration projects come in all shapes and sizes. In order to discuss the problems of organization of programs, I have classified programs into types, based on intent or goal, as follows:

Into the first group are put all programs that are limited only by availability of funding. Such programs are usually not dependent upon or limited by time. An example would be to utilize an exploration budget surplus on an area or property, which is not of immediate interest to a company's planning or production group. This is also called the "year-end special."

Another example is a resource inventory program wherein funds would be spread more or less equally among a group of prospects in order to further define the potential of each.

Cost limited programs do not present any special problems in terms of planning or monitoring, because no one expects that far reaching decisions will depend upon their results. The rules are simple: Spend so much. Get the most data for the money spent.

Programs of this group are also least significant from the standpoint of probability of occurrence. This is probably due to the fact that we are most often concerned with more scientifically challenging work than using up budget surplus. Others might argue that, as a group, we are almost never guilty of generating any budget surplus.

More important and much more likely to occur are all programs that are time constrained.

Until recently, an important type of time-constrained project would be one that was limited by climate or seasonal weather variation. More and more, companies are proceeding with exploration work through seasons of cold or wet weather at decreased efficiency, because not going forward might prove much more expensive than the increased cost of operations under poor conditions.

The situation most likely to occur involving the necessity of a time-constrained program would be the proposed acquisition of a coal property, which allows for an option period during which the prospective purchaser can for some consideration, conduct exploration to evaluate the potential of the property.

The idea, then, is to develop as much data as possible in the time available.

It would be nice if such option periods were always a year. Few are. Most range from 90 to 180 days. Shorter periods are for the gullible and the adventuresome.

The limiting factor in a time-constrained program is people. Even if cost were no object and the supply of drill rigs and steel was elastic, the ability to develop and use the results of exploration would be limited by the availability of qualified field geologists and experienced supervisors.

The emphasis of cost limited and time-constrained programs tends toward negative or pessimistic. This is because, in view of the constraints fatal-flaw analysis is liberally used in the evaluations.

Programs oriented toward a specific objective tend to be more constructive and optimistic in emphasis.

These programs can be subdivided into two types of targets: one, geographic, the other specific resource.

Geographic target types include all individual properties or tracts of land that are subjected to evaluation. An example of another type would be exploration on a regional basis within a limited area, say the service area of an electric utility.

Specific resource targets involve programs to define specific quantities of recoverable reserves or reserves having chemical qualities necessary for some specific utilization.

Both target types are well suited to a program organized into a multi-phased or step-by-step pattern of execution. The purpose of the phased approach is to minimize initial risk and develop data in an orderly fashion so that progressively increasing expenditures of funds are justified by commensurate increases in the level of confidence in the success of the project.

Such programs consist of two to four phases. The tasks assigned to each are designed to increase the data base, each time, to a point where a sound decision can be made, either to proceed to a further phase or to suspend operations.

If in a program having a specific objective, the primary emphasis is on quantity of reserves the chances for success can be reasonably well-determined in the initial phase.

If the emphasis is on quality, such as in the case of metallurgical coal, chemical feedstock or compliance thermal coal, it is likely that the success of the project may be in doubt until later phases when the results of more detailed analytical procedures on large scale samples are available.

The success of this type of approach depends upon having both sufficient data for interphase decision making and proper time to evaluate those data.

The goal of program management is to avoid all of three calamities:

1) The project runs out of time
2) The cost exceeds the budget
3) The primary objective is not achieved.

All of us are familiar with exploration budgets. When we make one, we are usually working with money. Time budgets need to be made as well, and they are usually made in reverse, that is, starting from a completion date and working back to time present.

Setting completion dates seldom is difficult. More often than not, one is imposed upon the project by circumstances beyond the control of the geologist. It's usually a "we absolutely have to have this done by ..." situation. The second most popular completion time is "as soon as possible."

One step backward in time from the desired completion date we face Common Pitfall No. 1 of time budgeting: failure to allow sufficient time to evaluate the data gathered in the field.

I am not suggesting that all of the data interpretation be left to a period which begins at the end of the field work phase. To the extent possible the information generated should be reviewed, evaluated, and used where possible to adjust program design or logistics.

Still, a time space is necessary beyond the conclusion of field activity to organize, interpret and evaluate physical data as well as laboratory analyses in order to ensure the best possible basis for decision making.

Electronic data processing and computer plotting are welcome newcomers to the task of evaluating ever increasing amounts of physical and chemical data. Still, the critical factor is the availability of personnel who understand the value of such data to the economics of coal mining and utilization.

At the opposite end of this time line is the planning period. That is the space between time now and the commencement of initial field activity. The end of the period is usually controlled by the availability of drilling equipment. The principal activities within the period are the development of the exploration plan, the analytical program and securing necessary regulatory approvals.

And here we have to consider what could be the end of our ability to plan, in an orderly fashion, programs to suit the needs of our industry. We are faced with the unpleasant prospect of Federal regulation of coal exploration through the abuse of Public Law 95-87, the Surface Mining Control and Reclamation Act,which created the truly awesome Office of Surface Mining in the Department of the Interior.

In part 776 of a draft of proposed regulations for the permanent regulatory program, it is stated that the authority of the regulatory agency is to establish minimum requirements for "coal exploration which substantially disturb the natural land surface" and substantially disturb is defined as having significant impact upon, such as blasting, mechanical excavation, etc. In other words, the intent is to regulate outcrop openings, test pits, etc.

However, in part 815 of the same document, the permanent program performance standards for coal exploration, exploration is defined as "the gathering of surface or subsurface geologic, physical or chemical data by mapping, trenching, drilling, geophysical or other techniques necessary to determine the quality and quantity of overburden and coal of an area." This section would require a filing with the regulatory agency in order to "gather ... coal exploration data."

We must remember that a supplement to the same document would require on Federal lands analyses of overburden of each stratum from samples of "not less than one hole on each acre."

To put this in perspective, please consider the following example:

A surface-minable coal deposit containing about 200 million tons extending over some 10,000 acres. After the explorationists get through with two or more phases of drilling, they end up with holes on an average of one per 40 acres. That works out to about 250 holes. However, they probably drilled about 40 percent more than that in order to have delineated the deposit. Call it 350 holes in all.

For the purpose of the example, assume an average depth to the top of the coal to be 100 feet. However, exploratory holes tend to be deeper in order to provide additional stratigraphic control. Let's say that the average depth of hole is 200 feet. This times our 350 holes equals 70,000 feet of drilling.

At a cost of $1.50 per foot, which would be a rare bargain today, we have run up a drilling bill of $105,000.00.

To make the point, we do not need to add on the cost of supervision, travel, subsistence, etc., but will add $40,000.00 for coal analyses and $10,000.00 for overburden testing. So, the total cost of the data base, which we are going to use to determine the economics of mining and utilizing this energy source is $155,000.00.

Now, the regulation drafters suggest that we analyze overburden samples on 1-acre spacing.

We said earlier that the average depth to the top of the coal was 100 feet. This will result in 11 samples per hole at a <u>minimum</u> if we sample each change in lithology or every 10 feet of overburden strata, plus one sample of the subjacent material.

The average hole needs to be 120-feet deep: 100 feet of overburden, plus 10 feet of coal plus 10 feet of floor.

Our drilling bill becomes:

120 x 10,000 x $1.5 = $1,800,000.00

Based on one quote I got, the cost of the suite of analyses required is $60.00. Let's assume $50.00, because we are certainly going to get a volume discount.

The cost of analyses is:

11 x 10,000 x $50 = $5,500,000.00

Again, excluding the cost of geologic supervision as well as the cost of 110,000 plastic sample bags, we end up with a drilling and analytical cost of $7,300,000.00, or roughly 47 times the cost of determining whether or not we ought to have a mine in the first place.

How all of this premining overburden sampling is going to enhance postmining reclamation and revegetation, I am not sure. I do think that the whole idea says a lot about the qualifications of the regulation writers.

The intent of the bureaucrats is clear.

They want to regulate your business.

This was certainly not intended by the Congress, because in the act, the only references to exploration are those which use the phrase "coal exploration operations which substantially disturb the natural land surface." The act does not require the issuance of a permit for routine operations of data gathering such as drilling, mapping or sampling. It does not require the filing of a notice of intention to gather exploration data by means other than those which could "substantially disturb the natural land surface."

If the public loses its battle with the bureaucrats over this issue of regulation of exploration, then we will have a situation similar to the one that now exists on Federal coal land but applying to all coal land.

It will mean that the only thing of which we can be sure about the period of time between the conception of an exploration project and the approval by a regulatory agency is that it won't be short.

If we were to assume a normal situation, without a lot of artificial delays, then the average program would probably have a period of evaluation approximately twice as long as the period of organization.

The period in between would normally be expected to be two to three times as long as the evaluation time. It is this space on the schedule where time, costs, equipment, and people must be employed to maximum advantage. Success here is dependent upon a little luck and a lot of good supervision.

The budget for a typical coal exploration project in North America would include the following:

Rotary drilling
Core drilling
Geophysical logging
Coal analyses
Topographic mapping and hole location surveys
Transporation
Communication
Staff supervision
Geologic mapping
Subsistence
Documentation

Special programs might also include the driving of sample adits or excavation of test pits to recovery of large samples for bulk testing.

Three important notes on exploration budgets:

First, in the average program the combined cost of drilling and logging usually accounts for about half of the total program cost.

Second, the most commonly overlooked program costs are the salary and fringes of regular full-time company staff.

And third, even if you do an incredibly astute job of estimating the costs of a project, the most appreciated item will be the contingency of 20 percent on all the other items.

Success is a relative term. A program is not necessarily unsuccessful if it fails to define a trouble-free fuel deposit having physical conditions conducive to low mining costs.

The success or failure of a program should be judged by how efficiently the solutions for unknowns in the subsurface were effected, regardless of the significance of those solutions.

Some programs are doomed from the outset by design. Contributing factors include trying to cram into one program what ought to be done in two or three phases, equipment that is inadequate in capacity or in poor repair and inexperienced or understaffed field supervision.

Exploration programs can be either controlled or uncontrolled. Our organization considers a controlled program to be one where qualified geologists are engaged in onsite supervision. The integrity of the data developed from such programs is guaranteed by the field staff.

Drilling programs designed to evaluate potentially surface-minable deposits usually consist of a lot of relatively shallow rotary drill holes. Rates of completions and rates of data flow per rig unit almost demand a geologist per rig throughout the working day.

Programs involving deep holes where the data flow tends to be more regular may allow a geologist to keep up with two or even three rigs assuming that their locations are by design scheduled to allow intermittent supervision.

In large programs, field staffs can be augmented by employing graduate students on a temporary basis. An experienced permanent staffer ought to be able to supervise the activities of two students. Given a reasonable effort by both sides, such arrangements are mutually advantageous and should be encouraged.

The effectiveness of field management is inversely proportional to the square of the distance between the project and the first level of responsibility, or the chief of party. This person must have the ability, resources, and authority to deal with all problems of logistics.

Exploration managers, usually located at some location remote from the project, cannot expeditiously deal with the day-to-day problems that occur during all projects. Emphasis at that level should be on monitoring of schedules and budgets.

The communications link between the chief of party and the exploration manager should be periodic progress reports. The main concern of such reports is to report only the activities whose course can be somehow altered by acts of management. The important facts need to be up front, such as:

Feet of advance
Feet drilled
Feet cored
Holes completed
Shifts worked
Downtime
Weather delays
Samples shipped

Simple forms, the simpler the better, can be employed to get the message up the ladder. Figure 1 presents the version used by the Paul Weir Company, although I am sure countless variations abound. The form is designed to be used for any period chosen: daily, weekly, monthly, etc. It is also used to report progress for a variety of modes, such as: per rig, per shift, per team, or even for the total effort.

Another advantage of forms is to provide uniformity of data reporting. Figure 2 shows the form used by Weirco to identify and describe laboratory samples. This form follows the sample from collection in the field through receipt of analytical results and interpretation by coal beneficiation and utilization engineers.

The results of exploration activities, interim interpretations of those results, and other geologic data should follow the same path, but use a different medium.

The party chief should not be burdened with unnecessarily tedious reporting procedures, and the paths of communication should not become clogged with paper.

From a management standpoint, the employment of novel or prototype procedures or equipment is looked upon with much disdain because the time and costs involved are difficult to predict. During this symposium, we will have the opportunity to discuss with each other both the new and the old, some good and some bad, and perhaps to dispel to some extent fear of sailing off the edge.

PAUL WEIR COMPANY

Date ____________

Rig No. ____________

Job No. ________

_____ Drilling Report

Page ____ of ____

Hole Number	Driller	TD		DTC	Remarks
		Drilled	Logged		

DRILLING

Drilling ________ Ft. ________ Hrs.

Coring ________ Ft. ________ Hrs.

Plugging ________ Hrs.

Moving ________ Hrs.

Waiting ________ Hrs.

Downtime ________ Hrs.

LOGGING

Logging ________ Ft. ________ Hrs.

Moving ________ Hrs.

Waiting ________ Hrs.

Downtime ________ Hrs.

SUPERVISION ________ Hrs.

CUMULATIVE PROJECT TO DATE

Feet Drilled ________

Feet Cored ________

Rig Hours ________

Rig Days ________

(Driller)

(Log Operator)

(Field Geologist/Technician)

Figure 1. Drill log form developed to cover any chosen time period

It wasn't a very long time ago that a meeting like this one just wouldn't have taken place. Then, there simply would not have been the interest.

But in the future, we have to employ more complex methods, equipment and techniques, and we will need to share our experiences to ensure our collective ability to evaluate future coal supplies.

PAUL WEIR COMPANY
SAMPLE DESCRIPTION

Job No. ____
Page ___ of ___
Date Sampled ____
By ____
Recovery ____ %
Seam ____

Hole No. ____ DTC ____
Date Drilled ____ Core Dia. ____
Core Condition ____

	Thick.	Cum.

Core Box	
Box No.	___ ___
GW	___ ___
TW	___ ___
NW	___ ___

Samples	Thickness	Weight	Material	Lab No.
Total ___				

Figure 2. Laboratory sample identification and description form

DISCUSSION

QUESTION: Do you feel that the inclusion of mapping and general data collection in the Federal regulations is an attempt by the government to obtain the raw data obtained for their own purposes like further regulation and more accurate strategic inventories?

ANSWER: I don't know why the government wants to get confidential exploration data. I only know they shouldn't have it. Politicians are fond of accusing the energy companies of a lack of competitiveness. The confidentiality of subsurface data is as important to that competition as private property is to free enterprise.

QUESTION: Can you comment, please: You discouraged the use of "new and novel" techniques. How do you reconcile this with the need for improved data and faster, lower cost operations?

ANSWER: I certainly don't mean to discourage advances in coal exploration technology or methodology. Today, geophysical logging for coal is accepted as a commonplace practice. In fact, it is considered a vital requirement of most drilling programs.

But, think back ten or twelve years ago. At that time, coal logging was in its infancy. We have some coal companies, utilities, equipment companies, and some geologists to thank for sticking their necks out in support of something new, which today we know is a contribution of greatest magnitude.

We are not going to move forward unless we have enough faith to try the untried.

Nevertheless, new methods and equipment need to be viewed with caution and suspicion from the standpoint of exploration program management. Very often, when we try out something new, we don't know at the front end exactly what the time or cost is going to be. And certainly, we can't be sure of the results.

In designing and implementing a coal exploration program, prudence needs to be exercised in application of new methods and equipment to ensure that the program's objectives are not unduly jeopardized.

Exploration Drilling: Onshore and Offshore

4.

Utilization of Oilfield Rig Equipment and Technology in Coal Exploration

by Ron W. Pidskalny

5.

Offshore Nova Scotia Coal Exploration Drilling

by Allan R. Beckering

4

Utilization of Oilfield Rig Equipment and Technology in Coal Exploration

By Ron W. Pidskalny
Managing Director
Kenting Drilling Services Limited
Wilford, Nottingham, United Kingdom

INTRODUCTION

In 1975, the National Coal Board mobilized oilfield equipment from Canada when they realized that additional equipment to that available in the United Kingdom was necessary, if they were to meet their target of drilling 150 deep exploratory boreholes every year until into the 1980's. Many reservations were expressed by senior mining people on whether oilfield practices could, in fact, produce coal core of acceptable quality.

As oilfield drilling contractors, we have always had to survive in a very competitive industry and "making hole" was everyones' objective. To do so has meant staying abreast of the latest technology and continuously up-dating equipment and training personnel. Although we were confident that our experiences in "making hole" could be complimented with the production of good core, the National Coal Board were not. The contract, therefore, carries a guaranteed recovery of acceptable 3-1/2 inch diameter core, with payment only on core cut and recovered. We also must produce 100 per cent coal seam recovery of acceptable quality. This specification meant first choosing the type and size of core barrel for the program. This, in turn, dictated the diameter of the cored hole, the necessary diameter of the open hole and finally the capacity of rig required to handle the program. The challenge was put before the oil drilling contractors and the results speak for themselves. Core recovery of 100 per cent is more the rule than the exception with most coal seams cut and recovered in excellent condition. Deep boreholes, 1000 metres - 1200 metres, which up to then took three months to complete, are now being routinely completed in three weeks or less.

A paper titled "The Development and Adaptation of Drilling Equipment to Coal Exploration" and presented by Mr. K. Shaw of the National Coal Board to the International Coal Symposium in 1976, highlighted the initial results of the oil drilling rigs. My intent here is to update the results and expand on various points revolutionary to British Coal Exploration, based on our experience as one of the three drilling contractors on National Contract to the National Coal Board for deep exploratory drilling.

PERSONNEL

The rigs were originally manned by Canadian crews operating three 8 hour shifts per day, 7 days per week as compared to the two 12 hour shifts 5-1/2 days per week, more familiar in the British Drilling Industry. A continuous and uninterrupted operation results in fewer problems and greater efficiency.

Over the course of three years, our training program to yield skilled British drilling personnel has been very successful. More than 90 per cent of our crews are now local employees and turnover is minimal compared to the "norm" in the drilling industry. This has been accomplished with "on-the-job" training, training in Canada, correspondence courses, formal training at the Drilling and Production Technology Training Centre at Montrose, Scotland and, of course, wage rates which recognize the special effort we expect from our employees.

RIG DESIGN

The rigs, a Failing Strat 90 (Fig 1.) which was brought over from Canada as a complete unit, and a Gardner Denver 3000 (Fig 2.) which was 50 per cent manufactured in England, were designed bearing two main concepts in mind; moving quickly with minimum rig-up and rig-down time and drilling with a low solids mud with the aid of efficient drill solids

FIGURE 1

FAILING STRAT 90 RIG

FIGURE 2
GARDNER DENVER 3000 RIG

removal equipment.

While a rig is on the move, it is considered on a non-revenue generating operation. Every effort is, therefore, made to design the rigs so that moves can be as fast and efficient as possible. The concept from Canada, whereby rig components are housed in steel buildings on skids was very successful. The compact buildings occupy a minimum of space on the very small sites, protect equipment from often very wet conditions and are ideal for loading or unloading in minimum time by the expensive cranes. (Fig 4.) - illustrates a typical rig layout on a threequarter of an acre site. The rig is kept basically to eleven loads - a rig trailer complete with motor, drawworks, rotary table and telescoping mast; mud tank complete with a shaker and desilter; a pump house; a combination building including two generating units, change room and storage; a combination building including fuel tank, workshop and doghouse; a rig managers quarters and five loads of cement silos, core storage shed, cat walk, drill pipe and drill collars. To simplify acquisition of permits and police routing, loads are kept to a maximum 9 foot 6 inches wide except for the rig trailer. Maximum weight of any skidded load is 25 tonne. All loading and unloading is done with two 25 tonne cranes. Normally only 12 - 16 hours are consumed from the time the rig is released on one borehole to being ready to spud on the next.

Novel to this industry was the investment in an expensive drill solids removal system. From previous experience in

FIGURE 3

DRILL SOLIDS REMOVAL SYSTEM

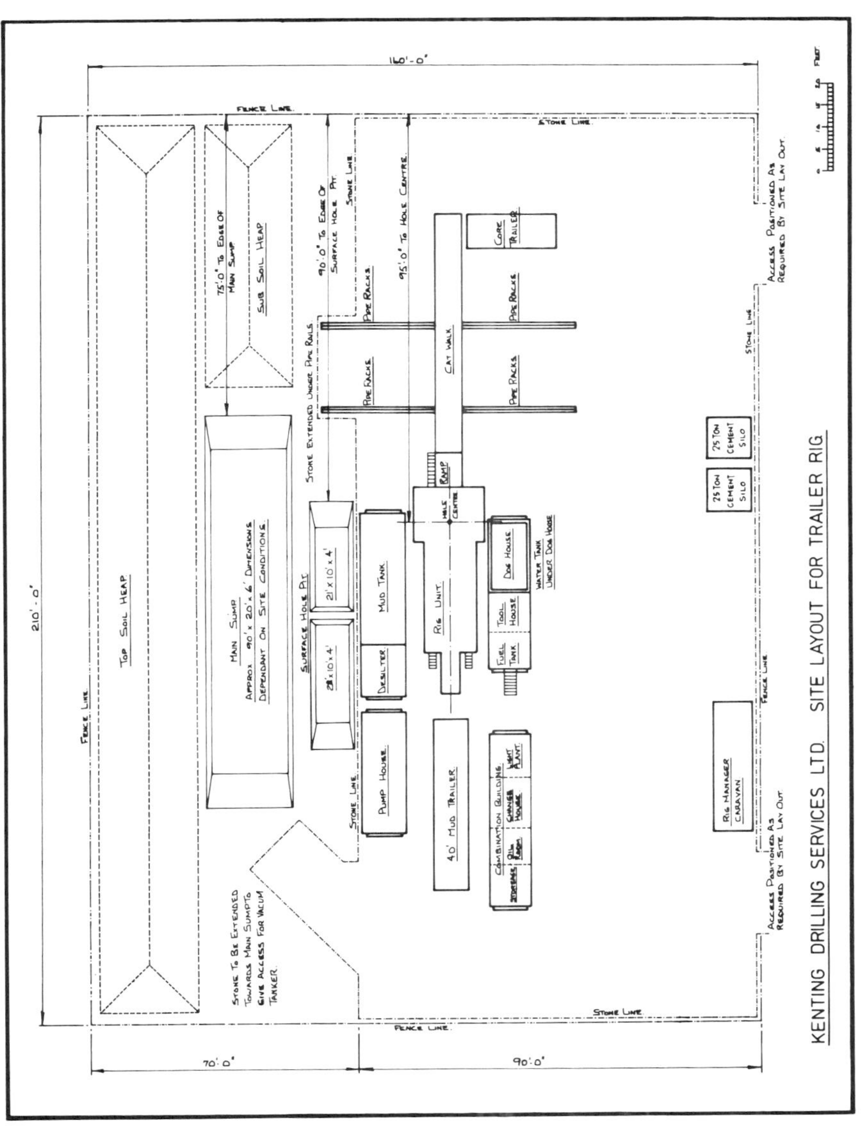
KENTING DRILLING SERVICES LTD. SITE LAYOUT FOR TRAILER RIG
160'-0"
210'-0"
70'-0"
90'-0"
FENCE LINE
STONE LINE
TOP SOIL HEAP
SUB SOIL HEAP
75'-0" TO EDGE OF MAIN SUMP
90'-0" TO EDGE OF SURFACE HOLE PIT
95'-0" TO HOLE CENTRE
MAIN SUMP
APPROX 90' x 20' x 6' DIMENSIONS
DEPENDANT ON SITE CONDITIONS
SURFACE HOLE PIT
21' x 10' x 4'
STONE EXTENDED UNDER PIPE RAILS
PIPE RACKS
CAT WALK
CORE TRAILER
RAMP
HOLE CENTRE
MUD TANK
DESILTER
PUMP HOUSE
RIG UNIT
40' MUD TRAILER
DOG HOUSE
WATER TANK UNDER DOG HOUSE
TOOL HOUSE
FUEL TANK
COMBINATION BUILDING
LIGHT PLANT
CHANGE HOUSE
OIL ROOM
STORAGE
25 TON CEMENT SILO
25 TON CEMENT SILO
RIG MANAGER CARAVAN
ACCESS POSITIONED AS REQUIRED BY SITE LAY OUT.
STONE TO BE EXTENDED TOWARDS MAIN SUMP TO GIVE ACCESS FOR VACUM TANKER.
FEET

Figure 4

the United Kingdom, we appreciated the importance of adequate mud conditioning equipment. A 200 barrel, three compartment mud tank was built (Fig 3.) with a high speed three screen shaker at the flow line, a 10 cone desilter at the suction end and a low pressure mud mixing system. This set-up has proved very efficient in removing drill solids from the circulating system. The mud weight can be controlled to a 9.0 - 9.2 pound per gallon with minimum dumping of tanks, thus minimizing the cost of drilling fluid materials and waste disposal.

Basically, the rigs are rated to 1500 metres and are equipped with 350 horsepower on the drawworks, a 17 1/2 inch rotary table and a 300 horsepower duplex mud pump. The drill string includes 3 1/2 inch O.D. 15.50 pound per foot internally coated drill pipe fitted with 4 3/4 inch O.D. flash welded and hard banded tool joints and a string of 5 inch O.D. drill collars. All drill pipe and drill collars are 31 foot in length enabling the 88 foot high telescoping rig masts to handle the drill string during trips in double lengths.

We found that rigs of this capacity were necessary for boreholes in the 1000 - 1200 metre depth range, if they were going to stand up to the vigours of as much as 800 metres of coring and the many associated trips which are required every 18 metres to recover the core. Considerable expense and effort still has to be made in preventive maintenance of equipment if 'down-time' is to be kept to a minimum. Because of the general unavailability of many of the equipment parts in the United Kingdom, it is necessary to maintain a sizeable stock of major equipment parts and, often necessary, to air freight equipment in an effort to minimize 'down-time'. At today's cost of running rigs, it is often found cheaper to replace a piece of equipment than waste time and money repairing it.

SITE PREPARATION AND RESTORATION

Most sites are situated such that very little, if any, access road is usually necessary. The sites leased from the landowner are kept to a minimum size, normally threequarters of an acre, seventy yards x fifty yards (Fig 4.). Because of design of the rigs and the utilization of cranes on rig moves, the previous use of railroad sleepers was both too time consuming and impractical. Instead, top soil is stripped and heaped on site. Pits are excavated for temporary waste mud and drill cuttings storage. An area of approximately

2400 square yards is then covered with 6,000-8,000 tonnes of 4 inch crushed stone plus another 150 - 200 tonnes of 2 inch crushed stone to act as binding material. A location of this sort allows moving in and out under any weather conditions without any loss of time. A normal site preparation by a sub-contractor can be completed within 2 - 3 days.

Occasionally, a site must be prepared with noise abatement as a prime consideration. Usually, orientating the site to direct the noise away from a residence and/or using straw bales or the top soil heap as a screen has been adequate. Noise levels are already reduced considerably since all the motors are in buildings and the rig floor is screened by asbestos canvas weatherproofing.

During the course of drilling operations, all waste mud is hauled to an approved tip by vacuum truck. After completion of the borehole, sludge and solid waste is removed with tip trucks and the hard core is taken up and either disposed of or stacked for the landowner. The sub-soil is then ripped to break up compaction and the top soil spread back over the site re-instating it to it's original condition. A new steel gate into the field is often the only evidence of a drilling operation having been present.

OPEN HOLE DRILLING

A standard program commences with drilling 25 metres of 11 inch diameter surface hole and then cementing in 2 joints of 8 5/8 inch casing. This is normally all that is required to provide a standpipe for mud return to the tanks, isolate the drift formations and accommodate a blow-out preventer. After 4 hours of waiting for the cement to set, a casing bowl and double gate mechanical blow-out preventer are installed. The blow-out preventer, a rarity on rigs previously utilized in coal exploration in the United Kingdom, is set up as standard equipment. It is present only to act as a diverter to protect men and equipment should any hydrocarbon flow to the surface be encountered during the course of operations.

For open hole drilling below the surface casing, the 6 3/4 inch diameter tri-cone bit was introduced for the normal coal exploration program. This was to accommodate the 5 3/4 O.D. core barrel chosen for coring operations. The attractive bearing design of this size of bit and the added clearance it provided between the down-hole drilling tools and the borehole wall, as compared to the more conventional 6 1/4 inch size, have yielded benefits so apparent that the 6 3/4 inch hole size is now a standard in the industry.

Top hole sections are drilled with water, mudding up only when hole conditions dictate or productive coal measure formations are being penetrated. The sophisticated solids control system does an excellent job of solids removal, allowing drilling at mud weights of 9.0 - 9.2 per gallon, which is far below those which were common in the past. With the reliable oilfield drill pipe and drill collar string, we can run 18,000 pounds weight on the bit at 120 revolutions per minute to attain maximum penetration. Penetration rates previously unheard of in the industry have been consistently attained ranging from 400 metres per day in a Bunter sandstone to 200 metres per day in the Coal Measures.

Borehole deviation is monitored every 150 metres, or more frequently if necessary, with a TOTCO drift recording instrument run on wire line. Any hole deviation tendencies are usually controlled or corrected with the aid of drill string stabilizers. A maximum of 5 degrees is allowed, although there have been exceptions under special circumstances. Special deviation control bits were found very effective in South Wales. The more conventional methods of stabilization and fanning were unsuccessful in controlling hole deviation in the hard, abrasive, faulted and tilted formations present in this area.

If there is any likelyhood at all of encountering old workings, larger diameter surface casing must be set to accommodate liner strings. A larger open hole is then drilled from under the surface casing shoe if we are certain that old workings will be encountered. If severe loss of circulation occur the interval is simply isolated by running and tag cementing the liner string 10 - 12 metres below the lost circulation zone. Normal operations are then resumed. If, on the other hand, old workings are not very likely, the drilling of the normal 6 3/4 inch hole below the surface casing shoe will be chanced. In this case, if severe loss of circulation is encountered and cannot be regained, the hole size must be reamed out to sufficient size to accommodate a liner string which would again allow the passage of 6 3/4 inch bits for drilling onward. Upon completion of the borehole, an attempt is usually made to recover the liner casing by mechanically cutting it at its free point.

CORING

Daily core recovery rates of as much as 50 metres per day have been achieved. Although a lot of this is due to the rig design and the trained rig personnel, a good part of the credit has to be given to the use of the oilfield core barrel and modifications in bit design.

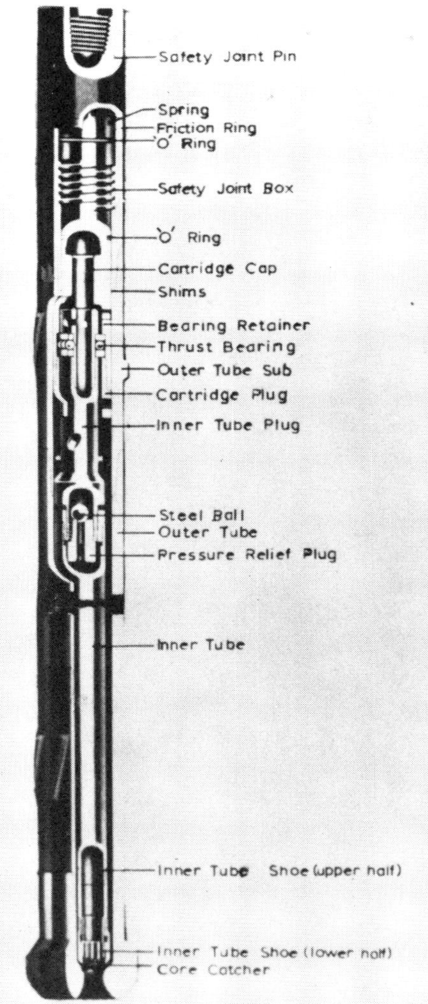

FIGURE 5

For the project, the Christensen Oilfield Products 250 P Series 5 3/4 inch O.D. core barrel - (Fig 5.) was chosen from satisfactory experience in the oilfield. This size of barrel also produced the National Coal Board requirement of a minimum 3 1/2 inch diameter core. A most important feature was the rugged design of the barrel which allows it to withstand the rigors of optimum oilfield drilling practices. The barrel, which can be run in multiples of 30 feet, is, due to available mast height being run as a 60 foot section at all times. This allows us to cut 18.35 metres of core per run. We have the most success in achieving a good break and often obtaining a root core with the "slip and knife" type of catchers. The barrel is stabilized with three integral blade tungsten insert spiral stabilizers. These, we find, have the longest wear life and provide the best core barrel stability.

The 6 23/32 inch diameter core head, instead of the more conventional 6 7/32 inch, was chosen for coring below the open holed section. We felt that the extra clearance between the outer barrel and the wall of the cored section of the hole would minimize any potential core barrel sticking problems. The option was always available to drop down to the conventional 6 7/32 inch size and still use the same core barrel should tight hole conditions develop in the 6 23/32 inch open hole. No such problems have been encountered to date. Experiments were run with a 6 7/32 inch core head with no noticeable increase in penetration rate. However, there were indications of tight hole problems developing. Also, the bit was being hydraulically lifted off bottom due to pump pressure created from the reduced anulus.

The Boards' several areas of exploration have presented their own types of coring problems with drillability and mixture of strata varying considerably in boreholes that are even only hundreds of yards apart. Bit designs have varied from step type with high penetration rate results, but poor bit life to the double and single cone bits shown in Figure 6.

FIGURE 6

Where the step bit has shown some success in coring the soft mud stones in Nottinghamshire, it is vulnerable to conglomerate, hard sandstone stringers and ironstone nodules which are very common in this area. The type of bit used in the end has become a compromise between the best penetration rate and the longest bit life.

The popular design now appears to be the single cone with a 15 millimetre inside guage and a .35 square inch total flow area for minimum fluid erosion of the core. The "Congo Round" diamonds appear least susceptible to fracturing with bit life excellent, ranging from 800 - 1500 metres, depending on formations being cored. Diamond size in the 3 - 5 stone per carat range is being used for coring in England. On the other hand, the difficult formations in South Wales have been cored with a "Premium" stone in the 10 - 12 stone per carat range with a life of often less than 200 metres. The single cone bit, regardless of stone size has given core recovery of excellent quality.

While coring the sandstones in the Nottinghamshire area, penetration rates of 4 - 6 metres per hour have been achieved. However, penetration rates as low as 1.5 metres per hour are still common in the mud stones and it is generally the mix in strata that is encountered that will determine the sort of average penetration rate one can achieve for the duration of the borehole. South Wales is a special problem with considerable amounts of hard mud stones often interbedded with pennant sandstone and quartzite which scrub bits in very short intervals. A bit is yet to be designed which, while achieving suitable rates of penetration in hard or soft mudstones, can also have a bit life that makes it an overall economic success.

Normal practice is to carry 14,000 weight on the bit, rotate at 120 revolutions per minute and circulate at 150 gallons per minute with a 550 - 700 pounds per square inch pump pressure. All four of these parameters are reduced before and while coring coal seams as a precautionary measure for good coal seam recovery. We have seen no evidence of the coal core being crushed from the weight of the core above it in a 60 foot barrel. The 60 foot barrel is run in every time and the coring will only be cut short if an expected seam, along with sufficient floor strata, are unlikely to fit into the barrel. It is what we call "lining up" for a seam and has become a standard and very successful procedure. There have been occasions where we have been caught with the full barrel or jammed off in the middle of a seam for one reason or another. However, by tripping out without any pipe rotation and carefully commencing coring with the pump off or at idle, we have been able to finish coring the seam with minimal

CORE HANDLING

FIGURE 7

SANDSTONE AND MUDSTONE

COAL SEAM CORE

or no loss of recovery. There have also been seams which are geologically faulted and required special precautions. It is often the ingenuity and skill of the men on site that has brought some powdered core to surface which has made one wonder how any of it was recoverable at all.

Recovery of the core from the barrel is also a very delicate operation. After cutting the core, the drill pipe and collars are tripped out and stood in the derrick. The core barrel is set in the rotary table slips. The safety joint is then backed off and the inner barrels retrieved and layed down on the catwalk in 30 foot sections. Empty inner barrels are picked up and reassembled into the outer barrel. The bit is then checked and run back to bottom. While coring continues the core is carefully pumped out of the previous inner barrels and recovered on trays. This is accomplished with the use of a rubber plunger in the inner barrel driven by water pressure from either a BAKER hand pump or the rig circulatory water pump. The core is washed, measured and racked in the core storage shed for examination by the geologist. Fig 7 illustrates recovery procedure and some typical recovered sandstone, mudstone and coal seam core.

Coring faulted and fractured formations has been the greatest challenge of all and has been met with limited success. The tilted formations often cause hole deviation problems which leave you helpless as there is not much one can do to reduce hole drift with a fully stabilized assembly already in the hole. Also, the steep dipping, fractured formations, characteristic of South Wales and Scotland and sometimes found in England, continuously jam off in the barrel. This occurs often only after a few metres of coring, necessitating tripping out of the hole. These extra trips naturally reduce on bottom time, thus reducing daily coring rates to 15 - 20 metres per day; that is 50 per cent of what we could expect if allowed to fill the barrel completely under normal coring conditions.

Also new to the industry are 3 pen automatic rate of penetration recorders. Unlike the manpower consuming and unreliable method of hand measurement used in the past, these provide a permanent record in 20 centimetre increments. The geologists have found these quite useful for correlating with the core itself as well as with electrical logs.

DIVERSIONS

Present diversion techniques are as described in Mr. K. Shaw's paper referred to previously, whereby mud motors with a bent sub are utilized instead of whipstocks. We have also found considerably better success using a diamond bit for the kick off rather than the conventional rock bit. The diamond bit appears to give a more positive kick off and can be kept in the hole as long as is necessary to get the desired distance away from the original hole. This is especially attractive when deviating in very hard formations such as South Wales or where a directional diversion is required.

ABANDONMENTS

The introduction of the continuous mix and displacement technique for sealing boreholes with cement was another area where considerable time and money is saved by the application of oilfield technology. Drill pipe is run to total depth and service companies familiar in the oilfield are contracted to mix and displace bulk Class 'B' cement with 3 per cent calcium chloride. The calcium chloride is used to accelerate the set time for the cement. Plugs 250 - 300 metres long are spotted through the drill pipe and then felt for in 4 hours to confirm placement. Each subsequent plug is run from the top of the previous plug until the entire borehole is sealed. No calcium chloride is run in the last plug to give sufficient time to lay down the drill pipe while withdrawing from the plug. Abandonment of a 1,000 metre borehole now takes only 18 hours compared to as much as a week when boreholes were sealed without the aid of specialist cementers and when the cement was batch mixed prior to displacement.

SUMMARY

In summary, Canadian oilfield rig equipment and technology were introduced successfully into the British Coal Exploration Program for several reasons.

1) The personnel came from a very competitive and progressive industry and with the attitude and expertise to make success inevitable.
   We now have virtually complete British Crews trained to these high standards.

2) The heavier duty oilfield equipment can cope with the considerable number of trips and rotary hours involved in coring operations.

3) Oilfield drilling practices apply heavier weights on the bit and use low solids mud systems for better penetration rates and less hole problems.

4) The 60 foot Christensen oilfield core barrel is capable of carrying higher weights and revolutions per minute to the bits; reduces time spent on trips and still produces core and coal seams of good quality.

5) Preventative maintenance reduces costly breakdowns and downtime.

6) All the aforementioned have now reduced the amount of time spent actually pulling and lowering the drill string to 20 per cent. Actual coring time averages 72 per cent with 4 per cent required to recover core and another 4 per cent to service equipment.

7) A minimum of time is required to mobilize and de-mobilize this design of rig allowing more time for actual drilling of the boreholes.

8) With stoned sites, less time is wasted moving in and out.

9) Introduction of the concept of using sub-contractors specialized in their own field for site work, rig moves, water haulage, waste disposal and abandonments. This, in the end, is not only time and money saving but frees the Rig Manager to concentrate his efforts on "making hole" and the "recovery of good core".

REFERENCES

1) SHAW K. The Development and Adaptation of Drilling Equipment to Coal Exploration. Proceedings of INTERNATIONAL COAL EXPLORATION SYMPOSIUM, London, May, 1976.

2) SHAW K. Drilling - A new Concept. COLLIERY GUARDIAN ANNUAL REVIEW, August, 1976.

5

Offshore Nova Scotia Coal Exploration Drilling

By Allan R. Beckering
Director, Contracts/Engineering Liaison
Global Marine Drilling Company
Houston, Texas, United States

INTRODUCTION

During the period 1 August 1977 through 16 November 1977 the GLOMAR CONCEPTION participated in a coring program offshore Cape Breton Island for the Nova Scotia Department of Mines. On 7 May 1978 the program was continued by a sister ship, the GLOMAR GRAND BANKS. The purpose of the program was to locate and determine the seaward extent of coal seams already being mined on land. Core samples and logs were utilized to determine the thickness of the seams and the quality of the coal. The purpose of this paper is to describe some of the equipment used and procedures followed while accomplishing this project.

DRILLSHIP DESCRIPTION

The GLOMAR CONCEPTION and the GLOMAR GRAND BANKS are ship shape drilling units owned and operated by Global Marine Inc. of Los Angeles, California. Both ships were built from the keel up as drilling vessels and are 400 feet long, 65 feet wide and have a 142 foot derrick. They are equipped to drill to 25,000 feet in 600 feet of water.

Since her completion in 1967, the GLOMAR CONCEPTION has operated in the severe environment of the Bass Straits off

Australia, in the South China Sea and in Nicaragua. During the summer of 1976 she participated in a scientific coring program off the East coast of the United States for the United States Geologic Survey. Since completing the program offshore Cape Breton, the CONCEPTION has worked in the Gulf of Mexico and is now off California.

The GLOMAR GRAND BANKS was completed in 1972 and operated in the Gulf of Mexico until her departure for Canada in May. While in the Gulf she participated in the Exxon Subsea Production System Project which involved setting a completely self-contained submerged drilling and production apparatus.

OPERATING PROCEDURES AND EQUIPMENT

The procedures followed in the conduct of a coring program are very similar to those followed by Global Marine when drilling an exploratory oil and gas well. In order to present these procedures and some of the equipment involved, a typical corehole will be described.

Mooring

The first procedure in any drilling program is to moor the vessel on location. A trisponder survey system, utilizing three onshore reference points, was used to place a marker buoy at the approximate site. The buoy is approached on a heading determined after a review of the local weather information and the prevailing winds and seas. Approximately 2,000 feet from location a stern anchor, Number 5, is dropped, which helps slow down the ship. (Figure 1) The anchor handling boat has a bow anchor, Number 10, already on board and it will be placed approximately 2,000 feet ahead of the vessel. While the remaining six 30,000 pound anchors are set in place, the trisponder determines the final location to the nearest

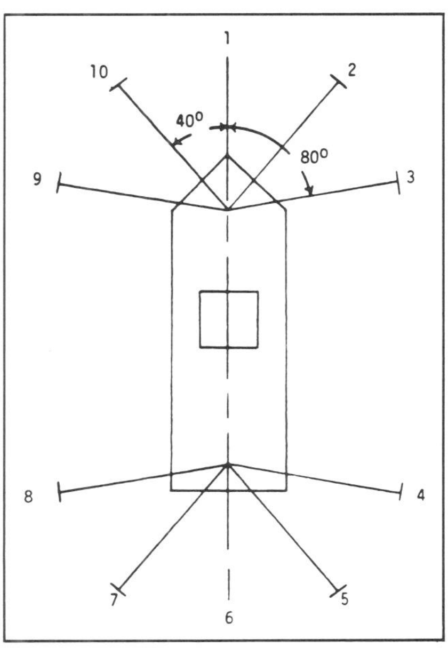

Standard Mooring Pattern
Figure No. 1.

thousandth of a degree of latitude and longitude.

The anchors are positioned in a 40-80 pattern, meaning they are 40 degrees and 80 degrees off the bow and stern. This pattern allows the Captain to adjust for variable weather and sea conditions by changing the vessel's heading up to 30 degrees without having to stop drilling operations to reanchor.

Bottom Hole Assembly

Once the vessel is determined to be on location and the anchors are holding, the bottom hole assembly is made up and drilling operations may begin. The GLOMAR CONCEPTION bottom hole assembly (Figure 2) consisted of a drilling bit, drill collars and a bumper sub. This assembly is typical for floating drilling units without motion compensators.

The design and composition of the bottom hole assembly satisfied two requirements for efficient drilling. First, it allows a constant, known weight to be kept on the bit which prolongs bit life and optimizes the penetration rate. Secondly, it keeps the relatively thin walled drill pipe from going into compression, where failure is likely to occur.

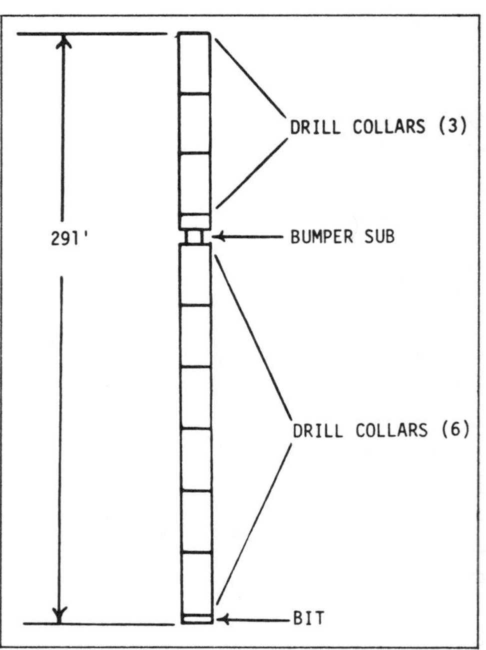

Bottom Hole Assembly
Figure No. 2.

Drill collars are thick walled, large O.D. pipe which have much higher compressive strength than drill pipe and are employed to provide the weight on the bit. When initially spudding in, we used the maximum size available on board - 8 inch O.D., 2-13/16 inch I.D., 150.5 pounds/foot. If these dimensions are compared with the drill pipe, which is 5 inch O.D., 4.276" I.D. and 19.5 pounds/foot, the strength difference becomes apparent. The number and size of drill collars used depends on the size hole, weight desired and the depth at which you are working.

The difference between drilling on land and drilling offshore is, of course, that the drilling platform heaves up and down with the natural movement of the sea. Something must be done to prevent the bit from moving up and down with the platform, as it is obvious if it does the bit will not be on bottom when the vessel heaves up and when it heaves down the impact force could be sufficient to put the drill pipe in compression. To compensate for the heave a bumper sub is included in the bottom hole assembly. (Figure 3).

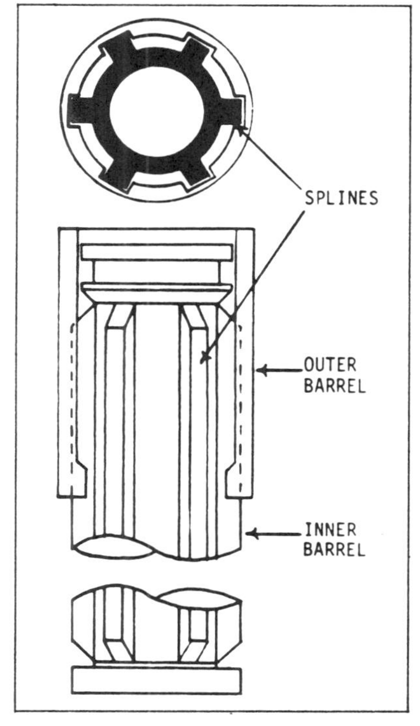

Bumper Sub. Figure No. 3.

The bumper sub isolates the portion of the bottom hole assembly below it from the vessel's motion, thus allowing constant weight to be maintained on the bit. This weight is not adjustable without changing the size or number of drill collars in this section of the bottom hole assembly. Very simply, the bumper sub consists of two barrels, or sleeves, one sliding inside the other. In order to transmit torque to the bit as the bumper sub strokes in and out, there are a number of splines on the inner barrel which fit into grooves in the outer barrel. Since each bumper sub has a 5 foot stroke, the number used is dependent on the amount of heave which must be compensated for.

Motion Compensator

The GLOMAR GRAND BANKS has replaced the bumper subs in its bottom hole assembly with a motion or heave compensator which is part of the hoisting equipment located in the derrick. (Figure 4) The pressure from the bank of high pressure air receivers supports the weight of the drill string, less the weight to be kept on the bit.* As the vessel heaves up or down the cylinder and traveling block move but the piston and rod, which are supported by air pressure, do not.

* The amount of weight on the bit may be changed by increasing or decreasing the air pressure.

Thus, the entire drill string is compensated for heave and the weight on the bit may be varied within reasonable limits without changing the number of drill collars in the drill string.

Motion Compensator System Figure No. 4.

Drilling Operations

Once the bottom hole assembly is made up and lowered to the sea floor, drilling operations are ready to begin. The bit is spudded in and rotated slowly until the drill collars are buried. When the depth of the first casing point is reached, the drill string is pulled out of the hole and preparations to run casing are made.

Since these coreholes were planned much shallower that the normal offshore exploratory hole, a complete offshore casing program of 30 inch, 20 inch, 13-3/8 inch and 9-5/8 inch was not used. The size of the first string was dependent on the anticipated total depth and whether one or two strings of casing would be required. The largest size casing used was 20 inch and the smallest was 9-5/8 inch. The running procedures, however, are the same, regardless of the size of casing run.

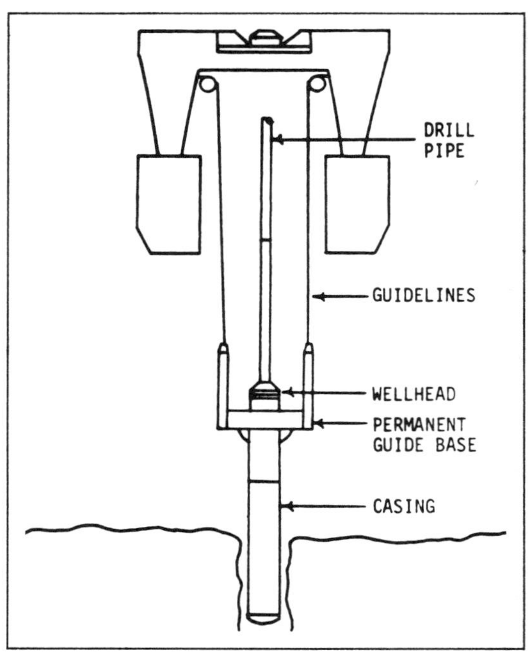

Running casing, guidebase and wellhead. Figure No. 5.

The required length of casing is made up with the

top joint being the wellhead, which is latched into the permanent guidebase. (Figure 5) The ship is held steady over the hole and the casing is blind stabbed in. The entire assembly is then lowered to the bottom and the casing is cemented into place.

The wellhead is the point where the BOP stack or diverter is attached in order to establish a means for returns to be brought to the vessel rather than left on the sea floor. The guidebase forms a foundation for the BOP stack and has four posts attached with wires running through them up to the vessel. These wires guide equipment, such as the BOP stack and TV camera, down to the wellhead.

Diverter System: During the 1977 program some of the core holes were shallow and since the history of the area did not indicate the presence of over-pressured zones, the Blowout Preventer Stack was dispensed with and the diverter was used instead. (Figure 6) The diverter system is sufficient for pressures up to 500 pounds per square inch and consists of a hydraulic connector, ball joint, marine riser, slip joint and flow diverter housing.

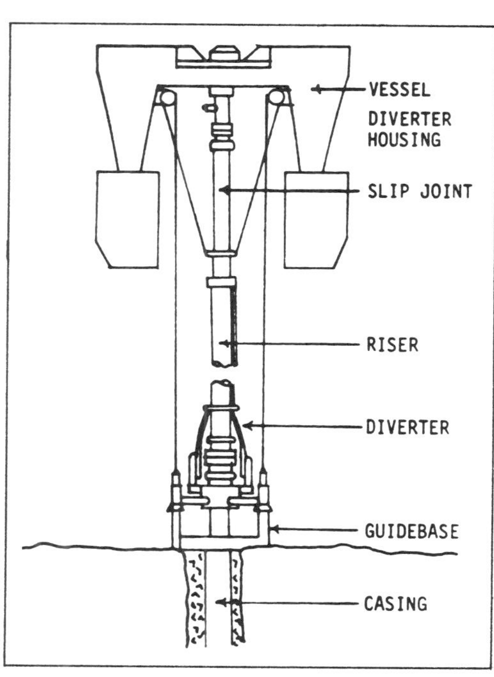

Running the Diverter Assembly
Figure No. 6.

The ball joint provides a flexible union between the BOP Stack and the riser and is used to compensate for the side to side movement of the ship. It can accommodate angles up to 5 degrees from the vertical. The marine riser is large diameter pipe which is used as a guide for drilling tools and to bring returns to the vessel. It also has two barrels, one connected to the riser and one to the diverter housing, but it has no splines to prevent the inner barrel from rotating if the heading of the vessel changes. The riser is kept in tension by wirelines attached to a hydraulic/pneumatic system similar to a heave compensator.

The flow diverter system is not designed to seal the well bore and control pressure. Instead, it keeps the gas off the rig floor by sealing around the pipe at the rotary table and diverting it out one of two 12 inch lines which are open to the atmosphere. The wind carries the gas away from the vessel.

<u>Blowout Preventer Stack:</u> The Blowout Preventer (Figure 7), or BOP Stack, was used on deeper holes during 1977 and on

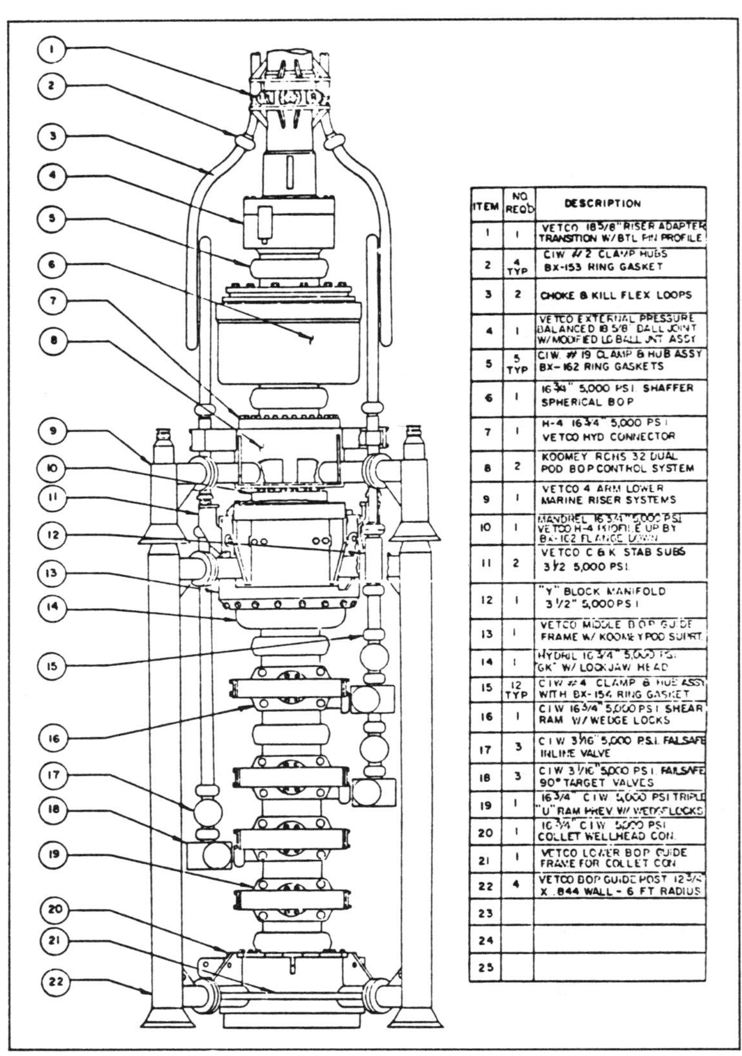

ITEM	NO REQ'D	DESCRIPTION
1	1	VETCO 18 5/8" RISER ADAPTER TRANSITION W/BTL PIN PROFILE
2	4 TYP	CIW #2 CLAMP HUBS BX-153 RING GASKET
3	2	CHOKE & KILL FLEX LOOPS
4	1	VETCO EXTERNAL PRESSURE BALANCED 18 5/8" BALL JOINT W/MODIFIED LG BALL JNT ASSY
5	5 TYP	CIW #19 CLAMP & HUB ASSY BX-162 RING GASKETS
6	1	16 3/4" 5,000 PSI SHAFFER SPHERICAL BOP
7	1	H-4 16 3/4" 5,000 PSI VETCO HYD CONNECTOR
8	2	KOOMEY RCHS 32 DUAL POD BOP CONTROL SYSTEM
9	1	VETCO 4 ARM LOWER MARINE RISER SYSTEMS
10	1	MANDREL 16 3/4" 5,000 PSI VETCO H-4 PROFILE UP BY BX-162 FLANGE DOWN
11	2	VETCO C & K STAB SUBS 3 1/2 5,000 PSI
12	1	"Y" BLOCK MANIFOLD 3 1/2" 5,000 PSI
13	1	VETCO MIDDLE BOP GUIDE FRAME W/ KOOMEY POD SUPRT.
14	1	HYDRIL 16 3/4" 5,000 PSI "GK" W/ LOCKJAW HEAD
15	12 TYP	CIW #4 CLAMP & HUB ASSY WITH BX-154 RING GASKET
16	1	CIW 16 3/4" 5,000 PSI SHEAR RAM W/ WEDGE LOCKS
17	3	CIW 3 1/16" 5,000 PSI FAILSAFE INLINE VALVE
18	3	CIW 3 1/16" 5,000 PSI FAILSAFE 90° TARGET VALVES
19	1	16 3/4" CIW 5,000 PSI TRIPLE "U" RAM PREV W/ WEDGELOCKS
20	1	16 3/4" CIW 5,000 PSI COLLET WELLHEAD CON.
21	1	VETCO LOWER BOP GUIDE FRAME FOR COLLET CON
22	4	VETCO BOP GUIDE POST 12 3/4" X .844 WALL - 6 FT RADIUS
23		
24		
25		

GLOMAR CONCEPTION - 16-3/4", 5,000 psi Blowout Preventer Stack. Figure No. 7.

all holes drilled this year. It consists of a series of valves which are designed to seal off or isolate the wellbore should an over-pressured zone be encountered. The GLOMAR CONCEPTION was equipped with a 16-3/4 inch bore, 5,000 pounds per square inch BOP Stack, which indicates well bore pressures up to 5,000 pounds per square inch could be controlled at the sea floor. This year, in order to simplify and speed up operations, the GLOMAR GRAND BANKS was outfitted with a smaller and lighter 20 inch, 2,000 pounds per square inch Blowout Preventer Stack.

On the bottom of the stack is a connector which is hydraulically latched to the previously mentioned wellhead. Next are three sets of pipe rams which may be configured to seal around various sizes of drill pipe, drill collars or casing. The top rams are shear rams which will cut the drill pipe in two in an emergency. Also installed are two annular preventers, which have a rubber element designed to seal around any size pipe, casing or tool joint or seal off an open hole. The opening and closing of all the BOP Stack equipment is controlled hydraulically from the surface by a hydraulic manifold or remote electric panel. The ball joint, riser, and slip joint described earlier are also used with the BOP Stack.

Once the casing has been cemented and the BOP Stack or diverter run, drilling or coring continues to the setting depth of the second string of casing, if a second string is run, or to total depth. Usually before setting the second string of casing, and always after reaching target depth, electric logs are run and sidewall samples are taken as requested by the Department of Mines. When logging is completed the hole is cemented to the surface and a mechanical cutter is run to cut the casing just below the wellhead. Thus, the guidebase and wellhead may be recovered for use on the next hole.

Coring Equipment

In order to maximize our chances for good core recovery, the GLOMAR CONCEPTION arrived on location prepared to core using two different methods - wireline and conventional.

The wireline core barrel system used was similar to that used by the GLOMAR CHALLENGER to recover cores in over 20,000 feet of water. The advantage of this method is speed. The inner core barrel is pulled through the drill string each time a core is cut, saving the time spent in pulling and running the drill pipe. This procedure is known as tripping and is required each time a core is cut conventionally.

The wireline system also has its drawbacks. In order to take a 2-3/8 inch core, which was considered a minimum size, the bottom hole assembly components had to be sized with a large enough I.D. to accommodate passage of the inner core barrel. This necessitated large O.D.'s also as the pipe wall had to be kept sufficiently thick to maintain compressive strength. The result was a large diameter bit - 9-1/2 inch - cutting a small core - 2-3/8 inch. Thus, the weight per square inch of cutting surface provided by the drill collars was small, the amount of formation to be cut was large and the penetration rate was slow. After various problems were worked out with the design of the bit and core barrel, the penetration rate reached 7.7 feet per hour and the recovery was 80 percent.

Due to the slow penetration rate, low recovery and preference for a larger diameter core, it was decided to switch to a 6-3/4 inch x 4 inch conventional Christiansen core barrel. The Hycalog diamond bit used was specifically designed for coring coal and was 8-15/32 inch in diameter and cut a 4 inch core. The increased weight per square inch of cutting surface increased the penetration to 9 feet per hour with recovery of almost 95 percent.

However, since the program started in August, late in the drilling season, and an objective was to drill as many sites as possible, it was again decided to switch barrels.

A 6-1/4 inch x 3 inch Christiansen Marine core barrel, which is stronger than the conventional barrel, was run with an 8-15/32 inch O.D. Christiansen C-20 diamond coring bit. The penetration rate with this assembly increased to 15 feet per hour, indicating increasing the rpm from 80 to 120 compensated for the decrease in weight per square inch of cutting area caused by taking a 3 inch core. Core recovery, however, decreased somewhat to 88 percent.

Overall, 2,388 feet of formation was cored at five sites and 2,058 feet recovered, a rate of 86 percent. Recovery was excellent in the sections directly above and below the coal seams, but inconsistent in the seams themselves.

Since the 1978 program commenced earlier in the drilling season the penetration rate while coring was not as critical as during 1977. The GLOMAR GRAND BANKS is using a 6-3/4 inch

Christiansen core barrel, which takes a 4 inch core, and an 8-15/32 inch Christiansen C-20 bit. Through the first five sites 1,240 feet have been cored and 1,194 feet, or 96.3 percent have been recovered. Recovery has also been consistently high in the coal seams.

PERFORMANCE

Information from last year's drilling program provided better geologic control for the program this year. Consequently, we have been able to drill deeper, faster because fewer cores have to be taken to find the coal. Figure 8 depicts the well performance curves for a typical hole drilled in 1977 and 1978. Each is an average for six core holes.

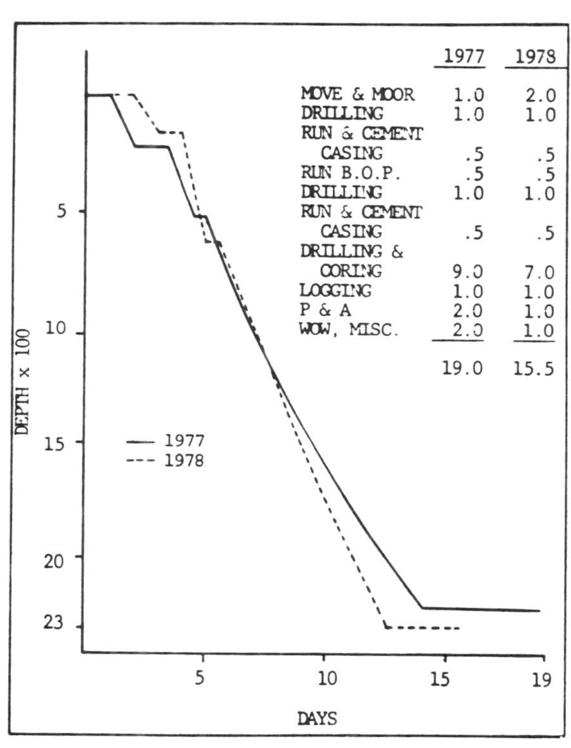

Drilling Performance Curves
Figure No. 8.

Table I is the time distribution for the 1977 and 1978 programs. The important numbers here are: (1) the amount of time spent doing productive work - drilling, coring, logging, (2) the amount of time spent on necessary activities - moving and mooring, abandoning and other, and (3) the amount of time spent on non-productive activities - waiting on weather and repair.

The amount of time spent waiting on weather during 1977 was anticipated due to our late start. We spent approximately 12 days waiting on weather, an average of two days per hole, but approximately 50 percent of the total time occurred during one storm which lasted 12 days in October. The storm struck at the worst possible time as the operations being attempted were those most affected by drillship or anchor handling boat motion. These operations include logging, cutting the casing and recovering the wellhead, picking up and running anchors and attempting to spud in. Although the

GLOMAR CONCEPTION was down part of the day over 12 consecutive days, an average of 12 hours per day was spent operating and the longest continuous down period was 29 hours. During that period, winds reached 67 miles per hour and the seas reached 36 feet. The downtime recorded during this 12 days would have been reduced had the GLOMAR CONCEPTION been drilling, as this type operation continued in 50 miles per hour winds and 15 foot seas. Weather downtime so far during the 1978 program has totaled 44 hours, or 7-1/2 hours per hole.

Table No. I
TIME DISTRIBUTION FOR 1977 AND 1978
CORING PROGRAMS

	1977		1978	
Drilling	52.0%		53.6%	
Logging	6.0		7.3	
Coring	10.6		12.8	
	68.6	68.6%	73.7	73.7%
Moving and Mooring	5.8		11.5	
Plug and Abandon	9.5		7.0	
Other	3.3		5.4	
	18.6	18.6%	23.9	23.9%
Waiting on Weather	12.4		2.0	
Repair	.4		.4	
	12.8	12.8%	2.4	2.4%
		100.0%		100.0%

RESULTS

The value of this program may be measured by more than percentage of core recovered. The results from last year's program proved the existance of 800 million tons of coal, double the amount earlier believed to exist. A virgin area of coal was found in the Donkin Region and feasibility studies are now underway to determine if a new mine should be opened. And, most significantly, the Canadian Government has announced a $265 million program for coal development in the Cape Breton area.

ACKNOWLEDGEMENT

The author appreciates the assistance provided by personnel of the Nova Scotia Department of Mines, GLOMAR CONCEPTION and GLOMAR GRAND BANKS. The information they provided was invaluable in the preparation of this paper. Special thanks are due to John Smith, Wynn Potter, Frank Shea and Danny Murray of the Department and Rusty Roberts and Gerry Gunter of Global Marine Drilling Company.

DISCUSSION

QUESTION: What water depth are we working in?

ANSWER: The water depth varied from 70 to 200 feet. I might say that the rig is capable of drilling in 600 feet of water, though.

QUESTION: What is the total depth of the basal coal seam?

ANSWER: The deepest coal found was from 3,500 feet on the east side of Cape Breton to 4,800 feet on the west side. Right now, they are limited to coal that's at 4,000 feet.

QUESTION: The nature of the coal deposits, multiple seams?

ANSWER: Yes, in the Donkin region there were three seams, but in the other coal beds there's five to 10 seams. The coal zones range over 2,500 feet, from 1,500' to 4,000'. The thickness ranges from about 4 to 14 feet per seam.

QUESTION: What is the drilling cost per foot in this environment?

ANSWER: The average cost of drilling figures out to about $300.00 per foot and that includes actually everything you can think of, including the drill ship, all the consumables, the shore base and everything that has to be set up by the client. That's the entire cost for the Department of Mines.

QUESTION: How far from shore will the mining eventually be extended?

ANSWER: Right now they are limited to five miles from shore.

QUESTION: What will the minimum mineable thickness be

in the same area?

ANSWER: About four to five feet.

QUESTION: What type of mining method is considered more suitable under these conditions? For instance, advanced longwall, and what is the thickness of overburden?

ANSWER: The mining method is advanced longwall. We try to keep 1,000 feet of overburden.

Review of Exploration Practices

6. A Critical View of Some Coal Exploration Programs

by Graham R. Wallis

6
A Critical View of Some Coal Exploration Programs

By Graham R. Wallis
Chief Coal Geologist
Robertson Research (North America) Limited
Calgary, Alberta, Canada

I. INTRODUCTION

The principal objective of this paper is, through constructive criticism, to illustrate a number of weaknesses observed in exploration programmes in various parts of the world by use of examples where possible. Since the scope of all elements normally involved in an exploration programme is too wide to cover in depth in one paper, selected aspects will be discussed.

The examples used are necessarily anonymous but are from actual situations and occurrences. The references to a specific country are not meant as a comparison of that country with another, rather that the examples are universally applicable. The word "critical" in the title of the paper however is used in the dictionary definition of "censorious, fault finding".

The hypothesis presented in this paper is that weaknesses in exploration programmes commonly are related to three elements, which for the ease of reference in the following paragraphs, I refer to as PRIME FUNCTIONS. These are:

A. MANAGEMENT CONTROL
B. EXPERIENCE OF PERSONNEL
C. FINANCIAL CONSTRAINTS

Expanding briefly on these functions the following observations are presented.

A. Management Control

Management is exercised by all levels of persons involved in a programme including, but not only, the exploration manager, the project geologist and the drilling foreman. Communication, and more importantly, understanding and support are essential to achievement of the objectives defined.

Further, it is management's responsibility to ensure that the objectives established have an industrial basis and that this objective is adhered to at all times. If a project doesn't have a chance at producing a profit in the then-operating economic climate, defer it for the present time.

B. Experience of Personnel

The transference of either base metal or petroleum geologist to coal geologist has been successful in the short term in only a limited number of cases. The medium to long term is a different matter, but there are programmes today being conducted by geologists inexperienced in coal exploration. Lest it be thought that the comments in this paper are directed only toward the "base metal or petroleum geologist", let me stress that they apply equally well to the "coal geologist". The situation is similar with drilling crews and foremen.

C. Financial Constraints

The imposition of financial constraints on exploration and development programmes can have and has had ramifications during the production stage of a coal mine. This is not to imply there should be no limitations on expenditure, rather that funds are adequate and wisely allocated and intelligently spent.

This may sound a rather facile remark, in much the same vein as the observation that can be drawn from later comments in this paper regarding the wastage of funds resulting from inadequate documentation of results. The question must be asked, however, if sufficient funds are not available to properly evaluate a prospect, even at reconnaissance level, should an inadequate job be done at all? Is 5¢ per ton of saleable coal for, say, 100 million tons, too much to allocate to exploration on a resource worth $3 to $5 billion?

Further, at what point does a professional geologist compromise himself by attempting to produce evaluations, and make judgements, based on data resulting from a restrictive budget.

Demands Placed on the Exploration Geologist

To be fair, the demands placed by the coal industry - and I speak here in the generic sense - on the exploration geologist are really quite extensive. From an exploration programme, which can extend over a number of years, he must produce data in order to satisfy:

(i) The Management: That there is a deposit of coal and that the stratigraphy, structure, quality and reserves are known;

(ii) The Mining Engineer: Who is to design, and later operate, the mine;

(iii) The Coal Preparation Engineer: Who may have to clean the coal;

(iv) The Marketing Manager: Who has to sell the coal;

(v) The Accountant: Who pays the bills.

2. REVIEW OF PROGRAMMES

Possibly the clearest manner of illustrating the contentions stated above is to review certain elements typical to an exploration programme in a sequential manner. It is not intended that all aspects will be discussed, only those considered to be of a critical nature which can have financial implications.

While it is not the intention of the paper to take a "how to do it" position, as opposed to "what should not have been done" position, it is inevitable that the former approach must be taken in part.

2.1 Planning

The foundation on which a coal mine may be designed and operated is based on the definition of objectives and the planning of a series of exploration programmes from their very inception to "completion". The weaknesses observed usually can be traced to the three PRIME FUNCTIONS above, as evidenced by any or all of the following:

(i) a lack of appreciation of the task;
(ii) the consequent failure to first clearly define the objectives of the exploration programme; and
(iii) a lack of technical capability of the various persons, at any level, involved in the programme.

While it might appear obvious to state that planning and the definition of objectives are important, the following examples will serve to illustrate how funds can be unnecessarily wasted by carrying out exploration and geological mapping in the wrong sequence.

Figure 1 is a diagrammatic geological cross-section showing a dip slope separated from a sequence of rocks containing a coal seam by an overthrust fault in British Columbia, Canada. During what was a rather late stage and frenetic search for strip coal to provide an early cash flow for a planned underground mine, it was stated that easily accessible coal was present on the dip slope. The statement was made from a promontory overlooking the dip slope 1 kilometre to the south. Coal was believed to occur to the east of the thrust fault, though its presence was not proven at the time.

Drill holes A, B and C, plus additional holes to the

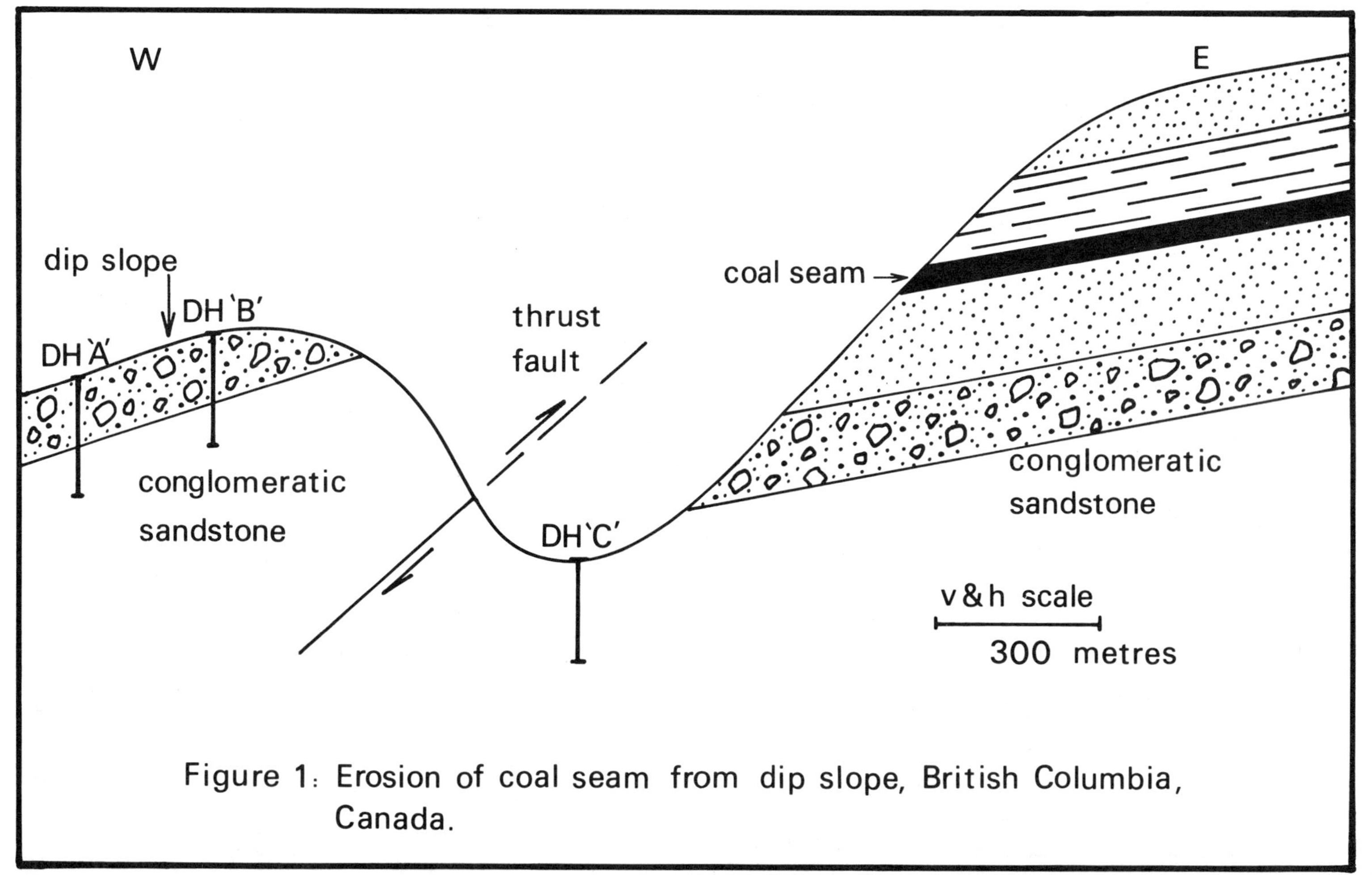

Figure 1: Erosion of coal seam from dip slope, British Columbia, Canada.

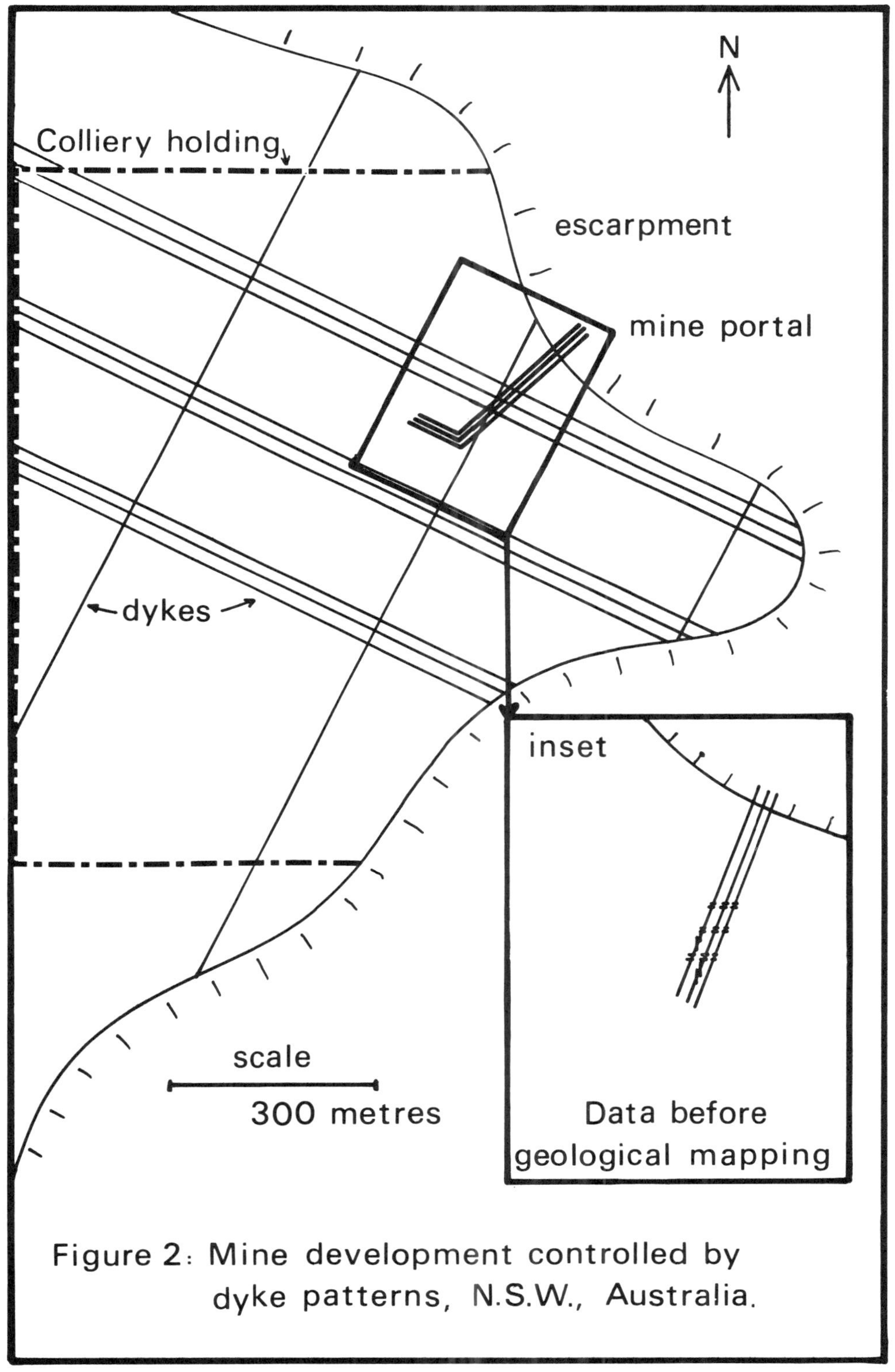

Figure 2: Mine development controlled by dyke patterns, N.S.W., Australia.

north, were drilled but did not intersect coal. Subsequent geological mapping and structural interpretation identified the presence of the fault upthrusting the western block. Thus it was deduced that the coal which was originally present above the dip slope is now part of a Recent sediment some hundreds of miles away, and that information was gained at the cost of five unnecessary drill holes.

An example in Australia where an operating coal mine incurred unnecessary and higher production costs is shown in Figure 2, illustrating the importance of geological mapping early in a planning stage. The initial mine layout was constrained by the frequency of an easterly trending dyke pattern, thus limiting panel development in the northerly direction. The geological data available at this time is illustrated in the inset in Figure 2. Subsequent geological mapping revealed that the northerly trending dyke pattern was more widely spaced thus allowing a more productive panel layout in the westerly direction, as was shown after the headings were turned to the west.

A drilling programme in the middle 1950's on Semirara Island in the Philippines "proved" the lenticular nature of two coal seams, since a number of drill holes did not intersect coal. However, surface mapping a decade later indicated faulting. A consequent drilling programme revealed a graben-like structure and showed that the coal seams were present throughout the drilled area. A lack of detailed stratigraphic documentation was at fault in this case.

Experience has demonstrated the practical necessity of having to conduct geological mapping in conjunction with a drilling programme. It is critical that, apart from scout or reconnaissance drilling, a basic geological map is available for planning of ongoing exploration programmes. It is believed that financial constraints, or indeed option agreements, should in no way result in a compromise to the production of a reliable geological map prior to more extensive exploration.

2.2 Operations

2.2.1 Drilling Programmes

The three Prime Functions of Management, Experience and Finance probably have the most impact in a drilling programme since it is common for sub-surface data to provide the basis of an evaluation of the potential of a coal deposit.

Management at all levels is operative from the calling of tenders or bids to final move-out of a drill rig, experience in gathering the necessary data in the required detail and finance in the ensuring there are sufficient funds.

Selection of the most suitable tender or bid should, in my opinion, be on the basis of price only after all technical factors have been satisfied. The lowest price is no justification for selecting a particular contractor if any doubts exist as to his ability to perform satisfactorily. Similarly, the use of non-core drilling should be avoided if important structural or engineering data is lost, even though it is more attractive to the bank account.

It is not unusual for the lowest tenderer to have inadequate spare equipment on site and so on. For example, a not uncommon occurrence is that of struck drill rods. In Canada one contractor had 300 metres of HQ rods stuck in a drill hole due principally to longitudinal splitting of the drill rods, stated to be second quality. He had on site a jack suitable for pulling the rods but no sub-assembly to connect to the rods. The loss of time and frustration to the geologists due to the onset of winter, were a natural corollary to this situation.

An example of good management is the education and supervision of the drilling crews. How often has a project geologist explained in detail why core recovery of the coal seam is critical, rather than just a hole in the ground? By this I mean showing and explaining to the driller what happens to the coal core: detailed logging, sampling, analysis and hopefully, the design of a coal mine.

Further, since in a multi-rig operation on site supervision during the drilling of all seams is rarely practicable, the driller should be provided, at least daily, with a prediction of when he should intersect the next coal seam, and as much information as possibly to allow his self education to proceed with respect to the rock sequence. (This does assume that the geologist knows the stratigraphy!) Only in this manner can good core recovery or chip sampling be expected.

2.2.2 Core Recovery

With the increasing sophistication of today's coal industry, and particularly market demands, 100% recovery of the coal seam intersection should be the prime objective of the explorationist. I am, even today, astounded at the acceptance of and the conclusions drawn from less than satisfactory core recovery; of course, this may attest to the flexibility of the coal industry and perhaps we should compliment its mining engineers!

It is suggested that low core recovery is a function of all three <u>Prime Functions</u>.

A. Management - Exploration manager, project geologist, drilling foreman.
B. Experience - A lack of both technical competence (drillers) and an awareness of the implications of core loss (geologists).
C. Financial - The use of NQ rather than HQ diamond drilling; stringent budget limitations.

The ramifications of low core recovery are, among other things,

(i) inadequate knowledge of seam structure and constitution;
(ii) incomplete quality data, at best, and misleading quality data at worst; and
(iii) incorrect seam thicknesses (in the absence of geophysical logs).

(a) <u>Triple Inner Tube and Recovery Clause</u>

In Australia, the use of the triple inner tube (TT) in diamond drilling for coal has been standard for about 15 years, as has a guaranteed recovery clause in drilling contracts. The triple tube has subsequently gained acceptance in North America, having been virtually standard in Canada for about 5 years and employed on about 10% of coal projects in the United States. I believe that a guaranteed recovery clause is common in the Appalachian region of the U.S.A. but this is not so in western Canada.

While the triple tube will not increase core recovery, per se, it allows,

(i) accurate determination of the core lengths recovered;
(ii) assists core recovery in difficult drilling conditions e.g. sheared coal; and
(iii) provides as undisturbed a sample as possible allowing for macro-petrographic logging, measurement of fracture patterns, x-raying, gas measurements, sampling and so on.

While there is no increase in cost for the use of the triple tube in Australia and Canada, I understand that in the United States a premium of 5% on the footage rate can be applied when the TT is used. There is, to my mind however, no alternative to its use if full value is to be gained from the exploration dollar.

To illustrate the point in dollar terms, if the base drilling rate is, say $60 per metre for NQ TT diamond coring, the overall cost per metre of core drilled can be as high as $150, when site logistics, geologists' time in field operations and data evaluation, and analytical costs are all taken into account. Thus the incremental cost, if applicable, for use of the triple tube is not significant.

A guaranteed recovery clause specifying that a coal seam will be re-drilled at the contractor's expense if less than 90% core recovery of the coal seam is produced, unless geological conditions mitigate against such action, is vital if good performance by the driller is to be maintained. It is axiomatic that the project geologist and the drilling contractor must be familar with the geological conditions and a state of trust or rapport must exist between both the contractor and client, if this clause is to operate fairly and effectively.

(b) <u>Core Diameter Size</u>

Larger core diameters assist in improving core recovery and the move to the larger HQ and

PQ sizes of 63.50 mm and 84.96 mm respectively, is becoming increasingly common, especially where difficult drilling conditions prevail. The Rocky Mountain region of western Canada is a particular case in point where HQ is now commonly used in the structurally disturbed sequences and the 20% increased cost of HQ over the NQ cost is justified.

(c) Determination of Core Recovery

Possibly the most neglected phase of an exploration programme is the accurate determination of core recovery and the apportioning of the core loss to its correct position in the coal seam. It should be understood that it has only been in the last 2 years, or 3 years at the most, that the development of detailed geophysical logs has reached a stage whereby accurate positioning of core losses has been possible. However, other methods have been available for accurately calculating core losses.

Detailed logging of coal seams always has been and still is the prime element in determining coal seam thicknesses. Drillers' marker blocks indicating depths were used but, in the best instance, were always to be regarded as questionable until proven correct.

Figure 3 shows the generally accepted symbols used for visual logging of coal seams in Australia Canada and the United Kingdom. This method provides a comparative and reproducible (by different geologists) graphic log of the varying brightness of the macerals in a coal seam. The log can be correlated visually on a seam to seam basis and also with both the various geophysical logs. Lithotypes should be recorded to a minimum of 1 centimetre, where practicable, to provide the necessary detail for correlation purposes.

Figure 4A illustrates a seam section as logged by the geologist on the basis of the core as it was placed in the core boxes by the driller. The correct seam thickness was calculated from the density log, and revealed a core loss of 12.1%. In attempting to match the graphic section with

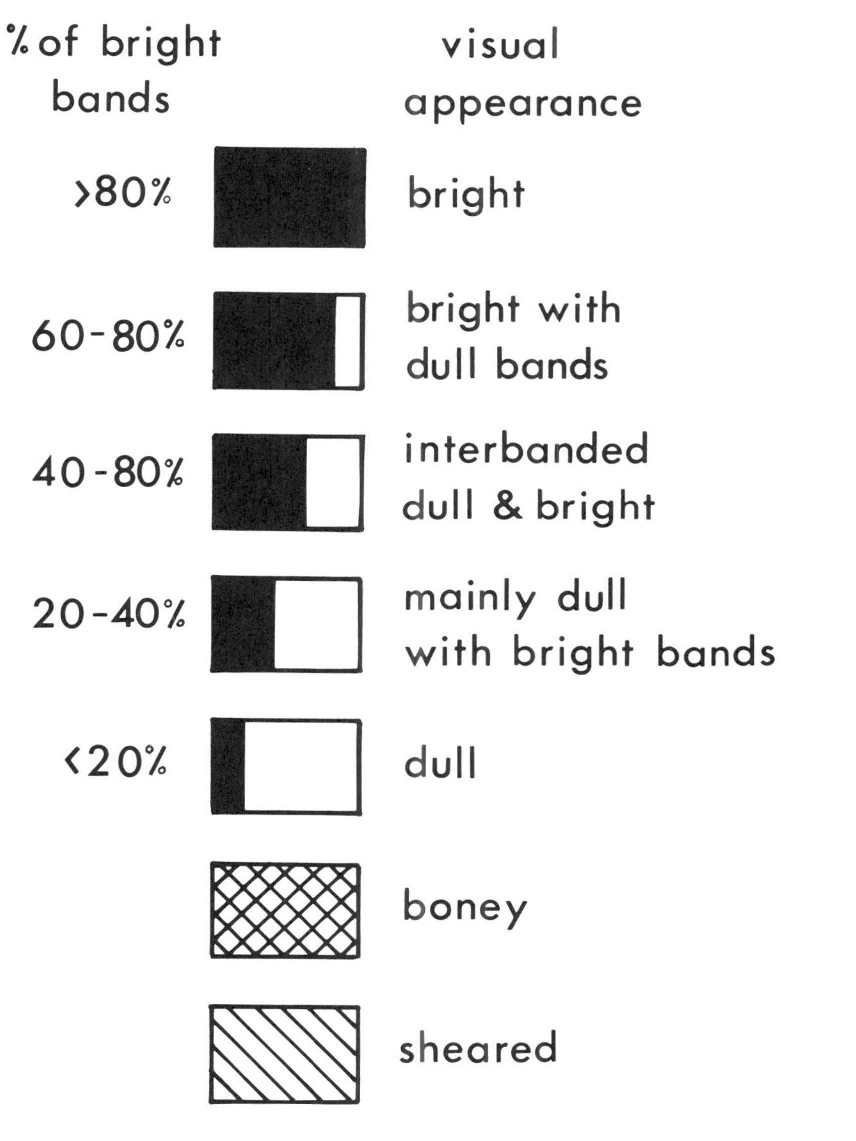

Figure 3: Lithological logging symbols for coal.

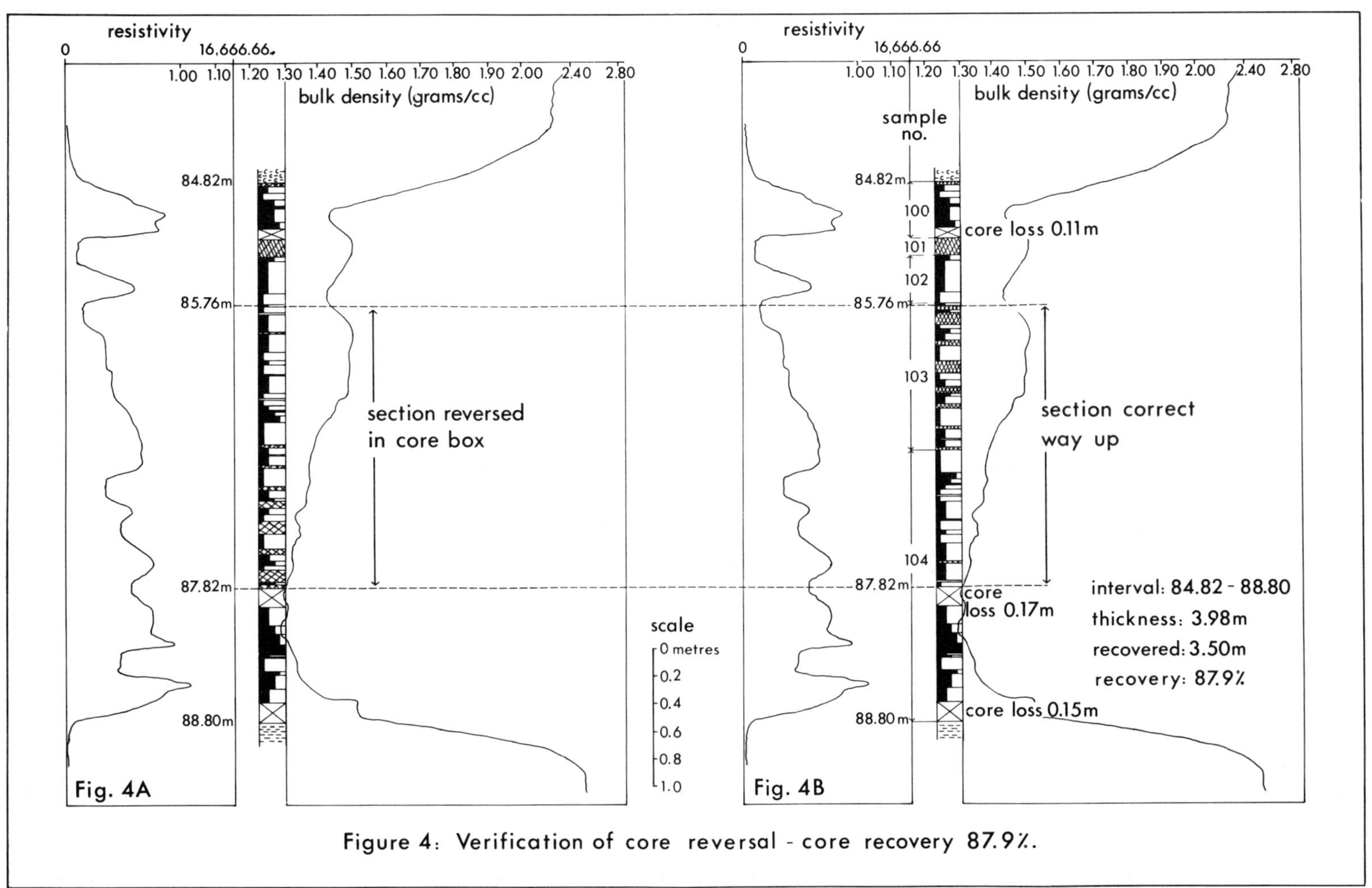

Figure 4: Verification of core reversal - core recovery 87.9%.

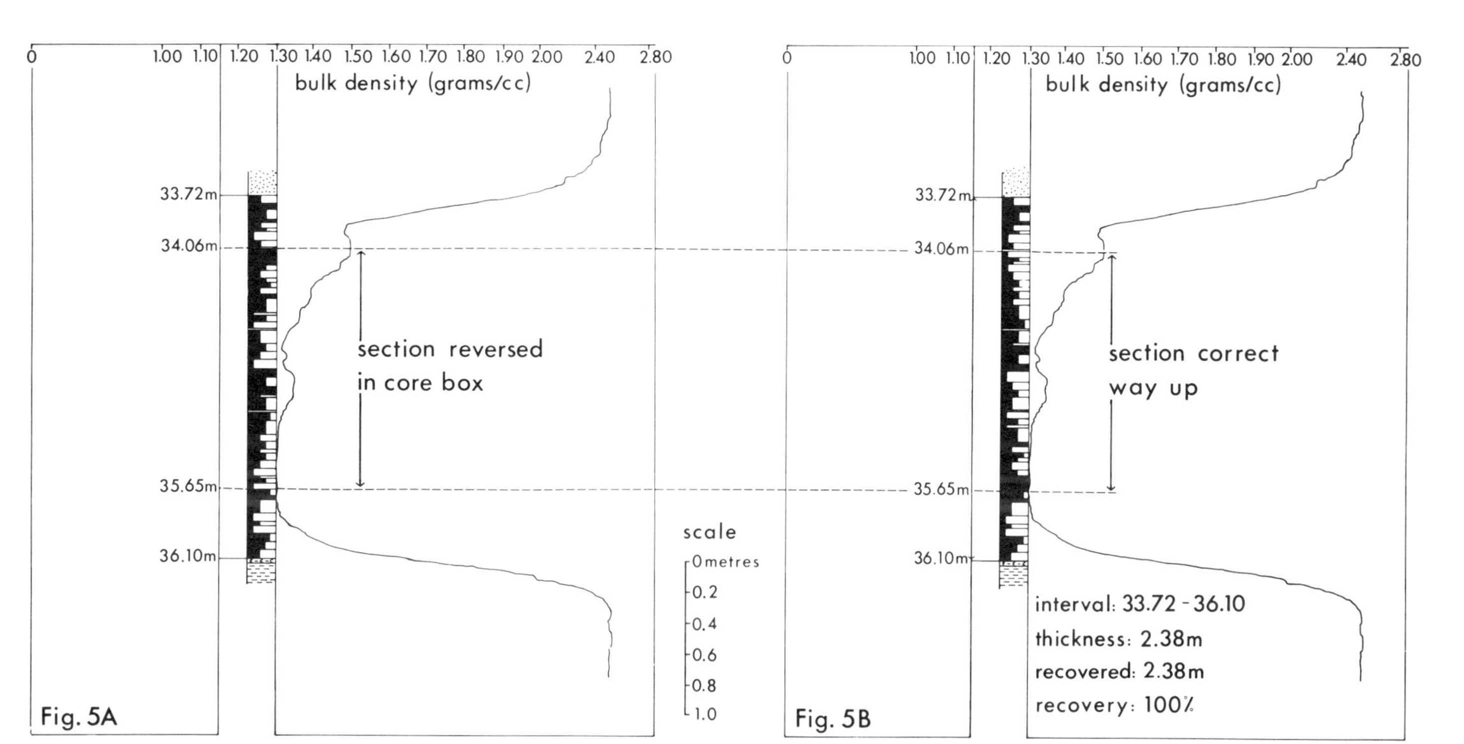

Figure 5: Verification of core reversal - core recovery 100%.

the density log and the focused beam (resistivity) log to determine the location of the core losses it became apparent that a problem existed between 85.76 metres and 87.82 metres. It can be seen that the density decreases in the higher ash zone at the base of the interval and vice versa at the top of the interval. Similarly, correlation of the resistivity log with the seam plies cannot be achieved.

Reference to Figure 4B illustrates how, by reversing the mismatched interval between 87.76 metres and 87.82 metres, a correlation is achieved and allows distribution of the core losses in three positions of the seam.

A second example is presented in Figures 5A and 5B where 100% of core recovery was achieved but the interval between 34.06 metres and 35.65 metres was reversed in the core box. This example also illustrates the correlation between the density log with the visual logging. Note that at the base of the reversed section (Figure 5A) the duller or higher ash coal corresponds to the low density (1.30) part of the curve and the bright coal at the top corresponds to the higher density (1.50) part of the curve. Both these portions are obviously incorrect; the correct correlation illustrated in Figure 4B shows the response of the density log to the varying degrees of brightness, and thus ash content, of the coal macerals.

Similar checks can be made using the bed resolution and high resolution density logs instea[d] of the resistivity logs.

2.2.3 Logging Detail

Basic to all coal exploration programmes is an understanding of the stratigraphy of the area. The necessity of producing a geological map was referred to above. The detailed logging of the core or cuttings recovered from a drilling programme is a basic tool in understanding the subsurface geology. Sufficient detail should be recorded such that discrete beds can be segregated but not such that the total picture is clouded by too much detail.

Only through a complete understanding of the stratigraphy can structural problems be solved. This aspect is illustrated in Figure 2, above and Figure 6, below. In the latter example, diamond drill hole (DDH 1) was prematurely stopped at 233 feet (71 metres) after it penetrated the first repeat of Unit A, designated A1 on Figure 6. Subsequent re-logging of the core a year later, owing to an absence of detail in the original log, defined a number of distinctive features in Units A and B. Stratigraphic and structural interpretation carried out using these and associated data necessitated deepening of the hole to 518 feet (157.9 metres) to confirm the presence of the repeated coal seams X and Y.

Had the stratigraphic sequence been documented in the earlier stage of drilling, thus allowing recognition of the repeated Unit A1 after the thrust fault F1 and thereby indicating that continuance of the hole was necessary, the additional expense of re-drilling would not have been incurred. In this case, the pebble band at 222 feet (67.6 metres) was a positive marker bed. The importance of stratigraphy in solving structural problems in the Rocky Mountain region of western Canada has been discussed by Wallis and Jordan (1).

It is important to note that the above remarks apply equally to chip logging in open-hole or non-core drilling programmes. Chip samples should be collected at 0.5 to 2.0 metre intervals, washed and described in detail under a binocular microscope.

A further consequence of inadequate logging of core is the loss of data required for engineering purposes - fractures, joints, slickensides and such like. Whenever underground mining is contemplated, some 10 to 15 metres of the roof and an adequate floor thickness must always be recovered. Similarly for open-cut operations, such parameters as clay materials, rippability etc. should be determined.

2.2.4 Sampling and Analysis of Coal Seams

As stressed in the preceeding sections documentation of the individual coal "plies" and "dirt bands" is essential to the understanding of the nature of the entire coal seam. It is on this information that the basic and most important technical problem of physical subdivision of the coal seam is based. Without this basis, all data

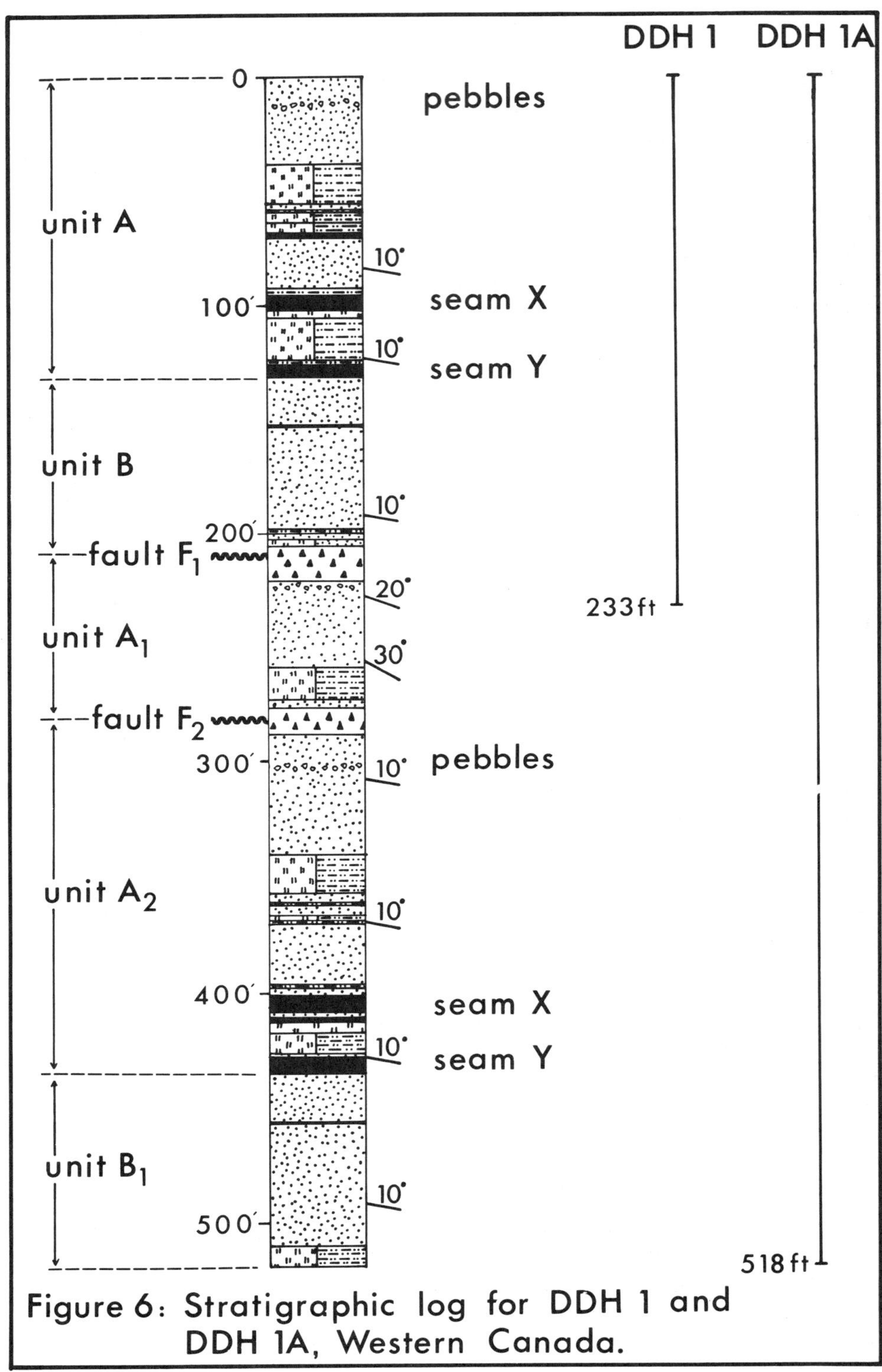

Figure 6: Stratigraphic log for DDH 1 and DDH 1A, Western Canada.

which follows is of questionable value. Once a bore core has been subdivided and sampled, it is not possible to effect further subdivision. A channel sample can be re-sampled but this should not be necessary.

Figures 4B illustrates the subdivision of the coal plies and dirt bands into samples for analysis, based on the visual and geophysical logs of the seam. Reconstitution of these samples into working sections can be achieved at a later date, if considered necessary. Longitudinal splitting of the coal core should not be undertaken since the volume of coal available for analysis is already at a minimum.

The following are some examples of sampling undertaken by professional geologists or engineers, and the results regarded as representative and meaningful!

(i) Sampling and combining two seams 18 metres (60 feet) apart. (Canada).
(ii) The collection of five (5) pieces of coal from a seam 15 metres (50 feet) thick and placed in separate bags with no log. (Canada).
(iii) An engineer, acting as the site geologist collected "a few pieces from here and there at about 1.0 foot intervals over a 6 foot seam to obtain a representative sample". (Philippines).
(iv) The collection of grab samples of a coal stockpile to comprise one sample. (U.S.A.)
(v) The sampling of coal only in a seam, excluding the dirt bands, indicated an ash content of 5%; subsequently, a complete seam sample revealed an ash content of 12%. (Philippines).
(vi) The sampling of 10 metres of coal seam (potential underground mine) as one sample with no regard for possible working sections. (Australia).

The analytical phase of a programme can be summed up by what might appear to be a sweeping statement that as full a testing programme as possible should be undertaken, provided the samples are representative. This policy should be adopted for at least a selected number of sample points in any property.

Quite obviously a full range of analyses is useless on a grab sample of "bright coal" from stockpile, a bore core with less than 90% recovery or chip samples contaminated by out of seam drill cuttings. All these have occurred.

As drilling costs far exceed analytical costs the apparent high cost of laboratory testing is not, in fact, real. The cost of re-drilling an area when market conditions change or new technological developments take place can and should be avoided by careful preparation and action at the outset. As an example of the shift in markets from coking coal to steaming coal, my colleagues in Australia tell me that some coal deposits originally explored for coking coal do not have one specific energy (calorific value) determination for the whole drilling programme. Similarly, there are numerous analyses of Appalachian coal seams which report only calorific value, ash content and sulphur content, but are believed to be attractive as blending coals for coke manufacture. No data is available to substantiate this however.

2.3 Reporting and Data Storage

The third broad phase of an exploration programme is that of synthesis of the data collected, reporting the results and storage of the raw data. The synthesis and reporting elements are natural corollaries to any programme, but the storage of raw data is commonly neglected.

2.3.1 Synthesis of Results

It is necessary that the various stratigraphic and structural problems encountered during a programme are solved since, apart from providing a base to the economic assessment of the project, the predictive value of those solutions cannot be determined until they have been resolved. Regrettably, there are still engineers today who regard these endeavors as "academic"; I heard that statement made 20 years ago and I know a coal mining engineer who has worked on three continents who made the statement in total seriousness that "geologists cause more problems than they solve".

In this context management, often at the operating level, has lacked an understanding of the function of applied geology in guiding mine design and operation. For example, the analysis of roof structures, both sedimentary and structural, to guide exploratory or trial mining operations can be critical to the early stages of development of a mining operation.

2.3.2 Reporting of Results

The reporting of results must have an industrial orientation and provide firm guidelines for management to make decisions. Regrettably however, financial constraints can place the professional geologist in a difficult position. How many geologists have placed themselves in a compromising situation by producing an evaluation on insufficient data?

Certainly the level of activity in Appalachia has produced volumes of "reports" recommending investment in coal properties based on data which has its origin in local folklore and the Keystone Coal Industry Manual. The data reproduced as Table I is from one such report.

Table No. I

Data from Property Report A

"These seams located at elevations 1560 and 1600 average 62 and 66 inches, respectively. Their analyses are similar and are as follows:

Moisture (%)	3.6 - 7.3
Volatile Matter (%)	33.8 - 37.2
Fixed Carbon (%)	51.2 - 57.5
Ash (%)	5.0 - 8.1
Sulfur (%)	0.6 - 1.3
Btu	12,580 - 13,480"

The basis for the analyses is not stated i.e. raw, washed, as received, etc. More importantly, the range of values is applied to two different, though adjacent seams on the property which, when sampled, ranged in ash content up to 23% on a raw coal basis. The seam elevations were based on local folklore.

The reserves of coal on this property were stated to be "18.6 million tons in place", but no reserve categories were given. Subsequent field investigations revealed in place, clean coal reserves of 8.1 million tons and 5.6 million tons in the proven and indicated categories, respectively, an overstatement of some 35%.

Standards exist for defining reserve categories and should be used without exception. However, while it is advantageous to have a regular drilling grid, it must not become the master of the situation. Geological analysis and judgement must always be the final arbiter of a drill site. There is no excuse for drilling a property on a grid pattern only because it was laid out in an office in Calgary or Denver. Nor should the dictates of a computer over-rule informed judgement.

Financial constraints can result in a more widely spaced drilling density than is acceptable under the various standards laid down by the various authorities. In all cases the reserve category of the figures quoted should be established and stated. In a lighter vein, perhaps a new reserve category of "speculative" or "hypothetical" could be created for some of the coal brokers in the eastern U.S.A.

As a second example, the following information was supplied relating to a property in a structurally disturbed region of Tennessee, U.S.A.

Table No. II
Data from Property Report B

Reserve estimate:	27,000,000 tons
Area:	6,000 acres
Drill holes:	6
Seam thickness:	28½ to 42 inches

The written logs which accompanied the report identified the stratigraphic sequences, contained sketchy description of the lithologies and identified the coal seams as "28 inches coal". In spite of the statement in the text that "the cores were studies and re-studied in detail", no seam sections were provided. The reserves were calculated using an "electronic planimeter", apparently implying that this instrument gives more veracity to the figures!

In commenting on this report the following observations are made:

(i) The absence of seam sections or even some attempt at precision makes verification of the reserves difficult, and gives no idea of what ratio of coal and rock is present in the seam.
(ii) The ash content of the seam ranged from 9% to 16%; but its distribution is unknown.
(iii) The reserve figures are not categorized as inferred or indicated - with 6 drill holes in 6,000 acres they must be inferred at best, given

their spacing along one line except for one hole which did not intersect the seam.

(iv) The seam thickness used in the reserve calculations was the maximum recorded interval which, additionally, was not corrected for the angle of dip.

(v) Structural features were virtually ignored, since it was assumed the seam was flat for most of the area; the structure contours were ignored.

(vi) No roof strata data were available even though it was recommended as a potential underground operation at depths of 500 feet to 1200 feet (152 metres - 366 metres).

The report referred to above, produced by a consultant geologist claiming to be versed in coal geology, is unfortunately all too common, even in today's sophisticated coal industry.

2.3.3 Data Storage

The third element in this section is that of data storage. Raw or untreated data, which must be objective in the first instance, is commonly not retained. The storage of core or cuttings is not common and even when detailed logs are produced, the availability of the core or cuttings for later evaluation and checking is important.

The New South Wales Geological Survey in Australia established a core library in the early 1960's and had been retaining core well before that date. In British Columbia, Canada, it has been mandatory to retain core since 1973. Possibly, the coal industry should take a leaf out of the petroleum industry's book regarding the sample retention.

The simple recording of information should be basic, but often is not adhered to. A colleague tells me of one property in Canada where holes have been drilled two and three times in the one locality because of a lack of awareness by each new geologist of the previous work carried out but not documented.

In carrying out an insurance survey of an underground mine in Queensland, Australia I learned that the mine geologist did observe that there had been roof failure but no joint measurements had been made. Nor had the roof falls been documented by the engineers in respect of mining practice, time relationship to stage of mining, water flow etc. No historic data were available in relation to this significant problem.

Data storage immediately conjures up the idea of a computer. Certainly for some information such as analytical results, a computer is fine, but there is great danger of loss of detail in condensing descriptive logging into computer compatible form for storage. It is almost inevitable that some data will be deleted, nuances of meaning lost, or the record shortened unnecessarily. While the bulk of the written record can be significant, microfiche storage is a most suitable medium for long retention of records.

3. CONCLUSIONS

An obvious conclusion to this paper is to ask how the coal industry can minimize the various weaknesses and reduce the wastage of funds which are occurring in exploration programmes being conducted today. It seems there are two avenues which present themselves - education and uniformity of approach.

The subject of education is being addressed by Professor Mathewson at this symposium. Further, the other papers being presented here are positive contributions to the education process, particularly in conjunction with the discussions, both formal and informal.

I have no knowledge of how many universities and colleges teach "coal geology" as a course, but have the feeling that there are relatively very few in North America. Now that the coal industry is attracting more attention, possibly some geology departments will feel there is something to be gained in introducing suitable courses.

In-house training in companies can be achieved by individual organization by way of seminars and visiting lecturers. The use of a "procedures manual" by companies for their staff is another mode by which a greater awareness can be gained. I wonder how many organizations have even considered the necessity of having such a volume available?

The logical extension of a manual is that a greater

uniformity in data collection and treatment will result. The following excerpt from a paper presented by Dick Sanders (2) at the symposium on Coal Borehole Evaluation in Australia a year ago is particularly appropriate and bears repeating.

> "What is needed for the future is a bold new standard on bore core evaluation. This standard should comprise three sections namely drilling, testing/analysis and evaluation of results. If founded on sound principles and agreed to by the majority of the industry, a document such as this would provide the basis for a rational, uniform approach to the problem and hopefully eradicate many of the expensive mistakes of the past".

Such a standard, entitled "Draft Australian Standard Code of Practice for the Evaluation of Hard Coal Recovered from Bore Cores" (3), is almost a reality in that country. A similar document is badly needed in both Canada and the United States, and I would go further to cover the wider field of exploration programmes.

There will be cries of restriction of professional freedom and complaints that it is not possible to cater for all possible contingencies. While the latter is true and there may be no unique solution to one set of results, the absence of any guidelines has only resulted in chaos. A set of codes or practices which have been found effective and realistic by usage should surely be the aim of the industry.

4. ACKNOWLEDGEMENTS

The preparation of this paper calls for sincere thanks to my many colleagues both here and in Australia who have provided thoughts, ideas and examples and to Geoff Jordan who critically reviewed the paper. Also regretful thanks should go to those geologists, even though anonymously, whose work I have used.

REFERENCES

1. Wallis, G. R. and Jordan, G. R., "The Stratigraphy and Structure of the Lower Cretaceous Gething Formation of the Sukunka River Coal Deposit in B.C.", Canadian Mining and Metallurgical Bulletin, March, 1974.

2. Sanders, Dick, "Optimizing a Bore Core Evaluation Test Programme, Symposium on Coal Borehole Evaluation", Australasian Institute of Mining and Metallurgy, October 1977, pp. 36-47.

3. Standard Association of Australia, "Draft Australian Standard Code of Practice for the Evaluation of Hard Coal Recovered from Bore Cores", Standards Association of Australia, DR78126, August, 1978.

DISCUSSION

QUESTION: Could the author offer advice on how core re-drilling clauses can be introduced into countries and areas where local contractors will not bid on tenders containing penalty clauses?

ANSWER: The inclusion, or otherwise, of core re-drilling or minimum recovery clauses in drilling contracts can present problems if they are approached in what might best be described as a "heavy handed" manner. While it might sound pompous, I believe I was the first geologist to introduce such a condition into a drilling contract for coal drilling in western Canada and, where I am managing a project today, am able to obtain such a clause with at least a number of contractors.

The heart of the problem lies in an understanding of the problem by both geologist and driller, and the establishment of a working relationship between client and contractor. These clients who insist on a re-drill regardless of geologic conditions does himself injustice and the contractor damage, in financial terms, as well as the industry as a whole.

It must be understood that such a penalty clause is principally directed toward ensuring that good drilling practice is maintained. However it has to be recognized that geologic constraints can often preclude the obtaining of high core recoveries in the coal seam. If the geologist is unable to recognize such a condition, the case for minimum recovery clauses is lost.

In summary, the potential problems which will be encountered must be fully understood by client and contractor, and the geologist must realize that all core losses cannot be attributable to the driller. A working relationship is the only sure way of ensuring such a clause is fairly applied. Idealistic, yes, but also workable.

The payment directly to the driller, not the contractor, of a bonus for "good" or "high" core recovery has been successfully employed by a number of companies, again with the stipulation of minimum core recovery. This avenue, in some instances is more conducive to increasing core recovery but negates the recourse to requiring a re-drill when bad drilling practice occurs.

QUESTION: Has the author any statistics on how increasing core diameter increases recovery in Western Canada? How much of the HQ and PQ core has had recovery factors greater than 90%? If this is not significant, the increased cost of greater core diameter surely cannot be justified.

ANSWER: Firstly I have no statistics which apply to PQ core. With respect to HQ core versus NQ core I am able to offer some comments below. As a background, however, it is now reasonably well established that core recoveries in the disturbed belt of western Canada generally average between 70% and 75% in the coal seams. While it is not uncommon to achieve higher core recoveries this situation cannot be regarded as the norm.

It must be remembered by non-Canadians, who might regard the acceptance of an average of 75% core recovery as satisfactory, that the drilling season in the Rockies is limited to about 4 months - not 12 months as in some countries, and structural deformation is virtually universal. Core drilling employing Christiansen and VTM barrels has not improved the situation significantly over the "conventional" wireline techniques.

With respect to the differing core diameters, i. e. NQ versus HQ, it is fair to say that recoveries from the two hole diameters are going to average between 70% and 80%. The HQ size might improve the recovery by 5% to 10%, and thus not enough to justify the increased cost of the larger diameter.

From a statistical viewpoint, of 115 coal seam intersections of NQ diameter 45% exceed 80% recovery and 14% exceeded 90% recovery on a particular property. In a structurally equivalent area, of 75 HQ seam intersections 53% exceed 80% recovery and 18% exceeded 90% recovery.

However, in defense of the larger diameter, if drilling or geologic problems are going to be encountered, the chance of success with HQ core is increased. To this must be added the greater mass of coal, and roof and floor rock, recovered for washability testing and engineering studies. This aspect is outside the realm of this discussion and subject to many personal opinions.

United States: Concepts and Practices

7

Coal Exploration Concepts and Practices in the Western United States

By Gerald E. Vaninetti
Director of Exploration
Department of Mining and Exploration
Utah Power and Light Company
Salt Lake City, Utah, United States

INTRODUCTION

Prior to the second world war, steam coal was used extensively as an energy source to fuel electric-generating power plants, homes, and steam locomotives. The coal fields of the eastern U.S.A. supplied the bulk of the steam coal although minor contributions were made from the coal fields of western U.S.A. However, the post-war expansion in the United States was primarily fueled by oil and gas derivatives because of greater ease in handling, cleaner burning, better availability, and lower cost advantages over steam coal. Accordingly, U.S.A. steam coal production has steadily decreased since the war.

However, the energy crises of the 1970's, when energy users became aware that oil and gas supplies were limited and available at radically escalated prices, has resulted in the revitalization of the coal industry of the U.S.A. The effect has been to expand the coal production of the eastern U.S.A. to old pre-war levels and unprecedented growth in the coal fields of western U.S.A.

The rapid expansion in exploration activities in the coal fields of the western U.S.A. in the states of Montana, Wyoming, Utah, Colorado, Arizona, and New Mexico has had a profound impact upon the exploration concepts and practices utilized in the Western coal fields. The intent of this report is to define the evolution of exploration concepts in terms of accumulation and interpretation of data and present case examples of the implementation of these concepts.

EXPLORATION CONCEPTS

Exploration concepts utilized in Western coal fields are based on a combination of three elements: (1) uses of data, (2) methods of gathering data, and (3) interpretations of data. Individuals conducting exploration efforts for coal in Western coal fields must be aware of the importance of and interplay between each of these data-oriented elements inasmuch as different goals for obtaining data profoundly impact upon the methods used to gather different types of data which ultimately reflect on the interpretation of these data.

Use of Data

Before gathering data, an investigator must first define the nature of the use of data to be collected. The three major end-member goals for gathering exploration data are academic knowledge, economic evaluations, and development of mining plans. Each goal defines the time frame in which an exploration program is to be conducted, the expenditures, the degree of accuracy desired, the methods used to obtain data, and the types of data obtained.

Academic investigations are usually time independent, relatively inexpensive, result in a moderate degree of accuracy, and are characterized by an emphasis on the interpretation of depositional environments, stratigraphy, and correlation of beds from the examination of outcrops.

Coal exploration programs geared to obtain information for economic evaluations are the most common type of investigation. Time constraints, moderate expenditures for drilling programs, a moderate degree of accuracy, and an emphasis on geometric and sample data are characteristic of this type of program.

Exploration programs to define the limits of a coal deposit for mine planning are characterized by extensive drilling programs on grid spacing patterns, significant expenditures, a high degree of accuracy, and an emphasis on coal seam correlations, lateral and vertical geometric variations, and quality.

Methods of Collecting Data

The methods used to collect data for coal exploration programs are phased and proceed, in order, from the collection of generalized regional information to the collection of specific detailed data. Data are obtained by (1) reviewing the information compiled by previous workers, (2) regional reconnaissance mapping, (3) detailed mapping and data verification, (4) drilling and down-hole geophysics, and (5) collection of samples for quality determination.

The phases of investigation noted above should be implemented in the order in which they are listed above. Studies for academic purposes usually proceed through the detailed mapping and data verification stages where outcrop exposures provide the bulk of exploration data. Studies conducted by industry personnel usually proceed through all five stages because of the need for three-dimensional data obtainable only from drill holes and the need for detailed fuel quality information.

Basic geologic information obtained in coal exploration programs should consist of data concerning the general geology, stratigraphy, coal seam correlations, lateral continuity of coal seams and enclosing units, geometry and quality of coal seams, and faults and folds of the deposit. Other topics such as depositional environments, sedimentation, petrography, coal reserves, and mineability should also be investigated. In most instances, particularly those concerning economic evaluations, the variations in thickness of coal seams and the intervals between and overlying the seams, the lithology and strength of the sediments that immediately overlie mineable coal seams, the rock splits or partings within coal seams, and the quality of coal seams are of particular importance.

Previous Workers. Geologic investigations concerning coal-bearing sediments of the western U.S.A. were conducted in three major phases. The first phase began about 1900 and was followed by the second phase which began about 1935. The most recent phase began about 1970 and is still in progress.

The first phase of geologic work was conducted primarily

by the United States Geological Survey (U.S.G.S.) for the purpose of locating and classifying valuable federal coal lands prior to leasing. The stratigraphy and general economic geology of the coal-bearing sediments of the western U.S.A. were defined during this phase. Early workers defined the location and boundaries of most of the coal fields of the West (Figure 1) and collected stratigraphic and paleontologic data that indicated that the coal deposits of the West were deposited along the western margins of a shallow seaway that covered much of the central part of the North American continent during the Cretaceous Period and into the Paleocene. A list of selected coal fields and geologists' reports concerning them are shown in Table I. Some of the major workers in the first phase of investigation were Spieker, Reeside, Lee, Veatch, Lupton, and Dobbin.

The second phase of investigation was more concerned with the oil and gas potential of coal-bearing sediments than the coal deposits but contributions made to the stratigraphy of Cretaceous sediments have been invaluable to the study of coal in the region. The second phase was implemented by petroleum geologists and geologists from state, federal (U.S.G.S.), and academic organizations. This "middle period" of investigation built upon the framework established by the first phase of workers and was concentrated on detailed stratigraphic correlations, early work on the interpretations of depositional environments, and paleontology. Regional and basinal studies resulted in a broader understanding of the tectonics (85), timing of the transgressions and regressions that characterize the gradual in-filling of the Cretaceous Seaway (86), paleoecology (87), and regional correlations, time equivalence, and paleontology of Cretaceous sediments (88).

The geologists of the "middle period" organized state and regional geologic societies for the purpose of exchanging geologic information and preparing guidebooks. Some of the guidebooks prepared by geologic societies in the Rocky Mountain states that are particularly useful in understanding the Cretaceous System in various regions are listed in Table II.

The U.S.G.S. collated existing data concerning the coal fields of the states of Colorado (118), Montana (119, 120), Wyoming (121, 122), New Mexico (123), and of the U.S.A. (124).

Figure 1. Major bituminous and sub-bituminous coal fields of the western U.S.A.

Table I
Selected Investigators of the Bituminous and Sub-bituminous Coal Fields of the Western U.S.A. during the Early Phase of Geologicl Mapping

State	Coal Field	Investigator
Utah	Wasatch Plateau	Spieker and Baker (1); Spieker (2,3,4,5,)
	Book Cliffs	Clark (6); Fisher (7); Richardson (8,9)
	Henry Mountains	Hunt and others (10)
	Emery	Lupton (11)
	Kaiparowits	Gregory and Moore (12)
Colorado	Book Cliffs	Erdmann (13); Richardson (8,9)
	Grand Mesa-Crested Butte	Lee (14,15)
	Grand Hogback-Danforth Hills	Gale (16,17); Hancock (18,19)
	Yampa	Fennemann (20); Campbell (21); Hancock (18); Bass and others (22)
	Durango	Schrader (23); Shaler (24); Taff (25)
	Trinidad	Richardson (26); Lee (27)
	Denver Basin	Martin (28); Goldman (29)
	North and South Park	Berkley (30); Washburne (31)
Wyoming	Hams Fork	Veach (32,33,34); Schultz (35,36,37)
	Rock Springs	Schultz (38,39)
	Great Divide Basin	Ball (40-41)
	Hanna-Rock Creek	Dobbin and others (42,43)
	Wind River Basin	Woodruff (44)
	Bighorn Basin	Fisher (45); Washburne (46); Woodruff (47)
	Powder River Basin	Gale and Wegemann (48); Wegemann (49,50); Dobbin and Barnett (51)
New Mexico	San Juan Basin	Schroder (52); Shaler and Gardner (53); Bauer and Reeside (54); Dane (55)
	Gallup-Zuni	Gardner (56); Sears (57); Hunt (58)
	Raton Basin	Lee (59,60)
Arizona	Black Mesa	Campbell and Gregory (61)
Montana	Blackfeet-Valier	Bowen (62,63,64,65); Collier (66,67); Baver (68,69); Beekley (70)
	Bull Mountain	Woolsley (71,72); Richards (73); Lupton (74)
	Great Falls	Barnett (75); Fisher (76)
	Lewistown	Calvert (77,78,79)
	Powder River Basin	Baker (80); Bass (81)
	South Central	Calvert (82,83,84)

Table II
Selected Guidebooks of the Geological Societies of the Western States Concerning Cretaceous Sediments

Geologic Societies	Year	Location or Topic of Guidebook
New Mexico	1950	San Juan Basin: North and East Sides (89)
	1951	San Juan Basin: South and West Sides (90)
	1958	Black Mesa Basin, Arizona (91)
	1977	San Juan Basin (92)
Colorado	1959	Cretaceous Symposium (93)
	1976	Geology of Rocky Mountain Coal (RS-1) (94)
	1977	Geology of Rocky Mountain Coal (RS-4) (95)
Wyoming	1958	Powder River Basin (96)
	1960	Overthrust Belt: Southwestern Wyoming (97
	1961	Late Cretaceous in Wyoming (98)
	1962	Early Cretaceous in Wyoming (99)
	1963	Northern Powder River Basin (100)
	1965	Rock Springs Area (101)
	1973	Green River Basin (102)
Utah	1946	Henry Mountains (103)
	1949	Colorado Plateaus - Great Basin (104)
	1965	South Central Utah (105)
	1966	Central Utah Coal Fields (106)
	1972	Plateau - Basin and Range (107)
IAPG*	1954	High Plateaus and Canyonlands (108)
	1955	Northwestern Colorado (109)
	1963	Southwestern Utah (110)
Montana	1972	Crazy Mountain Basin (111)
	1975	Energy Resources of Montana (112)
	1977	Thrust Belt of Western Wyoming (113)
Four Corners	1955	Paradox, Black Mesa, and San Juan Basins (114)
	1957	Southwestern San Juan Basin (115)
	1973	Cretaceous and Teriary of Southern Colorado Plateau (116)
R.M.A.G.**	1972	Geologic Atlas of the Rocky Mountain Region (117)

* Intermountain Association of Petroleum Geologists
** Rocky Mountain Association of Geologists

Some of the major workers in the "middle period" of Cretaceous investigations were Young, Spieker, Stokes, Hale, Gill, Cobban, Reeside, Dane, Hunt, Weimer, and Haun.

The most recent phase of investigation documents a resurgence in interest in coal and the sediments in which they are enclosed as well as a concentration on the interpretation of the environments in which these sediments were deposited. The current phase appears to show a movement toward a better understanding and application of various sub-disciplines within geology as well as outside of geology. This includes a better understanding of the climate, depositional environments, paleoecology, paleocurrents, depositional processes, etc., that have contributed to the geological features that characterize coal-bearing sediments.

Studies of modern active depositional sites in the Mississippi Delta region (125, 126, 127), The Texas Gulf Coast (130), the Atlantic Seaboard of the U.S.A. (131, 132), the Pacific Coast of Mexico (133, 134), the Pacific Coast of the U.S.A. (135), and the Nile Delta of Africa (136, 137), have contributed significantly to the understanding of depositional processes and sites where coal-bearing sediments accumulate. Publications by others (138, 139) have defined the nature and processes active in modern coal swamps.

Geologic workers of the third phase are generally adept at interpreting the environments in which Cretaceous coal-bearing sediments of the Rocky Mountain states were deposited (140-149). Studies by these workers are representative of the detail that can be gleaned from sediments utilizing the multi-disciplinary approach of investigation.

Studies by McGowan (150) and Kaiser (151) indicate that major coal seams of economic interest accumulate in swamps in deltaic and interdeltaic-strandplain settings (Table III). Minor thin coal seams of lesser quality are deposited in lacustrine and lagoonal swamps.

Coal seams that have accumulated in deltaic and interdeltaic environments are characterized by wide lateral continuity (several 1000 meters), mineable thickness (two meters), and high purity. Lacustrine and lagoonal coal beds are thinner, less pure, and lack the lateral continuity of deltaic and interdeltaic-strandplain coal seams. The characteristics that help differentiate the environment in which a given coal seam is deposited are summarized in Table III

Table III
Characteristics of Coals Deposited in Different Depositional Environments

Parameter	Deltaic	Interdeltaic-Strandplain	Lagoonal	Lacustrine
Geometry	Tabular; Laterally Discontinuous	Tabular; Laterally Continuous	Discontinuous Pods	Elongate; Discontinuous
Thickness	Three Meters +	Three Meters +	Less Than One Meter	Less Than Three Meters
Frequency	Two to Four Beds	Several Beds (5)	Numerous	Few
Quality	High	High	Low	Intermediate
Ash	Low to Intermediate	Low to Intermed-iate	High	Intermediate
BTU	High	High	Low	Intermediate
Plants	Grassy-Shrub	Grassy-Shrub	Salt Marsh	Woody
Enclosing Sediment	Sand-Rich	Mud-Rich	Intermediate	Intermediate

and are discussed in a later section of this report concerning interpretations.

Regional Reconnaissance. After reviewing the geologic literature to define the level of understanding that previous workers have developed, a field reconnaissance program should be implemented in order to confirm and understand their interpretations. This phase of investigation consists of measuring stratigraphic sections in various widely-spaced locations within the region of investigation. Traverses on easily accessible roads and trails in outcrop areas are also helpful in gaining a generalized understanding of the coal-bearing sediments.

Detailed Mapping and Data Verification. Information gained in regional reconnaissance mapping activities and by previous workers should be used as a framework to guide further investigations. Stratigraphic sections and drill hole data obtained by others should be reevaluated to determine accuracy and check others' interpretations.

Coal-bearing outcrops and enclosing strata should be mapped in detail. Stratigraphic sections should be measured at regular intervals and where access permits. Individual units of concern such as coal beds, marker beds, and other persistent horizons should be traced laterally to ascertain the equivalence of units and the correlation of coal seams. The environments in which the coal-bearing sediments were deposited should be interpreted during section-measuring and correlating activities.

Drill hole data obtained by previous workers should be examined to determine its reliability; promoters of coal properties are known to interpret raw drill hole data to their advantage. If at all possible, raw data such as cores and geophysical logs should be examined. Cutting logs by drillers and geologists alike commonly yield erroneous information. If raw data are not available for inspection, interpretations must be suspect.

Drilling and Down-Hole Geophysics. Exploration programs sponsored by industry for the purpose of obtaining data for economic decisions and/or mine development usually proceed beyond the mapping stage to drilling programs. The number and spacing of drill holes planned to aid decision making are determined by the goals of the exploration programs, the density and reliability of existing data, drill hole depths, geology, inferred depositional environments, and of course, budget.

Regardless of the parameters involved in the number and spacing of drill holes, the drilling program should be geared to obtain the maximum amount of raw data. If coal quality or rock mechanics are of major concern, cores should be taken through the coal-bearing portion of drill holes. If the thickness of coal seams, rock splits and partings in coal seams, and intervals between seams and the lithologies of enclosing sediments are of concern, down-hole geophysical data should be obtained.

All drill holes, regardless of the types of data desired from them, should be geophysically probed. A basic suite of geophysical logs consisting of nuclear (natural gamma), density, and resistance logs should be obtained for all drill holes. Caliper, neutron, spontaneous potential, and seismic logs also find application but are supplemental to the basic suite of logs noted above. Papers by Reeves (152, 153), Seimers (154), Archer and Warrick (155), and Bond and others (156) summarize the usefulness and interpretation of down-hole geophysical logs in coal exploration programs. Recent work by Fisher (129), Brown (157), and Seimers (154) indicate that environments of deposition of coal-bearing sediments can be interpreted from geophysical logs.

The three most useful geophysical logs used in exploration of coal-bearing sediments are the natural gamma (radioactivity), density, and resistance logs. The natural gamma log records total radioactive gamma rays for a given interval of strata and is useful in interpretation of lithology, grain size, and unit thickness. The density log is a modified natural gamma log that is useful in determining the quality and thickness of coal seams and the presence of limestones, dense iron-rich concretion zones, and caves in drill holes. The resistance log measures the electrical conductivity of sediments and is useful in determining lithologies, grain size trends, porosity, unit thicknesses, and coal purity.

A set of geophysical curves for sediments commonly encountered in coal-bearing strata is illustrated in Figure 2. As shown, the total natural gamma response of coal seams is usually very low due to the low content of radioactive material present in purely organic materials. Sandstone beds usually have only minute quantitites of radioactive materials.

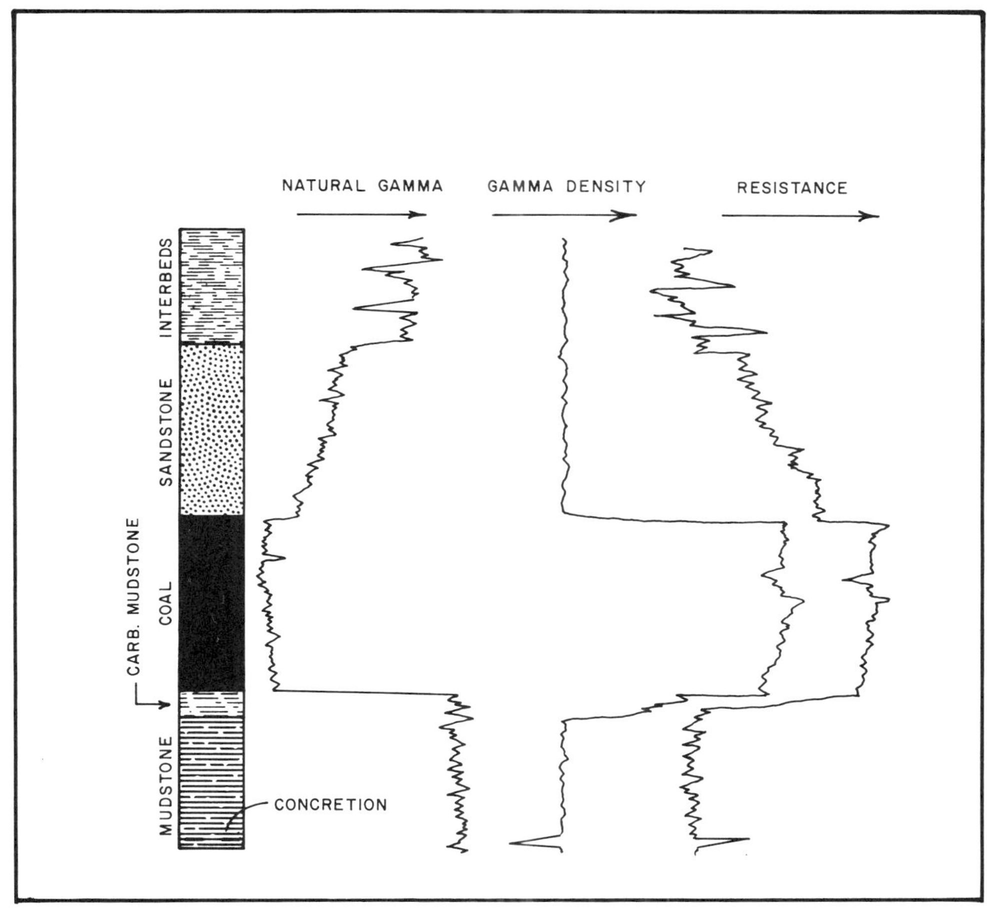

Figure 2. Typical Geophysical Responses for Natural Gamma, Density, and Resistance Logs run through Coal-Bearing Sediments.

mudstone beds have somewhat more. Interbed zones on natural gamma logs show a variability of gamma emitting materials. A gradual upwards increase in intensity of gamma particles emitted from a sandstone corresponds to a fining upwards change in grain sizes typical of fluvial and distributary sandstone beds.

The density log is a record of the capacity of a given rock type to retain induced gamma particles. The capacity to retain these particles is a function of the density of the earth material that is bombarded with gamma particles. Coal (1.4 specific gravity) and other low-density materials do not retain gamma particles and therefore permit gamma rays to return to down hole sensing devices (scintillometer). Higher density materials (2.5 specific gravity) such as sandstone, mudstone, and interbeds retain induced gamma particles. Very high density rocks (2.7 specific gravity) such as limestone and iron-rich concretions retain induced gamma particles more effectively than do sandstone, mudstone, and interbeds.

Resistance logs record the capacity of a given rock type to transmit induced electrons. Water-bearing sandstone and coal beds do not allow significant electron flow and yield high resistance curves (Figure 2). Variations in the water content of sandstone beds are sometimes a function of grain size and sorting, therefore, an upwards decrease in resistance can be interpreted as a decrease in water content and grain size. Mudstone beds are composed, in part, of clay particles and flakes deposited from suspension and allow a more efficient transmission of electrons than do sandstone and coal beds.

Coal Quality. Inasmuch as coal is used primarily as a fuel for power plant boilers, the quality of coal seams must be evaluated to determine optimum boiler design or amenability for usage in existing boilers. For purposes of determining coal quality, coal samples can be taken from outcrops, mine workings, or exploration core holes. The number, weight, distribution, and location of samples determines the reliability of coal analytical data.

Coal samples can be analyzed for ash, moisture, fixed carbon, volatile water, and sulfur content, heating value (BTU), mineral analysis of ash, ash fusion temperatures, ultimate analyses, surfur forms, grindability, and other coal quality parameters. The ash, sulfur, moisture, and BTU content of coal is of primary importance. The ash fusion temperatures and sodium content of coal ash materials are also of importance in determination of boiler performance.

The ash content determines the nature of ash handling equipment and boiler sizing. The sodium oxide content (Na_2O) of the ash, as well as other mineral analysis of ash data, determine electrostatic precipitator and boiler design and performance. Sodium oxide contents below 1.5% of the ash contribute to poor precipitation performance. Coal with greater than 3.0% sodium in ash content is likely to allow the formation of eutectic compounds which cause significant slagging problems in power plant boilers (158). Low ash fusion temperatures are commonly caused by high sodium content ash which results in slagging in boilers. Ash fusion temperatures (initial deformation) below 2300°F are problematic. The range between initial deformation and fluidization of coal ash also affects boiler performance. Close ranges (less than 150°F) are problematic and wide ranges are desirable (greater than 250°F).

The heating value of coal determines boiler design and sizing. Coal seams with high BTU's are preferred over those of lower BTU's. The sulfur content of coal is usually compared to the BTU content for purposes of determining the maximum allowable sulfur content of coal burned without power plant stack-gas removal equipment. Present federal guidelines do not permit more than 1.2 pounds of SO_2 per million BTU's for new power plants. Several states have imposed even more restrictive SO_2 emission regulations.

Other quality parameters such as mositure content, ultimate analysis, and grindability pertain to coal handling equipment, pulverizer sizing, air quality, and boiler design. Sulfur forms are determined for high sulfur coals in cases where washing may remove some of the sulfur. In most cases, the pyritic percentage of Western coals is too low to justify washing.

Interpretation of Data

The final goal of a coal exploration program is to interpret the data that have been assembled during investigations. Data can be interpreted to satisfy various goals depending on the nature of the investigation. Interpretations concerning the (1) general geology and correlations, (2) depositional environments, (3) geometry of coal seams, and (4) coal quality for a given tract of land are common.

General Geology and Correlations. Regardless of the goal of an exploration program, the general geology of a given coal deposit should be defined first. The structure, stratigraphy, and correlations of sedimentary units are of primary importance. A complete description of the sedimentary rocks which enclose the coal seams as well as the coal seams themselves is necessary. Bedding features, structural characteristics (cleats, joints, folds, and faults), rock splits in coal seams, grain size variations, lateral continuities and thickness variations of units, rock types, and paleocurrent analyses are among the important variables that should be noted.

General geologic data can be used to predict rock types that enclose coal seams, the location of rock splits in coal seams, the competence of sediments that overlie coal seams, overburden thicknesses, the size of fault blocks, and the hydrology of the coal-bearing sediments. These interpretations are of use in determining the most feasible mining method, the nature and locations of potential mining problems, areas that are well or poorly suited for future development, and in general, the value of the coal deposit.

Depositional Environments. An ability to recognize sediments deposited in deltaic, interdeltaic-strandplain and other environments within a given coal deposit can be used as a predictive tool in determining the value of the deposit.

Rock splits in coal seams that are present in drill hole data or outcrops are best analyzed in terms of depositional environments. Although some mining personnel would choose to believe that rock splits occur as pods completely surrounded and enclosed by coal, they are more commonly found in belts where dimensions can be predicted on the basis of a knowledge of depositional environments. Rock splits are usually formed by a rapid influx of clastic, riverborne sediment into coal swamps and originate from a definite source (crevasse splay, washover fan, etc.) (159). The trend, location, and lateral extent of many rock splits in coal can therefore be predicted based on a knowledge of depositional environments (160).

Rock splits that occur as pods completely enclosed by coal and seemingly without a source were probably deposited as ash falls from nearby volcanic eruptions. The alteration products of volcanic ash materials are sometimes diagnostic of ash fall origins.

Examples to be discussed later in this report indicate that the continuity and variations in thickness of coal seams are dependent to a large degree on the location of depositional environments previous to the growth of coal swamps (161). The effects of differential compactional subsidence play an important role on the location and development of coal swamps and other depositional environments.

The interpretation of major depositional environments

such as deltaic distributary channels and fluvial paleochannels within a coal deposit may indicate that significant portions of the deposit may have been eroded or never deposited. Examples to be discussed later in this report indicate the value of an awareness of the depositional settings in evaluating the potential coal reserves of a deposit.

The location of fluvial paleochannels in the sediments that immediately overlie coal seams is of interest to mining engineers in predicting potential mining problems. The location and quantity of water that is commonly found as perched water tables in fluvial sandstone beds in the roofs of coal mines can sometimes be predicted on the basis of an interpretation of depositional environments. Roof falls and other roof support problems can sometimes be related to lithology, slickensides, and abrupt lithologic changes in fluvial sandstone beds (162, 163).

Geometry. One of the more important interpretations to be developed in a coal exploration program concerns the geometry of the coal deposit. Economic evaluations of coal deposits invariably are highly dependent on the reliability and quantity of reserves that can be estimated. Other geometrical considerations have to do with the thickness of the overburden, interburden between mineable coal seams, and variations in the coal seams.

Raw data are usually converted to isopachous maps of coal seam, interburden, and overburden thicknesses. These maps are constructed from data gathered from drilling programs and mapping activities and are modified on the basis of depositional interpretations. The maps are planimetered to determine the average thickness and total reserves of a coal deposit.

Workers in the coal industry commonly utilize the three-part division of reserves as defined by the United States Geological Survey (164). This system is based on the relative reliability of reserve estimates and categories are referred to as (1) measured, (2) indicated, and (3) inferred in order of decreasing reliability. Coal exploration drilling programs designed to gain information for purchase of coal lands are usually geared to obtain inferred to indicated reserve estimates. Exploration drilling programs to facilitate mining plans consist of closely spaced drill holes that yield measured reserves. Outcrop mapping programs usually result in estimates of resources and inferred reserves.

Coal Quality. As noted earlier in this report, the quality of coal is a major factor in determining the value of a coal deposit inasmuch as boiler design and utilization in existing boilers is highly dependent on fuel quality. Desirable fuel quality characteristics are listed in Table IV.

An analysis of coal quality is not complete until the effects of dilution caused by mining, transporting, and stockpiling activities are assessed. In underground mining situations, the ash content of coal seams is commonly increased by as much as 2.5% as a result of normal mining practices. Underground coal mining practices commonly result in the inclusion of roof rock, roof falls, bottom rock, overcast rock, rock dust ($CaCO_3$), and rock splits as well as continuous miner bits, roof bolts, brattice cloth, cable, beverage cans, mine timbers, etc., in the run-of-mine coal.

The transportation of coal also results in contamination and degradation (165). Materials in the bottoms of vehicles that haul coal commonly become intermingled with the coal.

Coal oxidizes and is further contaminated in coal stockpiles. Oxidation decreases BTU content and increases moisture content in stockpiled coal. The movement of coal in constructing and depleting stockpiles results in a size sorting of coal which commonly concentrates impurities at certain levels in stockpiles. The foundation materials upon which stockpiles are constructed commonly become mixed with coal as the stockpiles are depleted.

The net effect of mining, transporting, and stockpiling coal is to increase the ash and moisture contents and ash fusion temperatures and to decrease the BTU's, sodium, and calcium content. The changes in quality that coal undergoes is exemplified by the data from a group of underground mines in Utah shown in Table V.

EXPLORATION PRACTICES

The concepts for conducting coal exploration programs discussed above have been used to advantage in several instances throughout the coal fields of the Western U.S.A. The two major depositional models in which coal seams accumulate, deltaic and interdeltaic-strandplain swamps, are represented in the two examples from Utah that are to be discussed here. The goals of the exploration program in both cases, the

Table IV

Applicability of Various Quality Characteristics to Boiler Design and Operation*

Parameter	Desirable	Undesirable
Proximate		
Moisture	Low (<7%)	High (>12%)
Ash	Low (<10%)	High (>15%)
BTU's	High (>10,000)	Low (<10,000)
SO_2/MMBTU	Low (<1.2 pounds)	High (>1.2 pounds)
Ash Fusion Temperatures (reducing F°)	High	Low
Initial Deformation (I.D.)	(>2300°)	(<2300°)
Fluid (F.)	(>2500°)	(<2500°)
Range Between I.D. and F.	Great (>250°)	(<150°)
Mineral Analysis of Ash		
Sodium Oxide (Na_2O)	Moderate (1.5-3.0)	High-Low (>3.0;<1.5)
SiO_2/Al_2O_3	Low	High
Base/Acid	Moderate (0.25-0.80)	High-Low (>0.80; <0.25)
Fe_2O_3/CaO	Moderate (0.31-3.0)	High-Low (>3.0; <0.31)
Grindability	High (>55)	Low (<47)

*Data in part, from Hensel and Skowyra (158)

Table V
Mean Averages and Changes in Fuel Quality in Various Locations within a Coal Utilization System for a Two Month Period

	In-Mine	Change (+ or -)	Mine Portal	Change (+ or -)	Power Plants	Total Change
Long Proximate						
No. of Samples	34		167		234	
Moisture (%)	5.33	+1.98	7.31	+0.35	7.66	+ 2.33
Ash (%)	7.33	+2.26	9.59	+2.34	11.93	+ 4.60
BTU/Lb.	12,785	- 891	11,894	- 310	11,584	-1,201
Sulfur (%)	0.54	+0.12	0.66	-0.08	0.58	+ 0.04
Ash Fusion Temp. (°F)						
No. of Samples	33		115		38	
Initial Def.	2,119	+ 93	2,212	+ 41	2,253	+ 134
Softening	2,215	+ 38	2,253	+ 65	2,318	+ 103
Hemis.	2,275	+ 31	2,306	+ 96	2,402	+ 127
Fluid	2,402	- 46	2,356	+ 112	2,468	+ 66
Mineral Analysis of Ash						
No. of Samples	25		116		34	
SiO_2	49.61	+4.24	53.85	+3.35	57.20	+ 7.59
Al_2O_3	20.45	-2.01	18.44	+1.68	20.12	- 0.33
Fe_2O_3	5.23	0.00	5.23	-1.25	3.98	- 1.25
CaO	9.08	+0.26	9.34	-2.84	6.50	- 2.58
MgO	0.97	+0.60	1.57	-0.15	1.42	0.45
Na_2O	6.74	-3.64	3.10	-0.34	2.76	- 3.98
K_2O	0.39	+0.74	1.13	+0.20	1.33	+ 0.94
TiO_2	0.25	+0.74	0.99	-0.06	0.93	+ 0.68
P_2O_5	0.14	-0.03	0.11	-0.01	0.10	- 0.04
SO_3	5.58	-0.35	5.23	-1.27	3.96	- 1.62
Si/Al	2.63	+0.42	3.05	-0.32	2.73	+ 0.10
Fe/Ca	0.59	+0.18	0.77	-0.18	0.59	0.00
Base/Acid	0.32	-0.04	0.28	-0.08	0.20	- 0.12

types of data obtained and methods used, as well as the interpretations of data will be discussed.

The first example is for the exploration of deltaic coal-bearing sediments in the Wasatch Plateau of central Utah (Figure 1). The example is indicative of the nature of exploration practices used in the Book Cliffs Coal Field of east-central Utah as well as the coal fields of Southwestern Wyoming.

The second example is for the interdeltaic strandplain coal-bearing sediments of the Kaiparowits Plateau region of south-central Utah (Figure 1). This example is applicable to understanding the exploration practices in the Black Mesa Coal Field of northeastern Arizona and the San Juan Basin Coal Field of northwestern New Mexico.

Example #1: Wasatch Plateau, Utah

The first example to be examined here is a coal deposit of about 100 square kilometers in size located in the central part of the Wasatch Plateau Coal Field. The deposit is bounded on its eastern and southern sides by erosional escarpments along which coal seams outcrop (Figure 3). The northern and western margins are defined by ownership boundaries.

The deposit to be examined here has been evaluated in terms of reserve potential so that an economic valuation of the deposit could be made. In order to show the procedure utilized to determine this valuation, the general geology, previous data, and present program will be summarized. The interpretation of these data will be used to evaluate the reserve potential of the deposit.

General Geology. The general geology and coal stratigraphy of the coal field were first mapped in detail by Spieker (2). His classic work defined the stratigraphic nomenclature that is, with minor modification, used throughout the region today. Stratigraphic units exposed along the margins of the coal deposit consist, in ascending order, of the Mancos Shale, Star Point Sandstone, Blackhawk Formation, Castlegate Sandstone, and Price River and North Horn Formations. The two major coal seams in the area, the Hiawatha (basal) and Blind Canyon (upper) Seams, are exposed at the base of the Blackhawk Formation and were extensively mapped by Spieker.

Although more recent studies add significantly to the understanding of the structure (166), regional correlations (167), geologic and depositional history (3, 4), depositional environments and coal seam correlations (168-173), and general coal geology (174, 175) of the region, none were specifically conducted on the subject deposit.

Previous Data. Data assembled concerning the deposit prior to the most recent exploration program were concentrated in outcrop areas and were available from Spieker (2), from the few drill holes completed by previous owners and the U.S.G.S., and from the mines developed within the deposit (Figure 3A). These meager data indicated that, for the most part, the two major seams were of mineable thickness over a majority of the deposit.

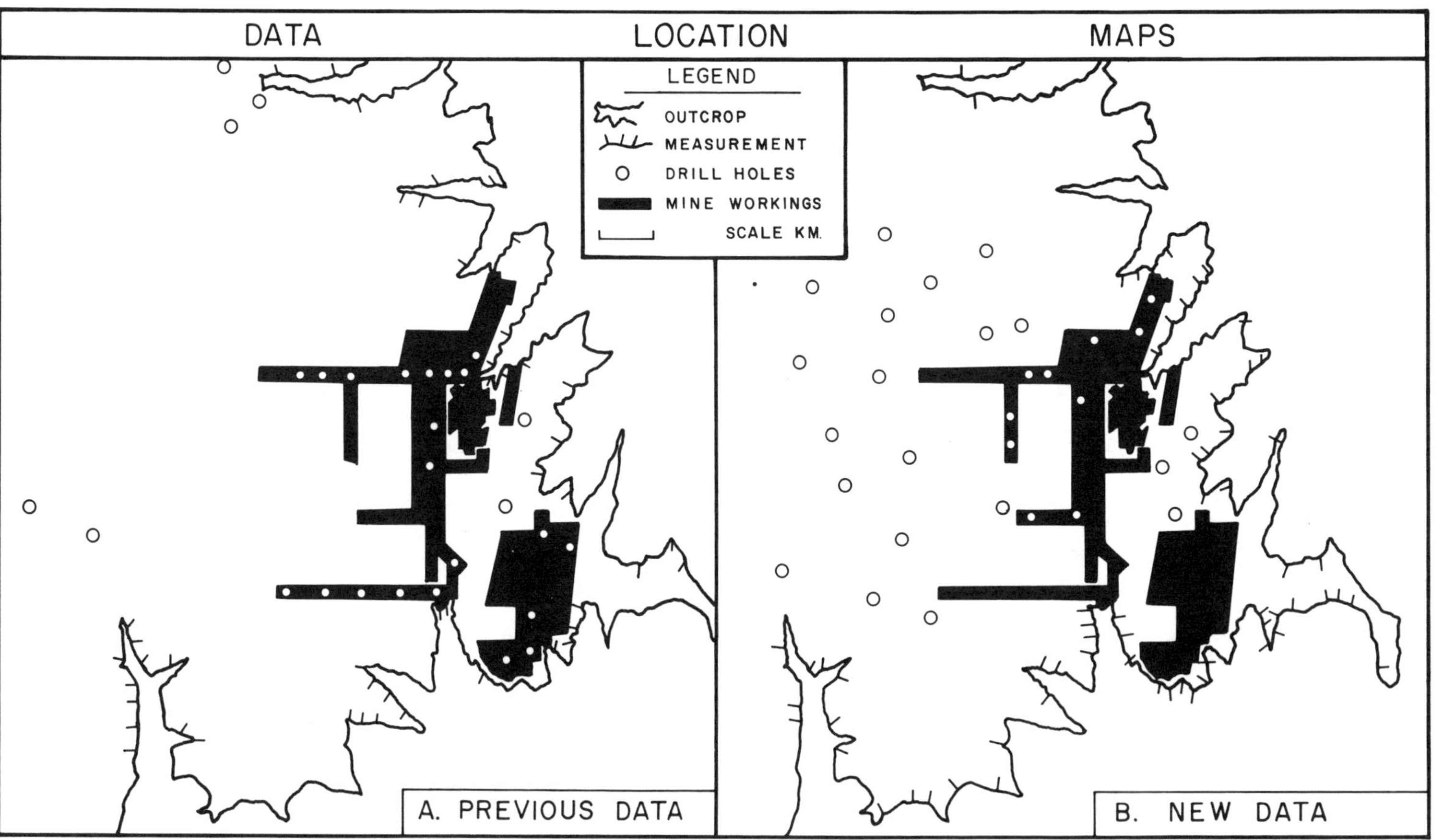

Figure 3. Data location maps of previous and present data for the Wasatch Plateau coal deposit example.

Available data indicated that there were areas in which each of the two seams were poorly developed. Outcrops and drill hole information in the central part of the area indicated that the Hiawatha Seam thins to less than one meter in thickness in that area. Poorly exposed outcrops in the southern part of the deposit indicated that the Blind Canyon Seam thins drastically in that area.

Present Program. The exploration program implemented on the deposit was geared to determine the presence and thickness of mineable coal seams in untested portions of the deposit and to verify previously assembled data. The program consisted of detailed outcrop mapping and a phased drilling and sampling program.

The outcrop mapping program consisted of verifying and supplementing coal measurement and correlation data originally collected by Spieker as well as an analysis of depositional environments. Original coal measurements obtained by Spieker were verified but his correlations of the Blind Canyon Seam in the southern part of the deposit, were not. Coal seams originally identified as the Upper and Lower Bear Canyon Seams were subsequently correlated with the Blind Canyon Seam. This correlation has been substantiated by the development of mine workings within the Blind Canyon Seam from the middle to southern part of the deposit.

Analysis of the Star Point Sandstone and the basal portion of the Blackhawk Formation indicated the presence of an anomalously bedded, lens-shaped sandstone unit at the contact of the two formations in the east-central part of the deposit. Depositional analysis indicates that the unit is a distributary channel sandstone that was deposited contemporaneous with the Hiawatha Seam. The characteristics that differentiate the unit from the deltaic beach sandstone of which the Star Point Sandstone is usually composed are listed in Table VI. Available data indicate a strong eastward component of flow for the anomalous sandstone unit; a direction that is perpendicular to regional shoreline trends.

Additional outcrop mapping and examination of mine workings indicated the presence of a northwest trending rock split within the Blind Canyon Seam in the southwestern part of the deposit. Depositional analysis indicates that the rock split was deposited in the upper reaches of a meandering distributary channel that temporarily traversed the Blind Canyon Coal Swamp during a short time in the

Table VI
Characteristics that Differentiate Beach from Distributary Channel Sandstone

Characteristic	Beach Sandstone	Distributary Sandstone
Bedding	Horizontal	Scoured; lensoid; convolute
Stratification	Horizontal in upper two meters	Trough cross-stratification; minor micro cross-stratification
Bioturbation	Common; *Ophiomorpha*	Rare
Petrography	Clean; well-sorted	Dirty; poorly-sorted
Grain Size Trends	Coarsens upward	Homogeneous to fines upward
Continuity of Beds	Very continuous	Discontinuous
Paleocurrents	Longshore transport	Perpendicular to shoreline
Carbonaceous Materials	Rare	Common

development of the Blind Canyon Seam. The channel may have originated as a crevasse splay deposit from a larger distributary channel located in the south of the deposit.

Other data assembled during outcrop mapping activities consisted of the measurement of the thickness of the interval between the major seams, structural characteristics of the sediments that enclose the seams, cleat orientations in coal, joint orientations in sandstone, detailed lithologic characteristics for eight meters above and two meters below each major seam, rock splits within coal seams, paleocurrent azimuths of cross-bedding in associated sandstone, and interpretations of depositional environments.

Exploratory drilling activities were conducted in two phases. The first phase consisted of two holes that were drilled in the center of the deposit where no previous data were available (Figure 3B). The second phase, the implementation of which was dependent on the results of the first phase, consisted of eighteen holes that were drilled to ascertain the reserve potential of the deposit.

Prior to the first phase of drilling, available data and mapping activities indicated two areas in which the development of coal seams to mineable thicknesses was suspect. Accordingly, two drill holes were positioned to test the presence and thickness of the two major seams in the suspect areas.

Both holes were "plug" drilled and geophysically logged inasmuch as core data for quality determination was of secondary importance to coal seam thickness data. The geophysical data for hole "A" (see Figure 3B) indicated the presence of a distributary channel sandstone at the stratigraphic position of the Hiawatha Seam and thick Blind Canyon coal. The data for hole "B" showed the Hiawatha Seam as two meters thick and the Blind Canyon Seam as 0.5 meters thick. These data confirmed projections of thin-coal trends as well as the presence of coal seams of mineable thickness in the subsurface.

Phase two exploration activities were designed to ascertain the reserve potential of the southern two-thirds of the deposit. Of the eighteen holes drilled, three were cored through the coal zone to permit coal quality analysis and rock mechanics testing. The drill holes were completed on a 1000 meters spacing grid so that identified reserves could be referred to as being in the measured category.

Drill hole data from phase two drill holes confirmed the projections of coal seam thicknesses into the subsurface and permitted the construction of isopach maps for the thickness of each of the coal seams as well as the interburden between and overburden over the seams. The isopach maps of the coal seams are illustrated in Figures 4A and 4B. The position of paleochannel sandstones in both seams was also shown by the assembled drill hole data (Figure 5).

Sample data were collected in three core holes that were drilled at locations to minimize drill hole depths in three of the four corners of the deposit. Samples taken on a 60 meter spacing pattern from within mine workings on the eastern side of the deposit completed the sampling grid.

Sample data indicate both desirable and undesirable coal quality characteristics. These data are tabulated in Table VII. Of particular concern is the high sodium oxide (Na_2O)

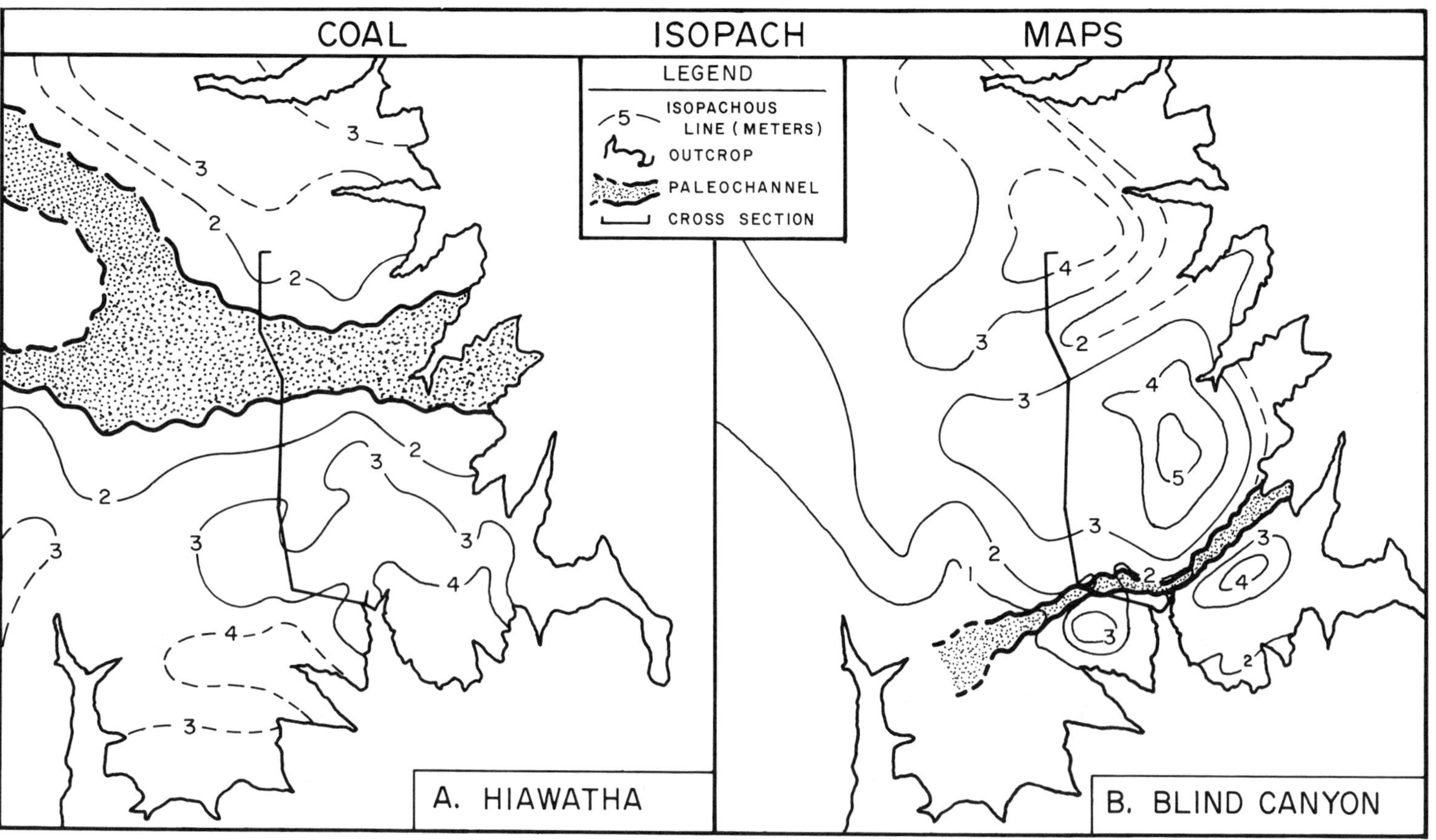

Figure 4. Isopachous maps of the Hiawatha and Blind Canyon Seams showing the locations of distributary channels.

Table VII
In-Place Coal Quality Data for
Wasatch Plateau Coal Deposit*

As Received	
Moisture (%)	4.6
Ash (%)	8.2
Volatile Matter (%)	40.2
Fixed Carbon (%)	44.3
B.T.U.	12,600
Sulfur (%)	0.56
Ash Fusion Temperatures (°F-reducing)	
Initial Deformation	2,230
Softening	2,269
Hemispherical	2,290
Fluid	2,320
Mineral Analysis of Ash	
SiO_2	50.4
Al_2O_3	22.1
Fe_2O_3	4.4
CaO	8.4
K_2O	0.6
MgO	1.2
Na_2O	5.4
P_2O_5	0.5
TiO_2	1.0
SO_3	5.6
Base/Acid	0.3
Fe_2O_3/CaO	0.6
SiO_2/Al_2O_3	2.6

*Based on 40 Samples

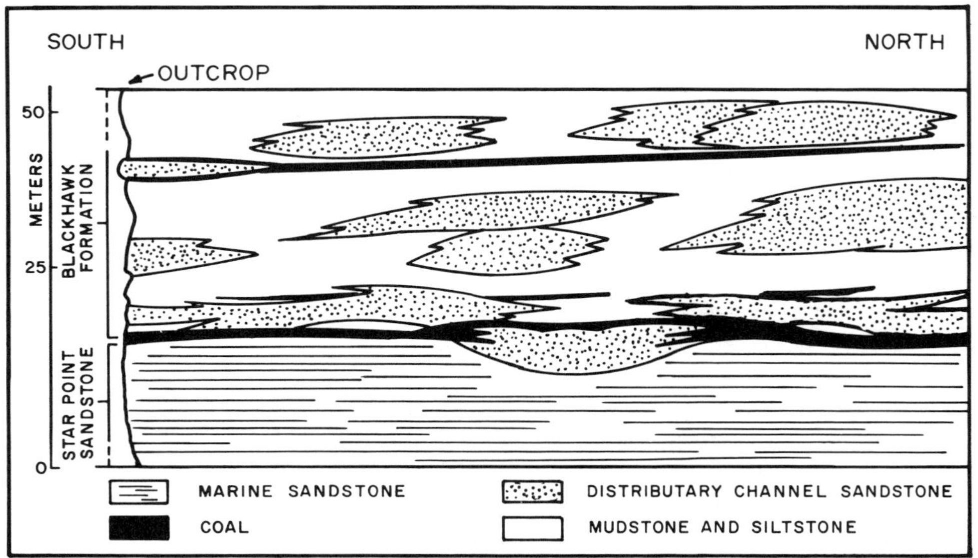

Figure 5. Cross-section through central part of Wasatch Plateau coal deposit example showing distributary channel sandstone at contact between Star Point Sandstone and Blackhawk Formation.

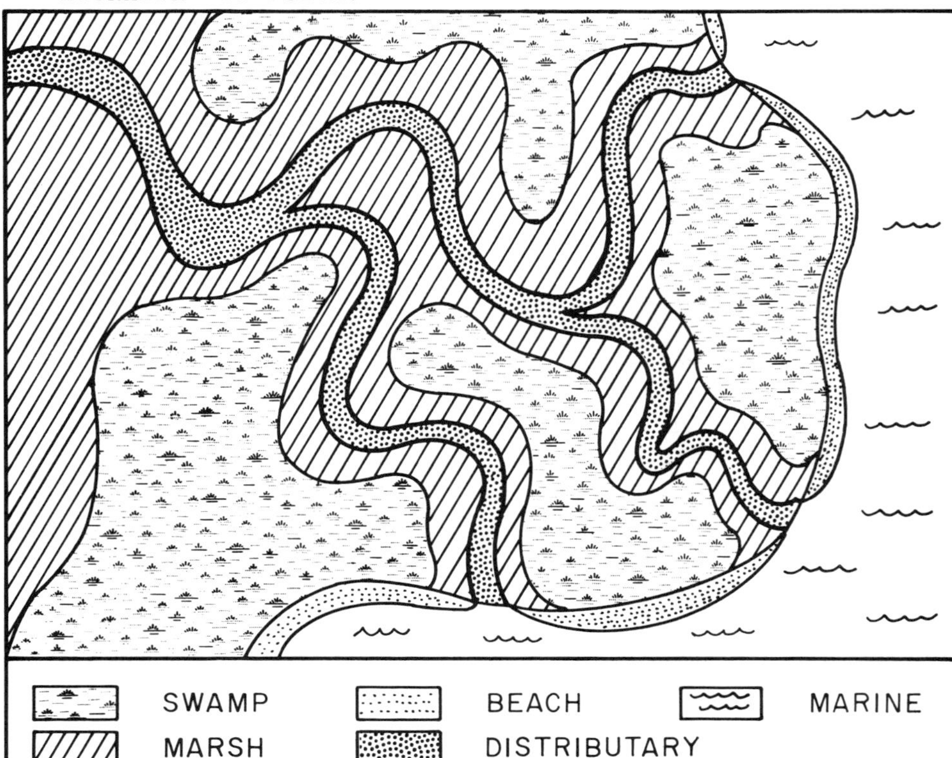

Figure 6. Major deltaic depositional environments.

content of the coal ash and low ash fusion temperatures of both seams. These data would indicate significant difficulties in preventing slagging in power plant boilers. Close grid spacing sampling in the mines indicates that the sodium oxide content is not distributed stratigraphically or laterally on a predictable basis.

Interpretations. Assembled data collected by previous workers and in the present drilling and sampling programs indicate that a significant coal reserve is present on the deposit. Although some negative factors do exist, the positive factors outweigh them and indicate that development of the reserve is viable.

Drill hole and outcrop measurements indicate that both major coal seams are well developed over large portions of the deposit. Depositional analysis indicates that coal accumulated in deltaic swamps that developed adjacent to and were locally traversed by distributary channels.

The Hiawatha Seam is in excess of two meters and commonly more than three meters in thickness over much of the area to the south of the distributary channel sandstone (Figure 4A). The seam appears to be equally thick to the north of the channel but additional drill hole information is needed for confirmation.

The Hiawatha Seam was probably deposited in coastal swamps behind beaches fringing the margin of a deltaic complex (Figure 6). The sharp depositional contact between the beach sandstones of the Star Point Sandstone and the coal of the Hiawatha Seam confirms this interpretation. The continuity of the swamp in which the Hiawatha Seam accumulated was broken by the local distributary channels that crossed the delta plain and emptied into the Cretaceous sea. The interfingering nature of the lateral contacts between the Hiawatha Seam and the distributary channel indicates that these units were deposited contemporaneously (Figure 5).

The Blind Canyon Seam attains thicknesses in excess of four meters and averages about 2.5 meters thick in a large pod that is located in the central part of the deposit (Figure 4B). The pod is traversed by a rock split that subdivides the seam into at least two seams of unmineable thickness in the southeastern part of the deposit.

The Blind Canyon Seam was probably deposited in the

landward portions of a deltaic swamp (Figure 6) inasmuch as the seam is enclosed entirely by sediments that are interpreted to have been deposited in marsh and fluvial environments (Figure 5). The rock split that subdivides the Blind Canyon Seam in the southeastern part of the deposit was probably deposited in a northeast-flowing crevasse splay channel that spilled from the upper reaches of a distributary channel

Coal quality information indicates potential problems in burning the high sodium coals in power plants. However, the diluting effects of mining, transporting, and stockpiling the coal from this deposit should lower the sodium oxide content two or three percentage points. Contamination caused by processing the coal from the mine workings to the power plants will also result in a decrease in BTU's and an increase in ash fusion temperatures and ash and moisture contents. The resulting coal that is delivered to power plants can be efficiently utilized if the boilers are designed and operated with an awareness of the delivered coal quality.

Example #2: Kaiparowits Plateau, Utah

The second example of a coal exploration program is of a 200 square kilometer area in the southern part of the Kaiparowits Coal Field of south-central Utah. The subject deposit has undergone two phases of outcrop mapping and two phases of drilling. The outcrop mapping programs were implemented by the U.S.G.S. The drilling programs were conducted for purposes of locating economic quantities of coal and defining the limits of coal of mineable thickness for the development of a mining plan.

General Geology. The outcropping Upper Cretaceous sediments of which the Kaiparowits Plateau is mostly composed were first mapped by Gregory and Moore (12). Their report notes the presence of local thick coal seams and burned strata but attention was focused mainly upon the general geology and stratigraphy of the region. More recent workers, notably Peterson (176, 177), Peterson and Waldrop (178), Zeller (179), and Doelling and Graham (180), have concentrated their efforts on mapping the coal geology of the region. This work has resulted in the definition of four members within the Straight Cliffs Formation: the Tibbet Canyon, Smokey Hollow, John Henry, and Drip Tank Members (176). The major coal seams of the region are found within the John Henry Member of the Straight Cliffs Formation.

Peterson (177) examined the stratigraphy, sedimentation, and coal seams of the southern part of the Kaiparowits Plateau. His analysis of the depositional environments defined the presence of an interdeltaic-strandplain beach complex behind which coal seams accumulated in coastal swamps. He also recognized that the several coal seams within the John Henry Member accumulated in superimposed swamp environments because of the static position of the shoreline of the Cretaceous Seaway during Santonian time.

Previous Data. Data concerning the development of thick coal seams in the Kaiparowits region, prior to the first exploratory drilling programs conducted in the early 1960's were meager. Information assembled by Gregory and Moore (12) indicated local coal beds in excess of four meters thick at various widely-spaced locations along poorly exposed outcrops in both the northern and southern parts of the region. They also noted the extensive areas in which coal-bearing strata had been burned and fused, presumably by the heat generated from burning coal seams. The excellent outcrops exposed along the eastern margin of the Kaiparowits Plateau, the Straight Cliffs Escarpment, which consist predominately of sandstone, were found to be nearly devoid of coal seams greater than one meter in thickness.

The first industry geologists to examine the coal-bearing outcrops in the area confirmed the presence of extensive areas of burned strata, the paucity of outcrops of coal, and the lenticularity of the coal seams. Their mapping activities indicated that the only method to determine the potential for mineable coal seams would be to drill in areas far removed from the burned outcrops and between the widely-spaced coal outcrops.

Exploration drilling programs conducted in the early 1960's in the southern part of the Kaiparowits Plateau identified the presence of several coal seams of mineable thickness. Drill hole spacing patterns of about 1500 meters between holes indicated that an 80 meter thick coal zone consisting of as many as six seams of mineable thickness is present in the lower portion of the John Henry Member of the Straight Cliffs Formation.

The original exploration holes were cored and were drilled on a wide spacing pattern (Figure 7A). Data from these holes identified the presence of several seams of mineable thickness and good coal quality. However, the drill holes

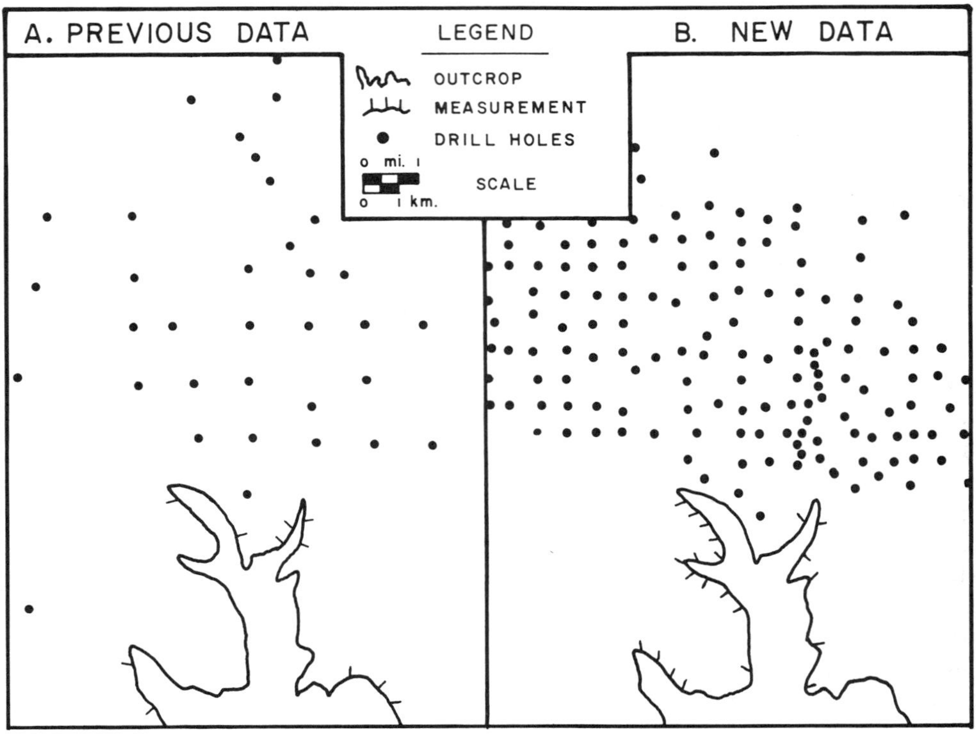

A. PREVIOUS DATA
LEGEND
B. NEW DATA
OUTCROP
MEASUREMENT
DRILL HOLES
0 mi. 1
SCALE
0 1 km.

Figure 7. Location of previous and new data for Kaiparowits Plateau coal deposit.

SEAMS AND
SUB-SEAMS
COAL SEAM
6
5b
5a
4
3b2
3b1
3a
2
1b
1a
6
5
4
3
2
1
WEST
EAST

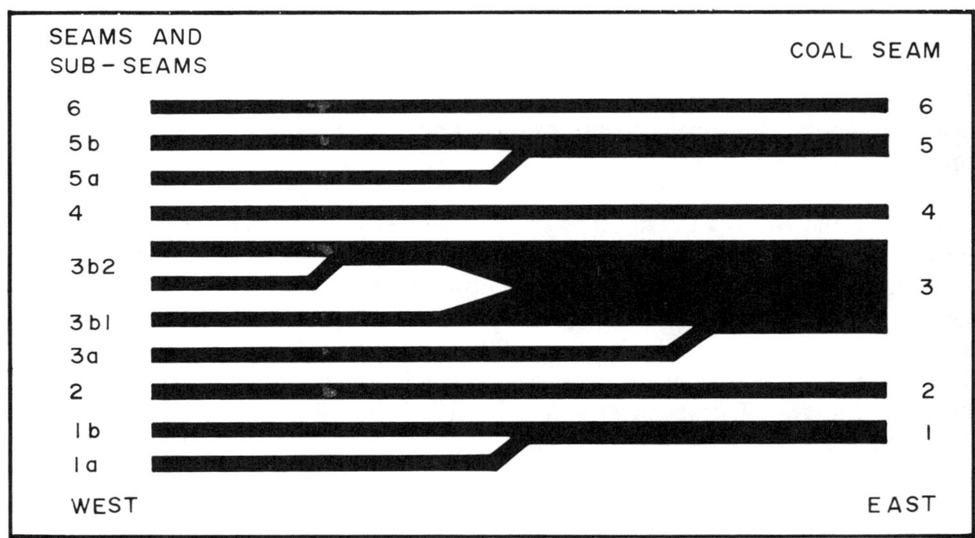

Figure 8. Coal seam correlation and nomenclature diagram for Kaiparowits Plateau coal deposit.

were too widely spaced to allow determination of the continuity and correlations of individual seams.

Additional stratigraphic and sedimentologic information compiled by Peterson (176, 177, 178) as well as the geologic maps prepared by U.S.G.S. workers (182-188) and Doelling and Graham (181), after the completion of the first exploratory drilling program also add to the general understanding of the geology of the deposit.

Present Program. Previous mapping and drilling activities on the deposit defined the presence of coal seams of mineable thickness and good quality. Additional data were needed to define the lateral continuity, correlation, and thickness variations of individual seams so that a mining plan could be developed and an accurate reserve estimate prepared. Accordingly, 165 additional drill holes were completed, outcrops were mapped, and a test adit was opened to obtain these data.

The most recent drilling program was conducted in two phases. The first phase was designed to determine the optimum spacing pattern and the second phase implemented the spacing pattern.

The first phase of recent exploration consisted of drilling 28 holes spaced about 400 meters apart in a north-south and east-west "cross" pattern in areas of shallowest drilling depths to ascertain the correlation of individual seams and define the optimum spacing pattern. The drilling program identified six major coal seams (Figure 8) that are of sufficient continuity to permit correlation on an 800 meter spacing pattern.

The second drilling phase consisted of implementing the 800 meter spacing pattern over a large portion of the deposit (Figure 7B). Drill holes completed in this phase consisted of an equal proportion of "plug" and core holes. All holes were geophysically probed. Data from the core holes were used to determine the lateral and stratigraphic variability of coal quality and to permit cross-checking of geophysical interpretations. Geophysical data from "plug" holes were used to determine coal seam thicknesses, general quality trends, depositional units, rock types, and stratigraphic relations. Geophysical data were also used to help determine sample intervals of coal cores. Occasional "plug" holes were drilled as deep control holes in advance of coring operations to ascertain the structure and thickness of the coal zone.

A test adit opened in a canyon in the southern part of the deposit provided information concerning the mineability of a thick coal seam and bulk samples for washability and other quality testing.

Outcrops were mapped along the southern margins of the coal deposit to ascertain previous interpretations, identify depositional environments, and to provide information concerning the lateral continuity of coal seams and enclosing sediments. These data were used to assist in the interpretation of drill hole data with the assumption that features commonly exposed on outcrops are more easily recognized in drill hole cores and geophysical logs if they have been recognized first in outcrop.

The data assembled above indicate the presence of six major coal seams locally subdivided by numerous rock splits. In instances where coal seams are subdivided by rock splits, resulting divisions are referred to as upper and lower sub-seams. Isopach maps of each of the six major coal seams and component sub-seams as well as the interval between seams were prepared. The total amount of sandstone in each interval between seams and sub-seams was also isopached to indicate the position of fluvial paleochannels in those intervals. Structure maps of the base of each coal seam were also prepared.

Outcrop mapping activities confirmed the presence of sediments deposited in fluvial channels, crevasse splays, swamps, marshes, and lakes in the deposit. Regional reconnaissance mapping indicated the presence of lagoonal and beach sediments to the east of the deposit. Geophysical logs confirmed environmental interpretations for fining-upward fluvial channels.

Interpretations. The drilling and mapping programs conducted on the subject deposit indicate the presence of a major coal reserve that consists of several seams of mineable thickness with lateral continuities that can be referred to in terms of 100's of square kilometers. These data indicate that the deposit is of significant economic value.

Coal seam correlations and isopach maps indicate that the

six major coal seams and component sub-seams are present in mineable thicknesses on the deposit (Figure 8). Depositional analysis indicates that all coal seams accumulated in the landward portions of an interdeltaic-strandplain swamp complex along the margins of the Cretaceous Seaway during Santonian time (Figure 9).

Of the six major seams, the No. 2 seam displays features common to most of the coal seams in the deposit and is used here as an example. The No. 2 seam was deposited on sandstone beds that accumulated in the interval between the No. 1 and No. 2 Seams. The isopach map of the thickness of this interval reflects the influence of the position of paleochannels inasmuch as the interval is thickest in the area in which paleochannels are present (Figures 10A and B). The area in which the No. 1 and No. 2 seams are separated by less than three meters of non-coal sediment is interpreted as a paleo-marsh area.

The development of the No. 2 Seam appears to have been influenced by the location of depositional environments that existed in the area prior to the accumulation of the vegetation that forms the No. 2 seam (Figure 10C). It is evident that the No. 2 coal seam developed to greater thicknesses in areas removed from previously-existing paleochannels. The process of differential compactional subsidence may have caused the variability in the development of the No. 2 coal seam. Sandstone deposited in paleochannels does not compact or subside as much or as rapidly as does mudstone deposited in marsh environments. Accordingly, differential compactional subsidence in marsh areas is sufficient to permit the accumulation of thick coal seams while the same process is retarded in fluvial channels and results in less well developed coal over paleochannels.

A diagram that illustrates the process of differential compactional subsidence and the effect it has on the development of overlying coal seams as well as other subsequently deposited sediments is presented in Figure 11. The example illustrated here is typical for the entire coal-bearing section within the study area. In panel A, a fluvial channel system and associated environments are flanked by marshes which in turn, are flanked by swamps. A crevasse splay subsequently forms along the outside of a meander bend in a lowland area in response to a seasonal flood (panel A) and eventually evolves into a fluvial channel (panel B). In time, the previous channel is abandoned, subsides into the

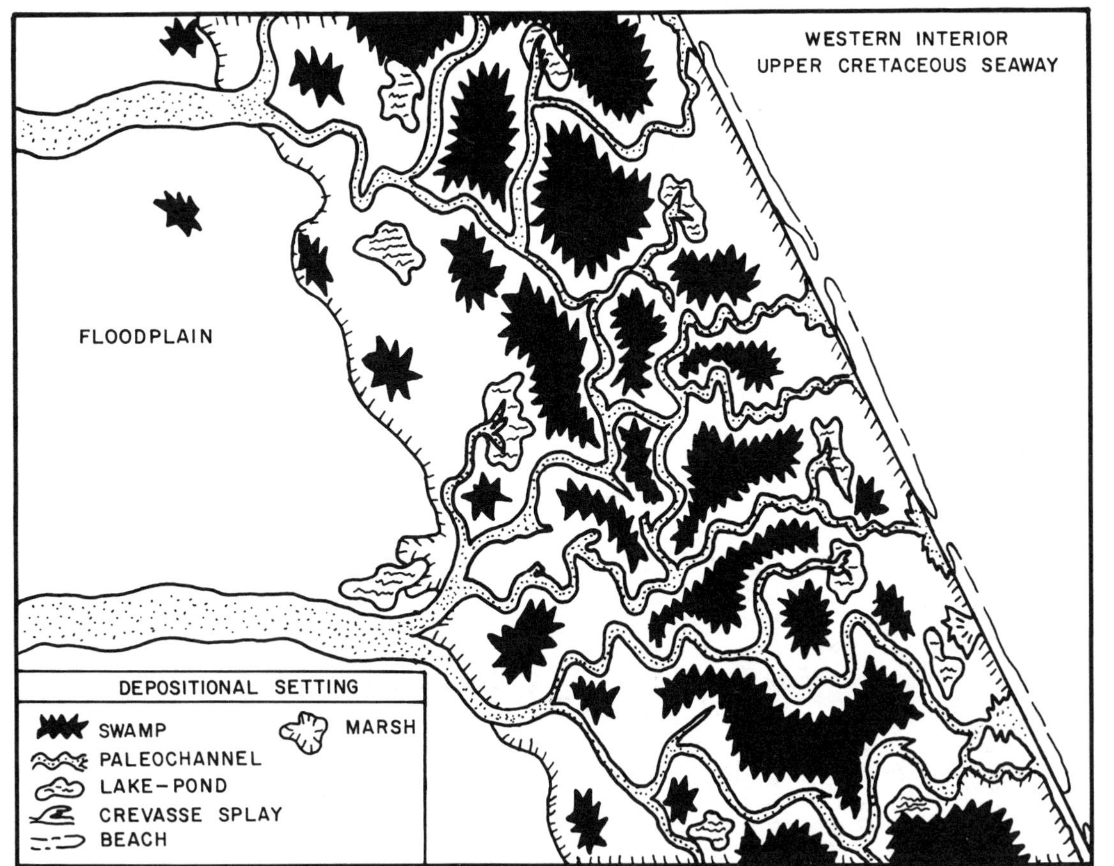

Figure 9. Idealized reconstruction of an interdeltaic-strandplain depositional setting in the Kaiparowits Plateau Region of south-central Utah.

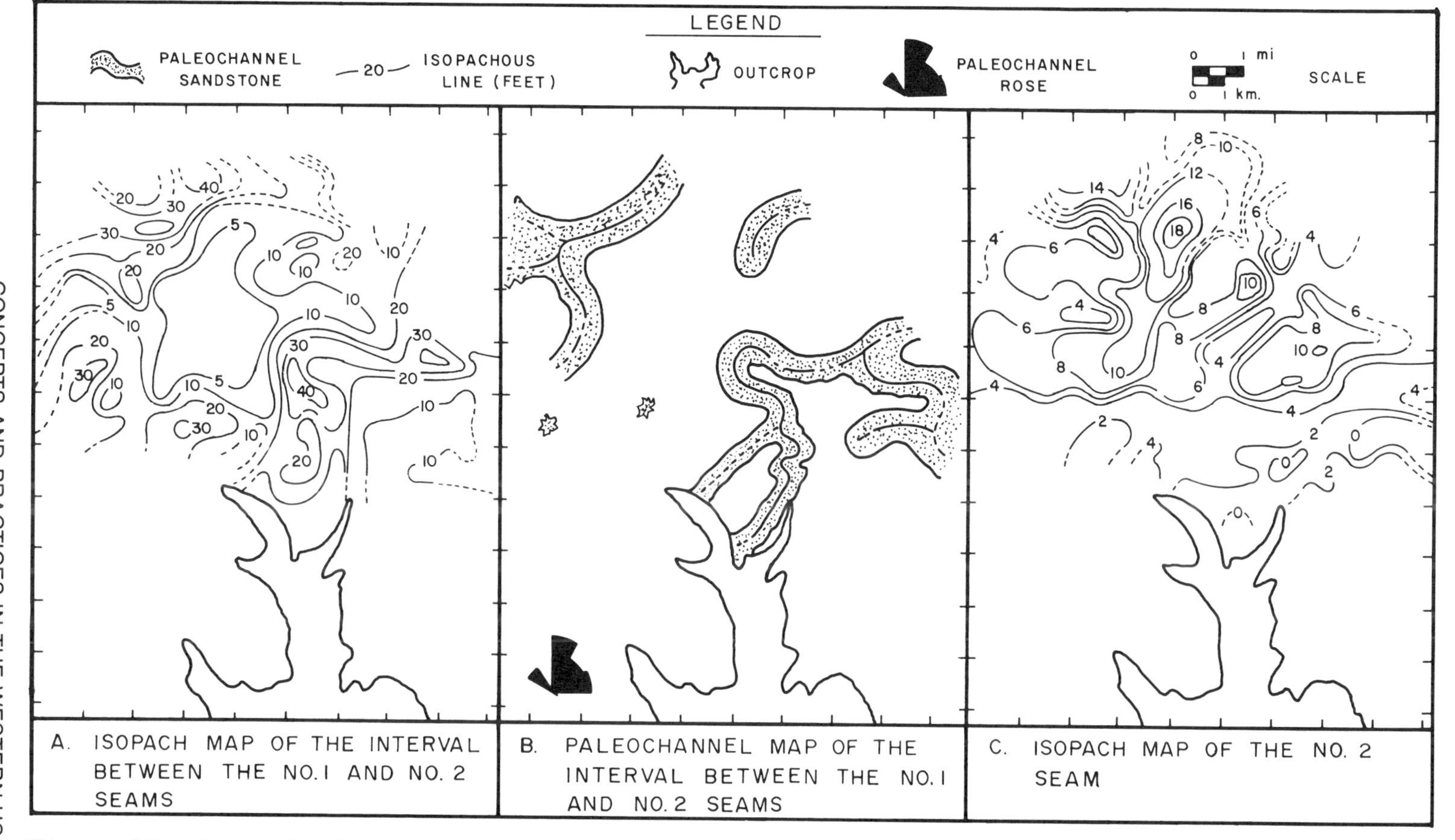

Figure 10. Maps showing the distribution of the No. 2 Coal Seam and the location and distribution of sediments deposited prior to the accumulation of the seam. Kaiparowits Plateau, Utah.

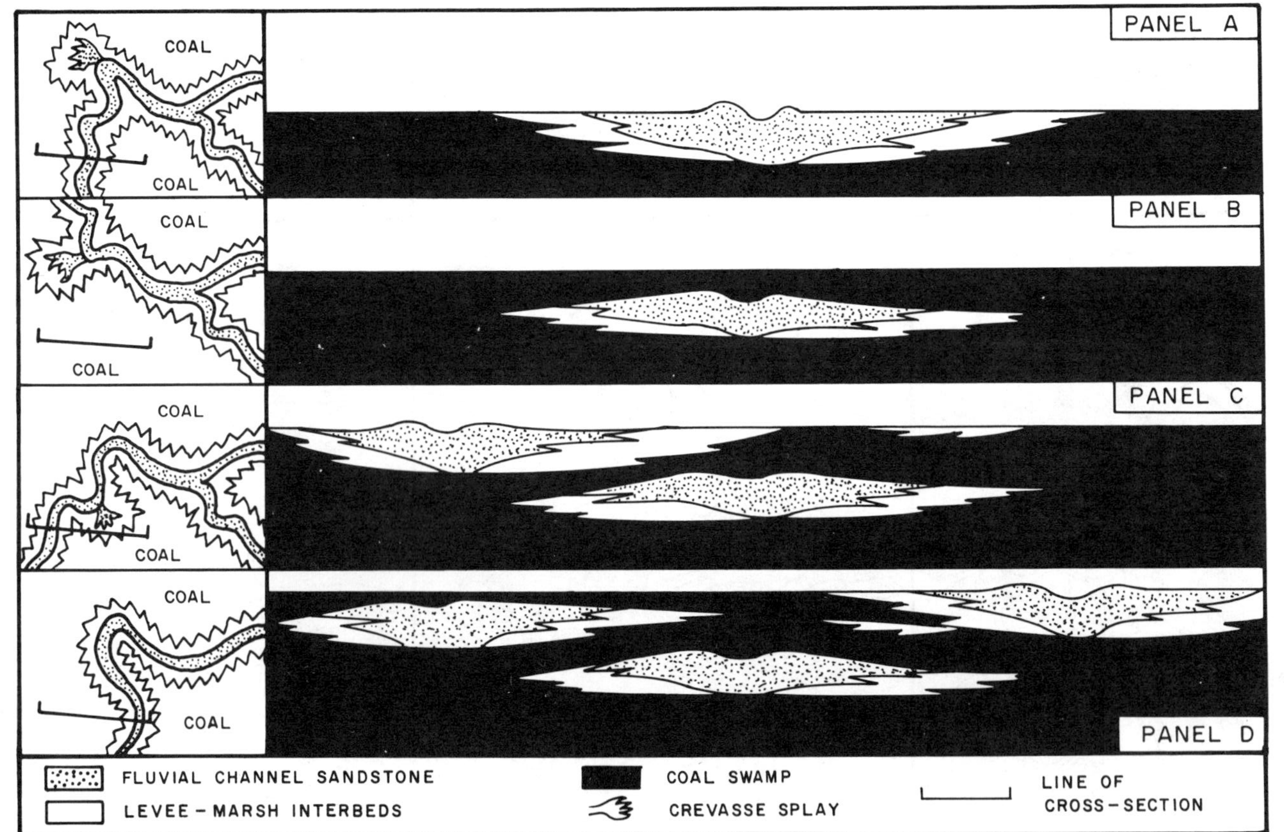

Figure 11. Four episodes of paleochannel deposition in an interdeltaic-strand-plain swamp complex as shown by "time-lapse" reconstructions.

substrate, and is overgrown by swamp vegetation. During the process of abandonment and subsidence, differential compaction results in greater subsidence and compaction along the margins of the paleochannel sandstone than in the center. The net effect results in the development of thick coal seams in the paleo-marsh areas and thin coal beds in the paleo-fluvial channel areas.

Other panels indicate that continued channel abandonment and the formation of crevasse splays that evolve into channel systems results in vertically offset paleochannel sandstone beds. These processes are inferred to have been an influencing factor in determining the location and persistence of the coal seams and enclosing strata within the Kaiparowits coal deposit.

The location of paleochannel sandstone in intervals between coal seams may exert an important influence on mining conditions when the deposit is developed. Differential compaction in sediments that laterally interfinger with paleochannel sandstone lenses may have caused the numerous microslickensides present in these sediments and may result in unstable roof conditions. Paleochannel sandstone beds commonly form perched water tables, that when tapped by mining activities, may cause water problems during mining.

SUMMARY AND CONCLUSIONS

Coal exploration programs in the western U.S.A. are usually conducted for (1) academic interests, (2) economic evaluations, or (3) definition of mining boundaries and plans. Data are best collected in phases that proceed from general to specific data gathering methods. These phases, in the suggested order of implementation are (1) review of information compiled by previous workers, (2) regional reconnaissance, (3) detailed mapping and data verification, (4) drilling and down-hole geophysics, and (5) sampling. Data to be collected during these phases consist of coal correlations and stratigraphy, structure, lateral continuity of coal seams and enclosing units, and the geometry and quality of coal seams. Other information concerning depositional environments, sedimentation, petrography, coal reserves, and mineability of coal should also be collected.

The interpretation of data collected on coal deposits should be carefully evaluated in terms of depositional environments. An ability to recognize and differentiate between coal deposits that have accumulated in deltaic and interdeltaic-strandplain swamp environments will influence the methods used in exploring the deposit. The characteristics of coal deposits that develop in each swamp environment are tabulated in Table III and are shown in Figures 6, 9, and 12.

Deltaic swamp coal deposits usually consist of only a few coal seams of mineable thickness (in excess of two meters thick). Lateral continuities are limited by the spacing of distributary channels. The speed of delta progradation and the frequency of lobe abandonment processes does not permit a vertical stacking of coal-forming depositional environments (Figure 12).

Interdeltaic-strandplain swamp coal deposits consist of several stacked coal seams of mineable thickness. These seams are characterized by extensive lateral continuities. The shorelines behind which interdeltaic-strandplain coal swamps develop do not fluctuate or prograde as rapidly as do deltaic shorelines which results in a vertical stacking of coal-forming depositional environments (Figure 12).

The mechanics of sedimentation in various depositional environments exercises an influence on the development of coal seam thicknesses and lateral continuities. An awareness of those depositional processes permits a more complete evaluation of a coal deposit. Mining engineers should be cognizant of the effects of differential compaction and hydrology of paleochannel sandstone beds in planning the most efficient extraction of coal seams.

Coal quality data are of significant import in determining the suitability of a coal deposit for use in power plant boilers. The presence of sodium oxide (Na_2O) in coal ash may result in a lowering of ash fusion temperatures of coal burned in boilers. Other coal quality parameters such as those listed in Table IV should be evaluated prior to designing new boilers or dedicating coal deposits to existing power plants. The effect of contamination on coal quality caused by mining, transporting, and stockpiling coal should be evaluated to permit proper boiler design.

ACKNOWLEDGMENTS

The data used and concepts expressed in this report have

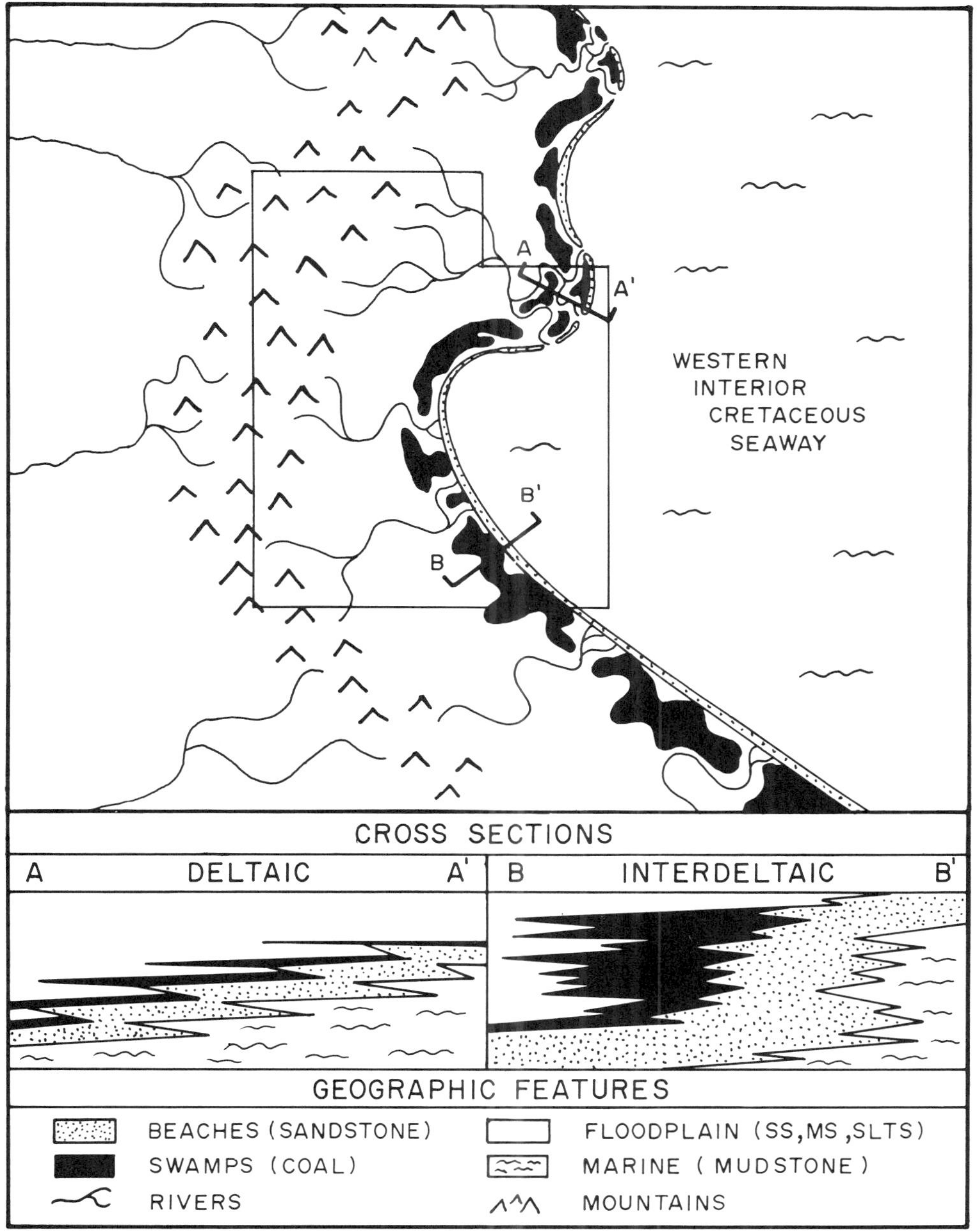

Figure 12. Regional diagrammatic reconstruction of the major depositional environments in which Cretaceous coal-bearing sediments accumulated.

become available through the hard work, enthusiasm, and intelligence of several technical as well as non-technical individuals among which are R. C. Townsend, Dr. W. L. Stokes, J. K. Balsley, J. M. Mercier, J. Bonaquisto, T. Lloyd, R. C. Fry, R. M. Thompson, J. S. Castleberry, H. L. Silkwood, O. Story, F. Peterson, R. S. Dewey, and R. E. Blackett. Special thanks go to M. Brewer and B. Wooley for their respective efforts in typing and illustrating the text.

REFERENCES

(1) Spieker, E. M., and A. A. Baker, 1928, Geology and coal resources of the Salina Canyon District, Sevier County, Utah: U.S.G.S. Bull. 796-C, p. 125-170.

(2) Spieker, E. M., 1931, The Wasatch Plateau Coal Field, Utah: U.S.G.S. Bull. 819, 210 p.

(3) Spieker, E. M., 1946, Late Mesozoic and Early Cenozoic history of central Utah: U.S.G.S. Professional Paper 205-D, p. 117-161.

(4) Spieker, E. M., 1949, Sedimentary facies and associated diastrophism in the Upper Cretaceous of central and eastern Utah: <u>in</u> Sedimentary Facies in Geologic History, Geol. Soc. Amer. Memoir 39, p. 55-82.

(5) Spieker, E. M., 1949, The transition between the Colorado Plateaus and the Great Basin in central Utah: Guidebook to the Geology of Utah, No. 4, 106 p.

(6) Clark, F. R., 1928, Economic geology of the Castlegate, Wellington, and Sunnyside quadrangles, Carbon County, Utah: U.S.G.S. Bull. 793, 165 p.

(7) Fisher, D. J., 1936, The Book Cliffs coal field in Emery and Grand Counties, Utah: U.S.G.S. Bull. 852, 104 p.

(8) Richardson, G. B., 1907, The Book Cliffs coal field, between Grand River, Colorado, and Sunnyside, Utah: U.S.G.S. Bull. 316-E, p. 302-320.

(9) Richardson, G. B., 1909, Reconnaissance of the Book Cliffs coal field, between Grand River, Colorado, and Sunnyside, Utah: U.S.G.S. Bull, 371, 54 p.

(10) Hunt, C. B., P. Averitt, and R. L. Miller, 1953, Geology and geography of the Henry Mountains region, Utah: U.S.G.S. Prof. Paper 228, 234 p.

(11) Lupton, C. T., 1916, Geology and coal reserves of Castle Valley in Carbon, Emery, and Sevier Counties, Utah: U.S.G.S. Bull. 628, 88 p.

(12) Gregory, H. E., and R. C. Moore, 1931, The Kaiparowits region, a geographic and geologic reconniassance of parts of Utah and Arizona: U.S.G.S. Prof. Paper 164, 161 p.

(13) Erdmann, C. E., 1935, The Book Cliffs coal field in Garfield and Mesa Counties, Colorado: U.S.G.S. Bull. 851, 150 p.

(14) Lee, W. T., 1909, The Grand Mesa coal field, Colorado: U.S.G.S. Bull. 341-C, p. 316-334.

(15) Lee, W. T., 1912, Coal fields of Grand Mesa and the West Elk Mountains, Colorado: U.S.G.S. Bull. 510, 237 p.

(16) Gale, H. S., 1907, Coal fields of the Danforth Hills and Grand Hogback, in northwestern Colorado: U.S.G.S. Bull. 316-E, p. 264-301.

(17) Gale, H. S., 1910, Coal fields of northwestern Colorado and northeastern Utah: U.S.G.S. Bull. 415, 265 p.

(18) Hancock, E. T., 1925, Geology and coal resources of the Axial and Monument Butte quadrangles, Moffat County, Colorado: U.S.G.S. Bull. 757, 134 p.

(19) Hancock, E. T., and J. B. Eby, 1930, Geology and coal resources of the Meeker quadrangle, Moffat and Rio Blanco Counties, Colorado: U.S.G.S. Bull. 812-C, p. 191-242.

(20) Fenneman, N. M., and H. S. Gale, 1906, The Yampa coal field, Routt County, Colorado, with a chapter on the character and use of Yampa coals, by M. R. Campbell: U.S.G.S. Bull. 297, 96 p.

(21) Campbell, M. R., 1923, The Twentymile Park district of the Yampa coal field, Routt County, Colorado: U.S.G.S. Bull. 748, 82 p.

(22) Bass, N. W., J. B. Eby, and M. R. Campbell, 1955, Geology and mineral fuels of parts of Routt and Moffat Counties, Colorado: U.S.G.S. Bull. 1027-D, p. 143-250.

(23) Schrader, F. C., 1906, The Durango-Gallup coal field of Colorado and New Mexico: U.S.G.S. Bull. 285-F, p. 241-258.

(24) Shaler, M. K., 1907, A reconnaissance survey of the western part of the Durango-Gallup coal field of Colorado and New Mexico: U.S.G.S. Bull. 316-F, p. 376-426.

(25) Taff, J. A., 1907, The Durango coal district, Colorado: U.S.G.S. Bull. 316-E, p. 321-337.

(26) Richardson, G. B., 1910, The Trinidad coal field, Colorado: U.S.G.S. Bull. 381-C, p. 379-446.

(27) Lee, W. T., 1922, Description of the Raton-Brilliant-Koehler quadrangles, New Mexico-Colorado: U.S.G.S. Geol. Atlas, Folio 214, 17 p.

(28) Martin, G. C., 1910, Coal of the Denver Basin, Colorado: U.S.G.S. Bull. 381-C, p. 297-306.

(29) Goldman, M. I., 1910, The Colorado Springs coal field, Colorado: U.S.G.S. Bull, 381-C, p. 317-340.

(30) Beekley, A. L., 1915, Geology and coal resources of North Park, Colorado: U.S.G.S. Bull, 596, 121 p.

(31) Washburne, C. W. 1910, The South Park coal field, Colorado: U.S.G.S. Bull. 381-C, p. 307-316.

(32) Veatch, A. C., 1906, Coal and oil in Southern Uinta County, Wyoming: U.S.G.S. Bull. 285-F, p. 331-353.

(33) Veatch, A. C., 1907, Geography and geology of a portion of southwestern Wyoming, with special reference to coal oil: U.S.G.S. Prof. Paper 56, 178 p.

(34) Veatch, A. C., 1907, Coal fields of east-central Carbon County, Wyoming: U.S.G.S. Bull, 316-D, p. 244-260.

(35) Schultz, A. R., 1907, Coal fields in a portion of central Uinta County, Wyoming: U.S.G.S. Bull. 316-D, p. 212-241.

(36) Schultz, A. R., 1914, Geology and geography of a portion of Lincoln County, Wyoming: U.S.G.S. Bull, 543, 141 p.

(37) Schultz, A. R., 1918, A geologic reconnaissance for phosphate and coal in southeastern Idaho and western Wyoming: U.S.G.S. Bull. 680, 84 p.

(38) Schultz, A. R., 1909, The northern part of the Rock Springs coal field, Sweetwater County, Wyoming: U.S.G.S. Bull. 341-B, p. 256-282.

(39) Schultz, A. R., 1910, The southern part of the Rock Springs coal field, Sweetwater County, Wyoming: U.S.G.S. Bull. 381-B, p. 214-281.

(40) Ball, M. W., 1909, The western part of the Little Snake River coal field, Wyoming: U.S.G.S. Bull. 341-B, p. 243-255.

(41) Ball, M. W., and E. Stebinger, 1910, The eastern part of the Little Snake River coal field, Wyoming: U.S.G.S. Bull. 381-B, p. 186-213.

(42) Dobbin, C. E., C. F. Bowen, H. W. Hoots, 1929, Geology and coal and oil resources of the Hanna and Carbon Basins, Carbon County, Wyoming: U.S.G.S. Bull, 804, 88 p.

(43) Dobbin, C. E., H. W. Hoots, C. H. Dane, and E. T. Hancock, 1929, Geology of the Rock Creek oil field and adjacent areas, Carbon and Albany Counties, Wyoming: U.S.G.S. Bull. 806-D, p. 131-153.

(44) Woodruff, E. G., 1907, The Lander coal field, Wyoming: U.S.G.S. Bull. 316-D, p. 242-243.

(45) Fisher, C. A., 1904, Coal of the Bighorn Basin in northwest Wyoming: U.S.G.S. Bull. 225-G, p. 345-362.

(46) Washburne, C. W., 1909, Coal fields of the northeast side of the Bighorn Basin, Wyoming, and of Bridger, Montana: U.S.G.S. Bull. 341-B, p. 165-199.

(47) Woodruff, E. G., 1910, The coal field in the southeastern part of the Bighorn Basin, Wyoming: U.S.G.S. Bull. 381-B, p. 170-185.

(48) Gale, H. S., and C. H. Wegemann, 1910, The Buffalo coal field, Wyoming: U.S.G.S. Bull. 381-B, p. 137-169.

(49) Wegemann, C. H., 1912, The Sussex coal field, Johnson, Natrona, and Converse Counties, Wyoming: U.S.G.S. Bull. 471-F, p. 441-471.

(50) Wegemann, C. H., 1913, The Barker coal field, Johnson County, Wyoming: U.S.G.S. Bull. 531-I, p. 263-284.

(51) Dobbin, C. E., and V. H. Barnett, 1928, The Gillette coal field, northeastern Wyoming: U.S.G.S. Bull. 796-A, p. 1-50.

(52) Schrader, F. C., 1906, The Durango-Gallup coal field of Colorado and New Mexico: U.S.G.S. Bull. 285-F, p. 241-258.

(53) Shaler, M. K., and J. H. Gardner, 1907, Clay deposits of the western part of the Durango-Gallup coal field, of Colorado and New Mexico: U.S.G.S. Bull. 315-I, p. 296-302.

(54) Bauer, C. M., and J. B. Reeside, Jr., 1921, Coal in the middle and eastern parts of San Juan County, New Mexico: U.S.G.S. Bull. 716-G, p. 155-237.

(55) Dane, C. H., 1936, The La Ventana-Chacra Mesa coal field (New Mexico): U.S.G.S. Bull. 860-C, p. 81-161.

(56) Gardner, J. H., 1909, The coal field between Gallup and San Mateo, New Mexico: U.S.G.S. Bull. 341-C, p. 364-378.

(57) Sears, J. D., 1934, The coal field from Gallup eastward toward Mount Taylor (new Mexico), with a measured section of pre-Dakota (?) rocks near Navajo Church: U.S.G.S. Bull. 860-A, p. 1-29.

(58) Hunt, C. B., 1936, The Mount Taylor coal field (New Mexico): U.S.G.S. Bull. 860-B, p. 31-80.

(59) Lee, W. T., and F. H. Knowlton, 1917, Geology and palentology of the Raton Mesa and other regions in Colorado and New Mexico: U.S.G.S. Prof. Paper 101, 450 p.

(60) Lee, W. T., 1924, Coal resources of the Raton coal field, Colfax County, New Mexico: U.S.G.S. Bull. 752, 254 p.

(61) Campbell, M. R., and H. E. Gregory, 1911, The Black Mesa coal field, Arizona: U.S.G.S. Bull 431-B, p. 229-238.

(62) Bowen, C. F., 1912, the Baker lignite field, Custer County, Montana: U.S.G.S. Bull. 471-D, p. 202-226.

(63) Bowen, C. F., 1914, Coal discovered in a reconnaissance survey between Musselshell and Judith, Montana: U.S.G.S. Bull. 541-H, p. 329-337.

(64) Bowen, C. F., 1914, The Cleveland coal field, Blaine County, Montana: U.S.G.S. Bull. 541-H, p. 338-355.

(65) Bowen, C. F., 1914, The Big Sandy coal field, Chouteau County, Montana: U.S.G.S. Bull. 541-M, p. 356-378.

(66) Collier, A. J., and C. D. Smith, 1909, The Miles City coal field, Montana: U.S.G.S. Bull. 341-A, p. 36-61.

(67) Collier, A. J., and M. M. Knechtel, 1939, The coal resources of McCone County, Montana: U.S.G.S. Bull. 905, 80 p.

(68) Bauer, C. M., 1914, Lignite in the vicinity of Plentywood and Scobey, Sheridan County, Montana: U.S.G.S. Bull. 541-M, p. 293-315.

(69) Bauer, C. M., 1925, The Ekalaka lignite field, southeastern Montana: U.S.G.S. Bull. 751-F, p. 231-267.

(70) Beekley, A. L., 1912, The Culbertson lignite field, Valley County, Montana: U.S.G.S. Bull. 471-D, p. 319-358.

(71) Woolsey, L. H., 1909, The Bull Mountain coal field, Montana: U.S.G.S. Bull. 341-A, p. 62-77.

(72) Woolsey, L. H., R. W. Richards, and C. T. Lupton, 1917, The Bull Mountain coal field, Musselshell and Yellowstone Counties, Montana: U.S.G.S. Bull. 647, 218 p.

(73) Richards, R. W., 1910, The central part of the Bull Mountain coal field, Montana: U.S.G.S. Bull. 381-A, p. 60-81.

(74) Lupton, C. T., 1911, The eastern part of the Bull Mountain coal field, Montana: U.S.G.S. Bull 431-B, p. 163-189.

(75) Barnett, V. H., 1917, Geology of the Hound Creek district of the Great Falls coal field, Cascade County, Montana: U.S.G.S. Bull. 641-H, p. 215-231.

(76) Fisher, C. A., 1909, Geology of the Great Falls coal field, Montana: U.S.G.S. Bull. 356, 85 p.

(77) Calvert, W. R., 1909, The Lewistown coal field, Montana: U.S.G.S. Bull. 341-A.

(78) Calvert, W. R., 1909, Geology of the Lewistown coal field, Montana: U.S.G.S. Bull. 390, 83 p.

(79) Calvert, W. R., 1912, Geology of certain lignite fields in eastern Montana: U.S.G.S. Bull. 471-D, p. 187-201.

(80) Baker, A. A., 1929, The northward extension of the Sheridan coal field, Big Horn and Rosebud Counties, Montana: U.S.G.S. Bull. 806-B, p. 15-67.

(81) Bass, N. W., 1932, The Ashland coal field, Rosebud, Powder River, and Custer Counties, Montana: U.S.G.S. Bull. 831-B, p. 19-105.

(82) Calvert, W. R., 1912, The Livingstown and Trail Creek coal fields, Park, Gallatin, and Sweetgrass Counties, Montana: U.S.G.S. Bull. 471-E, p. 384-405.

(83) Calvert, W. R., 1912, The Electric coal field, Park County, Montana: U.S.G.S. Bull. 471-E, p. 406-422.

(84) Calvert, W. R., 1917, Geology of the Upper Stillwater Basin, Stillwater and Carbon Counties, Montana, with special reference to coal and oil: U.S.G.S. Bull. 641-G, p. 199-214.

(85) Harris, H. D., 1959, Late Mesozoic positive area in western Utah: Amer. Assoc. Petroleum Geol. Bull., v. 43, No. 11, p. 2636-2652.

(86) Weimer, R. J., and J. D. Haun, 1960, Cretaceous stratigraphy, Rocky Mountain region, U.S.A.: Intern. Geol. Cong., 21st session, part XII, section 12 - Regional Paleogeography, p. 178-184.

(87) Reeside, J. B., Jr., 1957, Palecology of the Cretaceous seas of the Western Interior of the United States: Geol. Soc. Amer. Memoir 67, p. 505-542.

(88) Cobban, W. A., and J. B. Reeside, Jr., 1952, Correlation of the Cretaceous formations of the Western Interior of the United States: Geol. Assoc. Amer. Bull., v.63, p. 1011-1044.

(89) New Mexico Geological Society, 1950, San Juan Basin (north and east sides), New Mexico and Colorado: Kelley, V. C. (ed.), 152 p.

(90) New Mexico Geological Society, 1951, San Juan Basin (south and west sides), New Mexico and Arizona: Smith, C. T., and C. Silver (eds.), 163 p.

(91) New Mexico Geological Society, 1958, Black Mesa Basin: Anderson, R. Y., and J. W. Haushbarger (eds.), 205 p.

(92) New Mexico Geological Society, 1977, San Juan Basin III: Fassett, J. E., and H. J. James (eds.), 310 p.

(93) Colorado Geological Society, 1959, Cretaceous Symposium Guidebook

(94) Colorado Geological Survey, 1976, Geology of Rocky Mountain coal - a symposium: Murray, K. (ed.), Colo. Geol. Survey Resource Series 1, 175 p.

(95) Colorado Geological Survey, 1977, Geology of Rocky Mountain Coal: Hodgson, H. (ed.), Colo. Geol. Survey Resource Series 4, Proceedings of the second symposium, 219 p.

(96) Wyoming Geological Association, 1958, Powder River Basin: Strickland, J. (ed.), 15th Annual Field Conference, 341 p.

(97) Wyoming Geological Association, 1960, Overthrust Belt of Southwestern Wyoming: McGookey, D. P., and D. N. Miller, Jr. (eds.), 285 p.

(98) Wyoming Geological Association, 1961, Late Cretaceous Rocks: 16th Annual Field Conference, 350 p.

(99) Wyoming Geological Association, 1962, Early Cretaceous Rock: 17th Annual Field Conference, 340 p.

(100) Wyoming Geological Association, 1963, Northern Powder River Basin: Joint Field Conference with the Billings Geological Society, 205 p.

(101) Wyoming Geological Association, 1965, Late Cretaceous and Tertiary Sedimentation, Rock Springs Uplift: 19th Annual Field Conference, 240 p.

(102) Wyoming Geological Association, 1973, Greater Green River Basin Symposium: Schell, E. M. (ed.), 246 p.

(103) Hunt, C. B. 1946, Guidebook to the geology and geography of the Henry Mountain Region: Utah Geol. Soc. Guidebook to Geol. of Utah, No. 1.

(104) Spieker, E. M., 1909, The transition between the Colorado Plateaus and the Great Basin in Central Utah: Utah Geol. Soc. Guidebook to Geol. of Utah, No. 4.

(105) Utah Geological Society and Intermountain Assoc. of Petroleum Geologists, 1965, Geology and resources of south-central Utah: Goode, H. D., and R. A. Robison (ed.), Guidebook to the Geology of Utah, No. 19, 177 p.

(106) Utah Geological and Mineralogical Survey, 1966, Central Utah coals - a guidebook prepared for the Geological Society of America and associated Societies: Utah Geol. and Min. Survey, Bull. 80, 164 p.

(107) Utah Geological Association, 1972, Plateau-Basin and Range transition zone, central Utah, 1972: Baer, J. L., and E. M. Callaghan (ed.), U.G.A. Publication No. 2, 123 p.

(108) Intermountain Assoc. of Petroleum Geologists, 1954, Geology of portions of the high plateaus and adjacent canyon lands, central and south-central Utah: Grier, W. (ed.), 145 p.

(109) Intermountain Assoc. of Petroleum Geologists and Rocky Mountain Assoc. of Geologists, 1955, Guidebook to the geology of Northwest Colorado: Ritzma, H. R., and S. S. Oriel (ed.), 185 p.

(110) Intermountain Assoc. of Petroleum Geologists, 1963, Guidebook to the geology of Southwestern Utah: Heylmun, E. B. (ed.), 232. p.

(111) Montana Geological Society, 1972, Crazy Mountain Basin: Lynn, J. (ed.), 222 p.

(112) Montana Geological Society, 1975, Energy reserves of Montana: Twenty-second Annual Publication of the Montana Geol. Soc., 232 p.

(113) Montana Geologic Society, 1977, Rocky Mountain thrust-belt geology and resources: Heisey, E. I., D. Lawson, E. Norwood, P. H. Wach, and L. A. Hale (eds.), joint guidebook with the Wyoming Geological Assn. and the Utah Geological Assn., 787 p.

(114) Four Corners Geological Society, 1955, Geology of parts of Paradox, Black Mesa, and San Juan Basins: Cooper, J. C. (ed.), 217 p.

(115) Four Corners Geological Society, 1957, Geology of southwestern San Juan Basin: Little, C. J. (ed.), Second Field Conference, 198 p.

(116) Four Corners Geological Society, 1973, Cretaceous and Tertiary rocks of the Southern Colorado Plateau: Fassett, J. E., (ed.), 218 p.

(117) Rocky Mountain Association of Geologists, 1972, Geologic Atlas of the Rocky Mountain region: Mallory, W. W. (ed.), 331 p.

(118) Landis, E. R., 1959, Coal resources of Colorado: U.S.G.S. Bull. 1072-C, p. 131-232.

(119) Combo, J. X., D. M. Brown, H. F. Pulver, and D. A. Taylor, 1949, Coal resources of Montana: U.S.G.S. Circular 53, 28 p.

(120) Combo, J. X., C. H. Holmes, and H. R. Christner, 1950, Map showing coal resources of Montana: U.S.G.S. Coal Inv. Map C-2.

(121) Berryhill, H. L., Jr., D. M. Brown, A. Brown, and D. A. Taylor, 1950, Coal resources of Wyoming: U.S.G.S. Circular 81, 78 p.

(122) Berryhill, H. L., Jr., D. M. Brown, R. N. Burns, and J. X. Combo, 1951, Coal resources map of Wyoming: U.S.G.S. Coal Inv. Map C-6.

(123) Read, C. B., R. T. Duffner, G. H. Wood, and A. D. Zapp, 1950, Coal resources of New Mexico: U.S.G.S. Circular 89, 24 p.

(124) Trumbull, J.V.A., 1959, Coal fields of the United States U.S.G.S. Map.

(125) Coleman, J. M., S. M. Gagliano, and J. E. Webb, 1964, Minor sedimentary structures in a prograding distributary: Marine Geology, v. 1, p. 240-258.

(126) Fisk, H. N., E. McFarlan, Jr., C. R. Kolb, and L. J. Wilpert, Jr., 1954, Sedimentary framework of the modern Mississippi delta: Journal of Sed. Petrology, v. 24, No. 2, p. 76-99.

(127) Kolb, C. R., and J. R. VanLopick, 1965, Depositional environments of the Mississippi River deltaic plain - southeastern Louisiana: in Shirley, M. L. (ed.), Deltas in their Geologic Framework, Houston Geol. Soc., p. 17-61.

(128) Bernard, H. A., C. F. Major, Jr., B. S. Parrott, and R. J. LeBlanc, Sr., 1970, Recent sediments of Southeast Texas: a field guide to the Brazos alluvial and deltaic plains and the Galveston barrier island complex: Bureau of Econ. Geology, University of Texas, Austin, Guidebook No. 11, 200 p.

(129) Fisher, W. L., 1969, Facies characterization of Gulf Coast basin delta systems, with some holocene analogs: Trans., Gulf Coast Assoc. of Geol. Soc., v. 19, p. 239-261.

(130) Psuty, N. P., 1965, Beach-ridge development in Tabasco, Mexico: Coastal Studies Institute, Louisiana State University; Investigations in Tabasco, Mexico, Technical report No. 24, part A, contribution No. 65-2, 14 p.

(131) Howard, J. D., and H. E. Reineck, 1972, Physical and biogenic sedimentary structures of the nearshore shelf: Senckenbergiana marit., Band 4, p. 81-123.

(132) Hoyt, J. H., and R. J. Weimer, 1963, Comparison of modern and ancient beaches, central Georgia coast: Amer. Assoc. Petroleum Geologist Bull., v. 47, no. 3, p. 529-532.

(133) Curray, J. R., and D. G. Moore, 1964, Holocene regressive littoral sand, Costa de Nayarit, Mexico: in van Straaten, L.M.J.U. (ed.), Deltaic and Shallow Marine Deposits, Proceedings of the Sixth International Sedimentological Congress, the Netherlands and Belgium - 1963; Developments in Sedimentology, V. 1, p. 76-82.

(134) Curray, J. R., 1969, Shore zone and bodies: barriers, cheneirs, and beach ridges: in Amer. Geol. Inst. short course lecture notes, the new concepts of continental margin sedimentation-application to the geological record, November 7-9, 1969; Lecture 2, p. JC-II-1 to 18.

(135) Clifton, H. E., R. E. Hunter, and R. L. Phillips, 1971, Depositional structures and processes in the nonbarred high-energy nearshore: Journal of Sedimentary Petrology, V. 41, No. 3, p. 651-670.

(136) Allen, J.R.L., 1964, Sedimentation in the modern delta of the River Niger, West Africa: in van Straaten, L.M.J.U. (ed.), Deltaic and Shallow Marine Deposits: Proc. of the 6th International Sedimentological Congress - 1963 - the Netherlands and Belgium: Developments in Sedimentology, v. 1, p. 26-34.

(137) Allen, J.R.L., 1965, Late Quaternary Niger delta, and adjacent areas: sedimentary environments and lithofacies: Amer. Assoc. Petroleum Geol. Bull., v. 49, No. 5, p. 547-600.

(138) Dapples, E. C., and M. E. Hopkins, 1969, (eds.), Environments of Coal Deposition - Geol. Soc. America, Coal Geology Div., Ann. Mtg. 1964, Symposium: Geol. Soc. American Spec. Paper 114, 204 p.

(139) Spackman, W., A. D. Cohen, P. H. Given, and D. J. Casagrande, 1974, (eds.), The comparative study of the Okefenokee Swamp and the Everglades - mangrove swamp - marsh complex of southern Florida: Geol. Soc. Amer. Pre-convention field trip no. 6, 265 p.

(140) Weimer, R. J., 1973, A guide to uppermost Cretaceous stratigraphy, central Front Range, Colorado: Deltaic sedimentation, growth faulting and early Laramide crustal movement: Mountain Geologist, v. 10, no. 3, p. 53-97.

(141) Asquith, D. O., 1970, Depositional topography and major marine environments, Late Cretaceous, Wyoming: Amer. Assoc. Petroleum Geol. Bull., v. 54, no. 7, p. 1184-1224.

(142) Kauffman, E. G., J. D. Powell, and D. E. Hattin, 1969, Cenomanian-Turonian facies across the Raton Basin: Mountain Geologist, v. 6, no. 3, p. 93-118.

(143) Kauffman, E. G., 1969, Cretaceous marine cycles of the Western Interior: Mountain Geologist, v. 6, no. 4, p. 227-245.

(144) Kauffman, E. G., 1977, Cretaceous facies, faunas, and paleoenvironments across the Western Interior Basin: Mountain Geologist, v. 14, nos. 3 and 4, p. 75-274.

(145) Land, C. B., Jr., 1972, Strata of Fox Hills Sandstone and associated formations, Rock Springs uplift and Wamsutter Arch area, Sweetwater County, Wyoming: A shoreline-estuary sandstone model for the Late Cretaceous: Colo. School Mines Quarterly, v. 67, no. 2, 69 p.

(146) Fassett, J. E., and J. S. Hinds, 1971, Geology and fuel resources of the Fruitland Formation and Kirtland Shale of the San Juan Basin, New Mexico and Colorado: U.S.G.S. Prof. Paper 676, 76 p.

(147) Molenaar, C. M., 1973, Sedimentary facies and correlation of the Gallup Sandstone and associated formations, northwestern New Mexico: in Fassett, J. E. (ed.), Cretaceous and Tertiary Rocks of the Southern Colorado Plateau: Four Corners Geol. Soc. Memoir, Oct. 12-13, 1972, p. 85-110.

(148) Masters, C. D., 1966, Sedimentology of the Mesa Verde Group and of the upper parts of the Mancos Formation, Northwest Colorado: doctoral thesis, Yale University, 88 p.

(149) Hubert, J. F., J. G. Butera, R. F. Rice, 1972, Sedimentology of Upper Cretaceous Cody-Parkman delta, southwestern Powder River Basin, Wyoming: Geol. Soc. Amer. Bull., v. 83, p. 1649-1670.

(150) McGowen, J. H., 1968, Utilization of depositional models in exploration for non-metallic minerals: in Brown, L. F., Jr., (ed.), Proc., 4th Forum on Geol. of Ind. Minerals, Bureau of Econ. Geology, University of Texas at Austin; March 14-15, 1968, p. 157-174.

(151) Kaiser, W. R., 1974, Texas lignite: near-surface and deep-basin resources: Univ. Texas, Austin, Bur. Econ. Geology Report of Investigations, no. 79, 44 p.

(152) Reeves, D. R., 1971, In-situ analysis of coal by borehole logging techniques: Canadian Mining and Metallurgical (CIM) Bull., v. 74, p. 61-69.

(153) Reeves, D. R., 1977, Application of wireline logging techniques to coal exploration: in Coal Exploration; Proceedings of the first International Coal Exploration Symposium, p. 112-128.

(154) Siemers, C. T., 1977, Core and e-log analysis of a coal-bearing sequence: lower part of the Upper Cretaceous Menefee Formation (Mesaverde Group), northwestern New Mexico: Workshop - 1977 Rocky Mtn. Coal Symposium: Cities Service Company, Tulsa, Okla., 10 p.

(155) Archer, B. J., Jr., and M. Warrick, 1964, Better drill logging by gamma rays and resistivity: Coal Age, September 1964, p. 79-81.

(156) Bond, L. O., R. P. Alger, and A. W. Schmidt, 1969, Well log applications in coal mining and rock mechanics: Soc. of Mining Engineers of A.I.M.E. preprint No. 69-F-13, 19 p.

(157) Brown, L. F., Jr., 1969, Geometry and distribution of fluvial and deltaic sandstones (Pennsylvanian and Permian), north-central Texas: Trans., Gulf Coast Assoc. Geol. Soc., V. 19, p. 23-47.

(158) Hensel, R. P., and R. W. Skowyra, 1976, Properties of low-ranked coals and their influence on industrial boiler deisgn: Combustion Engineering Report TIS-4821, 9 p.

(159) Morris, D. A., 1968, Origin and significance of coal splits: in Geol. Soc. Amer. Spec. Paper 121 (Abstracts for 1968), 1969, p. 209.

(160) Donaldson, A. C., 1976, Origin of coal seam discontinuities: unpublished manuscript from talk given in Raliegh Co., West Virginia on Nov. 10, 1976.

(161) Vaninetti, G. E., 1976, Sedimentation and multiple coal seam correlations in Upper Cretaceous swamp complex, John Henry Member of Straight Cliffs Formation, southern Kaiparowits region, Utah: Amer. Assoc. Petroleum Geol. Bull., v. 60, no. 8, p. 1412-1413.

(162) Popp, J. T., and W. W. Carman, 1976, Mining problems related to geology in the Beckley coalbed: unpublished manuscript from talk given in Raliegh Co., West Virginia on Nov. 10, 1976.

(163) Dunham, R. K., 1976, The influence of discontinuities on mine operation: unpublished manuscript from talk given in Raliegh Co., West Virginia on Nov. 10, 1976.

(164) United States Geological Survey, 1976, Coal resource classification system of the U.S. Bureau of Mines and the U.S. Geological Survey: U.S.G.S. Bull. 1450-B, 7 p.

(165) Leonard, J. W., and D. R. Mitchell, 1968, Coal preparation: American Institute of Mining, Metallurgical, and Petroleum Engineers, Inc.

(166) Walton, P. T., 1955, Wasatch Plateau gas fields, Utah: Amer. Assoc Petroleum Geol. Bull., v. 39, p. 385-421.

(167) Fisher, D. J., C. E. Erdmann, and J. B. Reeside, Jr., 1960, Cretaceous and Tertiary formations of the Book Cliffs, Carbon, Emery, and Grand Counties, Utah, and Garfield and Mesa Counties, Colorado: U.S.G.S. Prof. Paper 332, 80 p.

(168) Young, R. G., 1955, Sedimentary facies and interfingering in the Upper Cretaceous of the Book Cliffs, Utah-Colorado: Geol. Soc. American Bull., v. 66, p. 177-201.

(169) Young, R. G., 1957, Late Cretaceous cyclic deposits, Book Cliffs, eastern Utah: Am. Assoc. Petroleum Geol. Bull., v. 41, p. 1760-1774.

(170) Young, R. G., 1966, Stratigraphy of coal-bearing rocks of Book Cliffs, Utah-Colorado: in Central Utah Coals - A guidebook prepared for the Geological Society of America and associated Societies: Utah Geol. and Mineralogy, Survey Bull. 80, p. 7-21.

(171) Young, R. G., 1976, Genesis of western Book Cliffs Coals: Brigham Young University Geology Studies, v. 22, part 3, p. 3-14.

(172) Balsley, J. K., 1977, Deltaic origin of the Sunnyside Coal, western Book Cliffs, Utah: in Hodgson, H. (ed.), Geology of Rocky Mountain Coal: Colo. Geol. Survey Resources Series 4, p. 219.

(173) Marley, W. E., R. M. Flores, and V. V. Cavaroc, 1978, Lithogenetic variations of the Upper Cretaceous Blackhawk Formation and Star Point Sandstone in the Wasatch Plateau, Utah: Geol. Soc. Amer. Abstracts with Programs, v. 10, no. 5, p. 233.

(174) Doelling, H. H., 1972, Central Utah coal fields: Sevier-Sanpete, Wasatch Plateau, Book Cliffs, and Emery: Utah Geol. and Mineralogy. Survey Monograph 3, 496 p.

(175) Maurer, R. E., 1966, The coal fields of eastern Sevier County, Utah: in Central Utah Coals, A guidebook prepared for the Geological Society of America and associated Societies: Utah Geol. and Mineralogy, Survey Bull. 80, p. 111-119.

(176) Peterson, F., 1969, Four new members of the Upper Cretaceous Straight Cliffs Formation in the southeastern Kaiparowits region, Kane County, Utah: U.S. Geol. Survey Bull. 1274-J, p. J1-J28.

(177) Peterson, F., 1969, Cretaceous sedimentation and tectonism in the southeastern Kaiparowits region, Utah: U.S. Geol. Survey, open file report, 259 p.

(178) Peterson, F., 1975, Influence of tectonism on deposition of coal in Straight Cliffs Formation (Upper Cretaceous), south-central Utah: Amer. Assoc. Petroleum Geol. Bull., v. 59, no. 5, p. 919.

(179) Peterson, F., and H. A. Waldrop, 1965, Jurassic and Cretaceous stratigraphy of south-central Kaiparowits Plateau, Utah: Utah Geol. Soc. and Intermountan Assoc. Petroleum Geologist, Guidebook to the Geology of Utah, no. 19, p. 47-69.

(180) Zeller H. D., 1973, Geologic map and coal resources of the Death Ridge quadrangle, Garfield and Kane County, Utah: U.S. Geol. Survey Coal Investigations Map C-58.

(181) Doelling, H. H., and R. L. Graham, 1972, Southwestern Utah Coal Fields: Alton, Kaiparowits Plateau and Kolob-Harmony: Utah Geol. and Mineralog. Survey Monograph Series No. 1, 333 p.

(182) Peterson, F., and H. A. Waldrop, 1966, Preliminary geologic map of the southeast quarter of the Gunsight Butte quadrangle, Kane and San Juan Counties, Utah, and Coconino County, Arizona; Utah Geol. and Mineralog. Survey Map 24-G (1967).

(183) Peterson, F., 1966, Preliminary geologic map and coal deposits of the northwest quarter of the Gunsight Butte quadrangle, Kane County, Utah: Utah Geol. and Mineralog. Survey Map 24-E (1967).

(184) Peterson, F., and G. W. Horton, 1966, Preliminary geologic map and coal deposits of the northeast quarter of the Gunsight Butte quadrangle, Kane County, Utah: Utah Geol. and Mineralog. Survey Map 24-F (1967).

(185) Waldrop, H. A., and F. Peterson, 1966, Preliminary-geologic map of the southeast quarter of the Nipple Butte quadrangle, Kane County, Utah; and Coconino County, Arizona: Utah Geol. and Mineralog. Survey Map 24-C (1967).

(186) Waldrop, H. A., and R. L. Sutton, 1966, Preliminary geologic map and coal deposits of the northwest quarter of the Nipple Butte quadrangle, Kane County, Utah: Utah Geol. and Mineralog. Survey Map 24-A (1967).

(187) Waldrop, H. A., and R. L. Sutton, 1966, Preliminary geologic map and coal deposits of the northeast quarter of the Nipple Butte quadrangle, Kane County, Utah: Utah Geol. and Mineralog. Survey Map 24-B.

(188) Waldrop, H. A., and P. L. Sutton, 1966, Preliminary geologic map and coal deposits of the southwest quarter of the Nipple Butte quadrangle, Kane County, Utah: Utah Geol. and Mineralog. Survey Map 24-D.

DISCUSSION

QUESTION: What are the physical characteristics of the coal seams associated with the interpreted "deltaic" and "interdeltaic" sediments? What is the frequency of splitting, thickness, and quality?

ANSWER: Western deltaic coals are characterized by the presence of only a few well developed coal seams (less than three) that offlap in the seaward direction and that are separated by sandstone-rich intervals, lateral discontinuities, local channeling of coal, distributary channel deposits that interfinger with coal seams, rock splits in areas of distributary channels, local high sulfur areas and higher ash coals.

Western interdeltaic coals are characterized by the presence of several well developed coal seams that are vertically stacked and separated by mudstone-rich intervals, lateral continuities especially well developed parallel to shorelines, rare channeling (if present at all) of coal, abundant crevasse splay deposits in areas marginal to the meandering channels that locally interrupt the continuity of the swamps, and very low ash and sulfur contents.

Both Western deltaic and interdeltaic coals are characterized by rock splits in swamp areas adjacent to channels. In that the channel density of deltaic environments is greater than in interdeltaic areas, one would expect to encounter more rock splits and higher ash coals there. The mobility and dynamic nature of a prograding delta is such that coals are less likely to develop to great thicknesses and stack vertically than the coals of the interdeltaic environment.

QUESTION: How does differential compaction affect the geometry and correlation of stratified units in coal-bearing sediments? Is there anything characteristic of certain geometries of units that overlie coals that can be related to depositional processes?

ANSWER: Differential compaction can result in profound geometric complexities in determining the correlation of coal seams. In many cases, the effects of differential compaction are interpreted as faults because individuals comparing drill hole data from hole to hole cannot comprehend the significant variations in thickness for the interval between seams that can result from differential compaction. A case in point is illustrated in a photograph (see p. 194) of a strip-pit wall at Kemmerer, Wyoming where a 32 meter thick seam is split, in the distance of 700 meters, into two seams with a 30 meter rock split (crevasse splay channel).

The effects of differential compaction are most pronounced in comparing sandstone with mudstone. As has been shown from the Kaiparowits Plateau example, sandstone does not compact to any significant degree while mudstone compacts quite significantly. The net result is to yield an interval between seams (or unit) that is characterized by great variations in thickness.

Sandstones deposited in stream channels are usually lens-shaped "ribbons" with sharp basal contacts and abrupt lateral interfingering contacts with mudstones deposited in marsh environments.

QUESTION: Compare and contrast the value of cores and geophysical well logs.

ANSWER: Geophysical well logs can be obtained in rotary drilled holes at from one-half to one-third of the cost of coring the hole. The information obtained is in many respects superior to just core information. In any event, geophysical logs should be obtained for all drill holes, regardless of whether they are cored or rotary drilled.

Geophysical logs permit the recognition and interpretation of unit thicknesses to within 2 to 3 centimeter accuracy, lithologies, general quality trends and rock splits and allow the recognition of gross features usually not identified in cores only. Geophysical logs must be thought of as raw data that can be used to verify the presence and thickness of coal seams and other units that may have been lost because of incomplete core recovery or are no longer available for inspection because of core sampling and testing. Carbonaceous rock splits and bony zones that are sometimes not identified in megascopic analysis of wet coal cores are easily seen in geophysical logs. Depositional units and events (such as fining upward fluvial units) are more easily recognized on geophysical logs than by inspection of cores.

The three major reasons for collecting cores in coal-bearing sediments are for (1) determination of detailed coal quality, (2) rock mechanics testing, and (3) gaining confidence in geophysical log interpretations. Usually a few cores will suffice for satisfying these reasons and the majority of drill holes can be rotary drilled so that drilling costs may be minimized.

If drill hole cores or geophysical logs are not available for inspection when evaluating the merits of a coal deposit, one must realize that he is not evaluating raw data and therefore any conclusions arrived at from examining available "data" must be considered tentative.

QUESTION: Other than the three main logs (gamma, density, and resistance) what logs find application in evaluating coal-bearing sediments?

ANSWER: Velocity, sonic, and induction logs have been used to advantage to determine the engineering properties (strength and competency) of the sediments that enclose coal seams. Neutron and S. P. logs are useful in determining lithologies and locating water-saturated and fresh water units. Caliper logs are of use in determining the diameter of the drill hole and locating units which erode or disaggregate easily. Deviation logs permit the determination of the inclination of a drill hole from vertical and the direction of the deviation.

QUESTION: Are the illustrations of the geophysical logs for coal actual? If so, how are such "square-shouldered" logs obtained (particularly so on the gamma log)?

ANSWER: The illustrations are based on actual logs obtained throughout the coal fields of Utah. The "square-shouldered" nature of these logs is characteristic of most coal logs I've seen. The upper contact of a log in a coal seam is usually very sharp because of the sharp contact with overlying sediments that is characteristic of coal-bearing sediments. This is because coal swamps are usually terminated by an instantaneous event such as a flood or development of a crevasse splay which results in the deposition of non-coal material directly on the coal vegetation.

The basal contacts of coal logs are usually not as sharp because of the gradual development of a coal swamp during its early stages of growth. The early stages of swamp development is characterized by a gradual takeover of an area by vegetation with the incorporation of much of the substrate (mudstone) by the action of growing roots. As later vegetation "roots" itself in earlier vegetation, the swamp vegetation becomes purer and has less ash content.

The ultimate sharpness of geophysical curves in coal seams is quite dependent upon the speed at which the hole is logged. The faster a log is run, the poorer the quality of the log because the probe equipment does not have sufficient time to react to all of the subtleties commonly present in coal seams and enclosing sediments. I've found that a speed of two meters per minute is the optimum logging speed.

The speed at which a log is run determines the accuracy of measurements of unit thicknesses taken from that log and the point in the log where the thickness is measured. If logs are obtained at slow speeds, measurements will tend to be nearer the maximum deflection side of the curve than the base-line side.

QUESTION: Was the drill spacing pattern in the case of the Wasatch Plateau example determined by rough topography and difficult access?

ANSWER: To a certain degree, most drilling program spacing patterns are determined by access. In this particular case, access problems helped determine the initial spacing pattern but the main concern was to intersect the distributary channel that was projected into the subsurface so as to ascertain its presence in untested portions of the deposit. In this case, more interest was placed on where the coal wasn't present than on where the coal was present. Later drill holes were spaced so as to permit the calculation of proven reserves so that accurate mining plans could be developed.

QUESTION: Which are the depositional environments that produce high sodium values?

ANSWER: No data are presently available that indicate that high sodium values for coal are influenced by or are a function of depositional environments. Available data indicate that the sodium is bound up within the mineral structure (molecular level) of the plant materials that form coal. It is reasonable to assume that certain plants' metabolic functions encouraged the concentration of sodium from the waters in the coal swamp at the time of growth but no associations with depositional environments have yet been identified.

QUESTION: Are sodium concentrations related to the structure of the deposit?

ANSWER: We've tried to relate sodium concentrations and the overall high sodium levels of all the coals in the Wasatch Plateau example to structure, stratigraphy, groundwater, jointing, faulting, and lateral continuity but have been unable to find a cause-effect relationship. We've taken detailed and closely-spaced samples both vertically and laterally to determine trends and none has yet been identified. There are no special concentrations of sodium along faults or in the synclinal areas. We've noticed that the gypsiferous fracture coatings on some of the coal cleats is somewhat enriched in sodium but not to the extent that it can be considered significant.

I know of only a few other high sodium coal deposits in the West (Montana and New Mexico) but no data are available that indicate the origin of those concentrations. Obviously, additional research is needed to determine the origin and reason for concentration of sodium in the ash of certain coal deposits. I'd be very interested in hearing of others' problems and other deposits where abnormal concentrations of sodium have been identified.

QUESTION: Please expand on the adverse mining conditions that are associated with the differential compaction of channel sandstone.

ANSWER: Differential compaction on the margins of channel sandstone lenses where sediments are thinly-bedded, rapid variations in lithologies, and sharp contacts between different lithologies are common results in the development of compactional slickensides. The slickensides in combination with the variations in lithologies noted above present difficult roof control problems in areas where an underlying coal seam has been extracted because of a lack of competency and a tendency to slack readily.

Differential compaction in the center of paleochannel sandstones commonly results in "rolls" in coal beds where the overlying sandstone has been pushed downward into the coal seam and deformed the seam so that it does not conform to the regional structure. The location of rolls in coal seams is predictable with a knowledge of the trend and location of paleochannel sandstone bodies in the roof of coal mines where it has been previously exposed and from outcrop and drill hole information.

Recent studies by the U. S. Bureau of Mines in the Beckley Coal Bed in West Virginia have shown numerous instances where roof conditions in coal mines can be directly related to the effects of differential compaction near paleochannel sandstones.

Figure 13

8

Exploration for Gulf Coast United States Lignite Deposits: Their Distribution, Quality and Reserves

By James A. Luppens
Area Geologist
Phillips Coal Company
Tyler, Texas, United States

INTRODUCTION

Cyclic deposition has been the fundamental pattern of deposition from the entire Paleocene/Eocene lignite-bearing strata through the modern sediments of the United States Gulf Coast. The thick clastic wedge of sediments which has accumulated in the Gulf Coast geosyncline consists of alternating sequences of regressive, lignite-bearing fluvial-deltaic units and transgressive marine units.

Within these regressive sequences, three major lignite-bearing depositional environments occur. These are fluvial, deltaic, and strandplain-lagoonal. Because the mode of lignite deposition enables one to predict the range in thickness, geometry, areal distribution, and quality of the lignites, a basic understanding of these environments is essential for implementation of a sound exploration program.

The geology of the Gulf Coast lignites is somewhat unique in that all the principal ancient lignite-bearing environments have, as a modern Gulf Coast analogue, one of the most studied modern peat-forming environments in the world - namely the Mississippi delta and associated Louisiana and

Texas coastal deposits. To gain better insight into the nature of the lignite formation and distribution, an ancient example of each of the three principal lignite-bearing environments is compared to a modern analogue. The ancient examples are all taken from Texas Eocene strata due to the recent excellent and thorough studies of Fisher and McGowen (1) and especially Kaiser (2,3); however, these examples can be applied to the entire Gulf Coast region.

STRATIGRAPHY AND ENVIRONMENTS OF DEPOSITION

General Stratigraphy

The Wilcox, Claiborne, and Jackson Group are the principal lignite-bearing strata in the Gulf Coast. The uppermost strata of the Midway Group in portions of Mississippi and Alabama also contain lignite of significant thickness. A map showing the distribution of these groups is found in Figure 1. The strata generally dip gently towards the Gulf Coast at 100 feet per mile or less and are relatively void of any significant structural complexities. A correlation chart from state to state is provided in Figure 2. Due to the scope of this paper, the sections are somewhat generalized; however the important lignite horizons are noted and each unit is designated either transgressive or regressive to emphasize the cyclical nature of the depositional history.

Environments of Deposition

Figure 3 shows the typical distribution of primary depositional systems for the three principal lignite-bearing groups in Texas and a comparison to the modern Gulf Coast depositional systems. This figure clearly demonstrates the similarities in the Gulf Coast depositional systems through time. While other units in Figure 2 may differ in areal extent and geometry, almost all of these units represent all or portions of the systems shown in Figure 3. Of all the depositional environments shown in Figure 3, only three contain lignite as a component facies. These are the fluvial, deltaic and lagoonal environments.

Fluvial The Wilcox group of northeast Texas exhibits sand-body geometries and vertical fining-upward sequences characteristic of fluvial systems, Kaiser (3). Figure 4 demonstrates the dendritic, dip-oriented sand geometry associated with this depositional system.

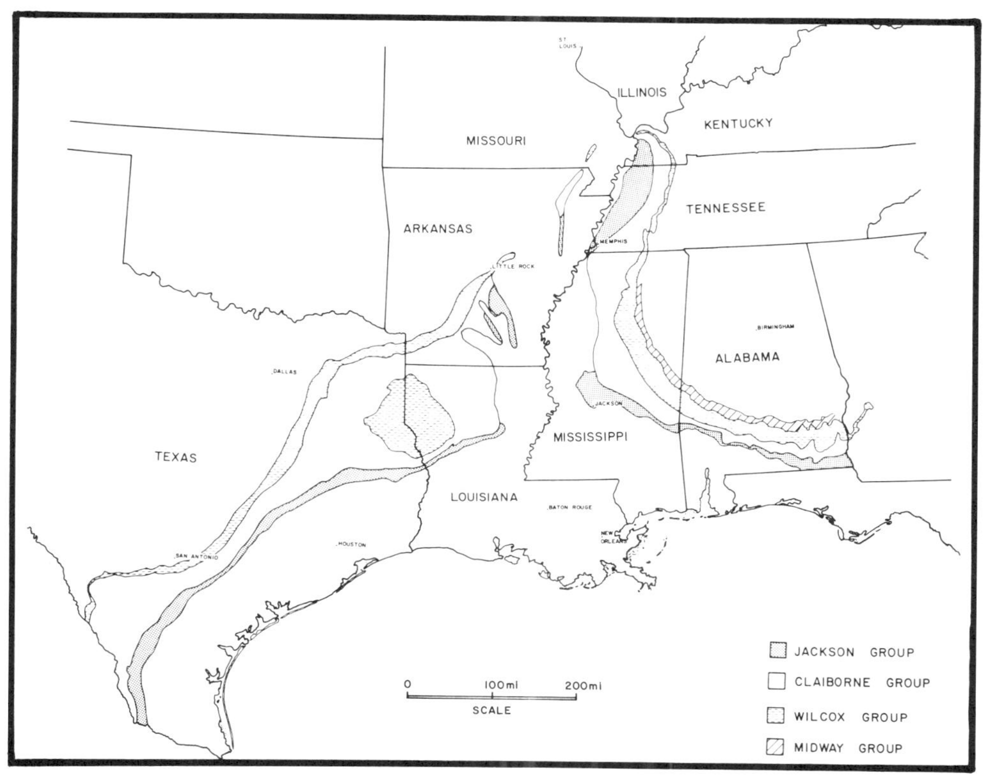

Figure 1. Distribution of principal Tertiary lignite-bearing Groups in the Gulf Coast region of the United States.

SYSTEM	GROUP	EAST, SOUTHEAST, CENTRAL TEXAS	SOUTH TEXAS	LOUISIANA	ARKANSAS	TENNESSEE	MISSISSIPPI	ALABAMA
EOCENE	JACKSON	WHITSETT ★MANNING WELLBORN CADDELL	UNDIVIDED ★	DANVILLE LANDING YAZOO MOODYS BRANCH	UNDIVIDED	UNDIVIDED	YAZOO MOODYS BRANCH	YAZOO MOODYS BRANCH
EOCENE	CLAIBORNE	★YEGUA COOK MOUNTAIN STONE CITY SPARTA WECHES QUEEN CITY RECKLAW CARRIZO	★YEGUA LAREDO EL PICO CLAY BIGFORD CARRIZO	COCKFIELD COOK MOUNTAIN SPARTA CANE RIVER CARRIZO	★COCKFIELD COOK MOUNTAIN SPARTA CANE RIVER CARRIZO	★ UNDIVIDED	★COCKFIELD COOK MOUNTAIN ★KOSCUISKO ZILPHA WINONA TALLAHATTA	GOSPORT LISBON TALLAHATA
EOCENE	WILCOX	★CALVERT BLUFF SIMSBORO HOOPER	★INDIO	★SABINETOWN ★PENDLETON ★MARTHAVILLE	★ UNDIVIDED	★ UNDIVIDED	★HATCHETIGBEE ★TUSCAHOMA ★NANAFALIA	★HATCHETIGBEE ★TUSCAHOMA ★NANAFALIA
PALEOCENE	MIDWAY	WILLS POINT KINCAID	KINCAID	PORTERS CREEK KINCAID-CLAYTON	PORTER CREEK KINCAID	PORTERS CREEK CLAYTON	★NAHEOLA MATHEWS PORTERS CREEK CLAYTON	NAHEOLA ★ CLAYTON

PREDOMINANTLY REGRESSIVE — PREDOMINANTLY TRANSGRESSIVE — ★ SIGNIFICANT LIGNITE OCCURRENCES

Figure 2. Generalized geologic column for the principal lignite-bearing Groups.

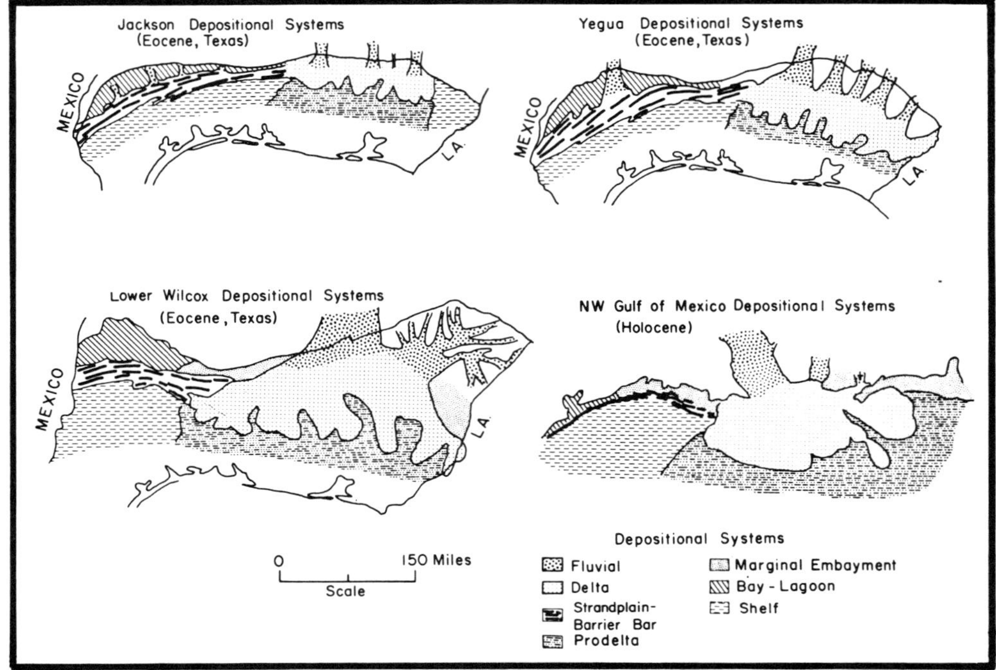

Figure 3. Distribution of principal depositional systems, contrasting Eocene Lower Wilcox, Yegua, and Jackson and the Holocene of northwestern Gulf of Mexico. From Fisher et al (4).

This sand geometry is consistent with that found in the alluvial plain region of the modern Mississippi delta shown in Figure 5. The peats in the modern alluvial plain are accumulating in hardwood back-swamps between natural-levee ridges (Figure 5). Swamps persist because peat accumulation keeps pace with subsidence and are far enough removed from the active channels so that periodic influxes of sediments due to flooding do not inhibit growth of vegetation.

As shown in Figure 4, areas of known lignite accumulation are also found in these sand-poor, interchannel areas. Commercial lignite deposits formed in the fluvial environment typically have one or more seams from 2 to 13 feet thick and range in size from 25 to 400 million tons, Kaiser (3). Larger deposits are commonly cut by one or more sand channels.

Lignites from the fluvial system characteristically have a high percentage of vitrinite (coalified woody fragments), low sulfur contents, moderate ash contents, and moderate BTU values.

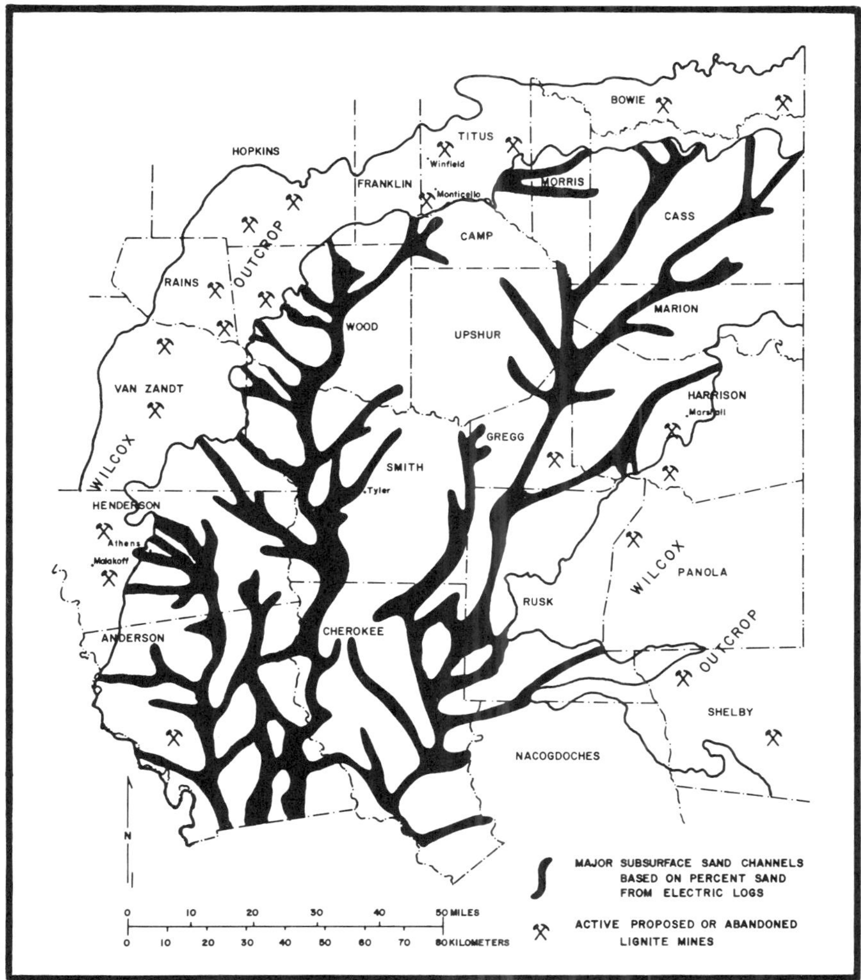

Figure 4. Wilcox sand geometry and lignite occurrences in a fluvial depositional environment of East Texas. Modified from Kaiser (3).

Deltaic The Calvert Bluff Formation of the Wilcox Group in central Texas exhibits a straight or slightly dendritic sand channel geometry updip that merges downdip with a bifurcating channel geometry (Figure 6). As shown in Figure 6, significant lignite deposits are found between the major sand trends.

Analogues of the Calvert Bluff interchannel basins are the Des Allemands-Barataria and Atchafalaya basins shown in Figure 5. Gulfward, these interchannel basins increase in number as trunk streams bifurcate into distributary networks which enclose increasingly smaller interdistributary basins.

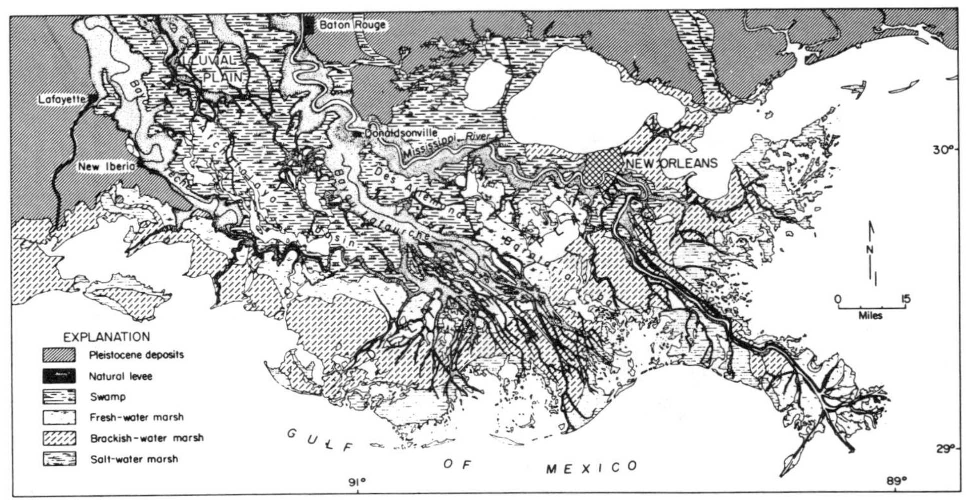

Figure 5. Mississippi delta physiography. From Kaiser (3) and Frazier and Osanik (5).

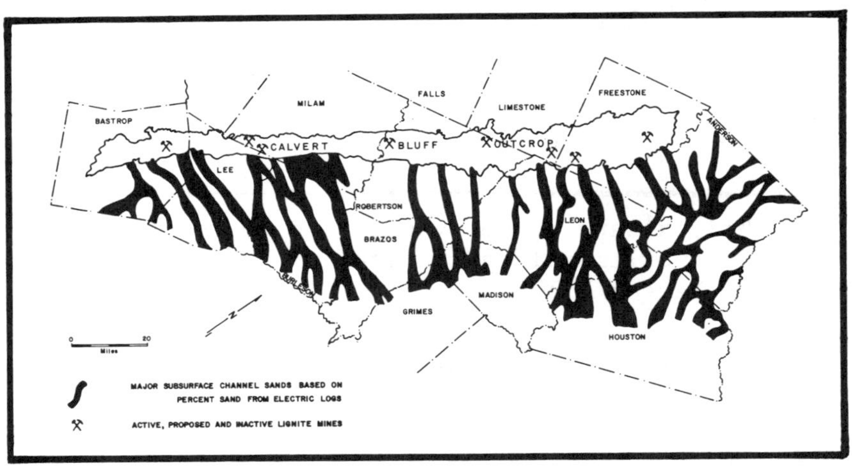

Figure 6. Calvert Bluff sand geometry and associated lignite occurrences from a deltaic environment of deposition in Central Texas. Modified from Kaiser (3).

Maximum peat development is found away from active distributary channels and associated sedimentation and inland from the destructive effects of the Gulf at the junction of the delta and alluvial plain. The blanket-like peats of the upper delta plain and inland interchannel basins are thicker and more widespread than are coastal marsh peats, Frazier and Osanik (5). Coleman and Smith (6) observed blanket peats with areal extents up to 200 miles. These blanket peats commonly cover several abandoned distributary channels.

Lignites in the Calvert Bluff Formation and all delta plain deposits throughout the Gulf Coast are generally the thickest and have the largest areal extent. Commercial lignite seams formed in this environment are typically 5 to 10 feet thick and range from 2 to 25 feet in thickness, Kaiser (3). Individual seams can be traced along strike up to 32 miles. Areally, a seam may cover up to nearly 100 square miles a depth less than 200 feet and contain up to 600 to 700 million tons in place.

Blanket lignites that accumulated in deltaic environments have lower vitrinite contents than fluvial lignites and are characterized by low ash contents, moderate sulfur contents, and high BTU values. In general, sulfur and ash contents increase gulfward with a corresponding decrease in BTU values as the marshes change from fresh to brackish and as the destructional effects of the marine environment become closer.

<u>Lagoonal</u> The Yegua Formation/Jackson Group of South Texas offers an excellent example of lagoonal lignites. The pattern of sedimentation is one of strike-oriented, multistacked, progradational or coarsening - upward, barrier and strandplain beach sequences in which the lignites are associated with inland or updip lagoonal muds Kaiser (2). Regionally, the lignites lie updip (landward) of the axes of the ancient barrier bar - strandplain systems reflecting their lagoonal origin (Figure 7).

Peat is found as a very minor component facies of the modern linear, regressive clastic shorelines of the Texas Gulf Coast barrier bar-lagoon system, Kaiser (3). These modern lagoonal environments also exhibit the coarsening-upward barrier-strandplain sequence topped with lagoonal sediments including peat.

Generally, lagoonal lignites are the poorest in terms of

overall quality. They usually have a high sulfur content reflecting a salt marsh origin. They also have high ash contents and low to moderate BTU values. Lagoonal lignite deposits of more than 200 million tons are present in south Texas in seams 2 to 13 feet thick and extending for up to 13.5 miles, Kaiser (3).

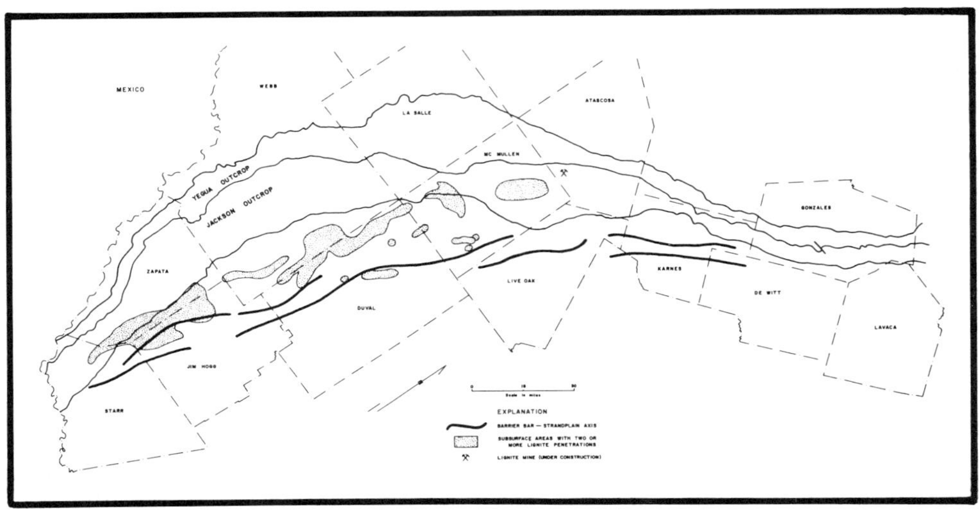

Figure 7. Occurrence of lignite in the Jackson Group-Yega Formation strandplain-lagoonal system of South Texas. Modified from Kaiser (3).

Karst One other type of lignite deposition of very minor importance is worth noting because of its marked contrast to the previously discussed modes of deposition and the extraordinary thickness displayed. In eastcentral Alabama, lignites in the Gravel Creek Sand Member of the Nanafalia Formation were deposited in depressions caused by collapse over solution channels in a karst plain developed on top of the Clayton limestone (Midway). A modern analogue would be the karst region of Florida where there is an extensive zone of closely-spaced solution basin lakes, White (7). The lignites are highly lenticular with thicknesses exceeding 50 feet locally. The seams, however, are discontinuous and very limited in areal extents rarely exceeding one mile in width, Self and Williamson (8).

EXPLORATION METHODS

Initial exploration target selection requires an investigative stage to define the previously discussed lignite-bearing environments. This stage involves a thorough literature search, and contacting appropriate government agencies for available information. One source of particular value is the number of water well lithologic logs available in many areas. The vast amount of oil and gas well logs and related information from the Gulf Coast can also be helpful. Because of the lack of sufficient outcrop data for the most of the Gulf Coast region, such subsurface data may be the only source of information suggesting the presence of, or at least the potential for lignite-bearing environments.

After the target area has been selected, a reconnaissance drilling program is recommended. An initial drill hole grid spacing pattern of 2 miles has proven satisfactory for most areas of the Gulf Coast. This 2 mile grid system may need to be modified occasionally depending on what the particular depositional environment dictates.

Mobile drilling and geophysical equipment as shown in Figure 8 is generally used. A typical drilling crew consists of two drilling rigs and one geophysical logging vehicle. Each drilling crew should be capable of drilling on the average of six to eight 300 feet holes per day. Geophysical logging is essential as it provides very quick and accurate depth and thickness information as well as facilitating coal seam correlations and providing lithologic information used in studying depositional environments. Generally, natural gamma, resistance, and gamma-gamma density logs are the most common suite of curves used in coal exploration.

Upon penetration of a coal seam(s) in one or more holes, the hole spacing is reduced to better define the extent of the seam(s). Several cores are usually taken to determine coal quality. It is at this stage where an understanding of the particular depositional environment becomes especially important.

Awareness and prediction of the complexities that may occur in a particular depositional model not only aids in correlations but also is very helpful in laying out a coring progran. Since coal quality exhibits more variability in the vicinity of seam complexities, prediction of these complexities allows more intelligent placement of cores and biasing of seam quality contours.

In summary, from the initial selection of exploration targets through the detailed prospect evaluation, a thorough understanding of the particular depositional environment is invaluable.

Figure 8. Mobile drilling and geophysical equipment used in Gulf Coast lignite exploration.

QUALITY

Typical as-received proximate analyses are shown in Table 1. Two conclusions are readily apparent from the analyses. First, the relative quality of the lignites is a function of the environment of deposition. Secondly, in general, the overall rank of the lignites decreases from the western to the eastern portions of the Gulf Coast. The lignite rank also decreases with progressively younger lignites with similar depositional environments. The decrease in rank is demonstrated both in the decrease in calorific value and an increase in moisture contents.

Table I. Typical as-received proximate analyses.

	TEXAS				LOUISIANA
	Northeast Wilcox	Central Wilcox	Central Claiborne/Jackson	SW Wilcox Claiborne/Jackson*	Wilcox
M	32.00	31.66	40.36	30.00	30.22
A	14.85	11.42	13.40	28.40	12.05
VM	26.43	29.95	26.82	23.20	29.05
FC	26.72	26.97	19.42	18.40	28.68
S	0.59	0.93	1.20	1.67	0.71
BTU	6584	7038	5837	5000	7073

	ARKANSAS			TENNESSEE
	Southwestern Wilcox	Northeastern Wilcox	Southwestern Claiborne	Claiborne
M	33.91	41.70	37.13	45.41
A	15.44	13.06	10.96	11.89
VM	25.83	25.40	30.97	26.56
FC	24.82	19.84	20.94	16.14
S	0.53	0.60	0.98	0.59
BTU	6171	5465	6733	5379

	MISSISSIPPI			ALABAMA	
	North Wilcox	South Wilcox	Claiborne	West Wilcox	Central Wilcox
M	43.51	43.19	41.69	45.87	49.54
A	12.08	11.84	13.89	8.45	9.48
VM	25.26	24.33	30.11	23.60	21.94
FC	19.15	20.64	14.31	22.08	19.04
S	0.54	1.17	0.54	2.65	2.08
BTU	5396	5509	5855	5738	5156

*From Sondreal et al (9)

The Texas Wilcox analyses from Table I clearly illustrate the relationship between environments of deposition and quality. From the northeast to central and finally to south Texas, the environments change from fluvial, deltaic, and lagoonal respectively. In general, there is an overall decrease in lignite quality southwestward along the Wilcox outcrop trend. Trends in relative ash, sulfur and BTU values are quite predictable. Sulfur increases progressively southward as the ancient depositional environment becomes distinctly more marine. Ash varies from moderate to low to high southward and is a function of the proximity of the area of peat accumulation to a source(s) of contaminating sediments. Correspondingly, the BTU values mirror the ash variations.

Similar trends can be seen throughout the Gulf Coast lignite region. For example, from the Wilcox in northern Mississippi to the Wilcox of western Alabama, the units change from dominantly fluvial and deltaic to deltaic and shallow marine, and is reflected in the marked increase in sulfur contents southeastward.

Essentially all the Gulf Coast lignites are classified as Lignite A according to the ASTM coal classification. Locally, seams may be classified as Lignite B in a few instances in the eastern Gulf Coast region.

Examples of complete analyses from Texas, Mississippi, and North Dakota lignites are listed in Table 2 for comparison between the two lignite provinces.

RESERVES

The following state by state in-place lignite reserve figures are preliminary estimates based on both published information and drilling information of Phillips Coal Company. The figures represent coals 3 feet thick or greater and generally at depths less than 200 feet. For Mississippi, Tennessee and Louisiana, these figures represent the first known published total lignite estimates.

Texas The total estimated lignite reserves for Texas are 11.5 billion tons which ranks Texas first among the Gulf Coast states in total lignite reserves. Over half of these reserves are found in the Wilcox Group. Individual seam thicknesses for the Wilcox lignites range from 2 to 25 feet but are typically 5 to 10 feet thick. In general, lignite seams in the Claiborne and Jackson Groups are thinner with typical thicknesses of 4 to 8 feet.

Louisiana The total estimated lignite reserves for Louisiana are 1.1 billion tons. All of these reserves are in the Wilcox Formation. Seam thicknesses range from 2 to 17 feet but are typically 3 to 8 feet thick. One thick, continuous seam referred to as the Chemard Lake Lentil contains over one-third of the total reserves for Louisiana, Roland et al (11).

Arkansas Most of the lignite is found in the southwest and southcentral portions of the state. Total reserves are estimated to be 2.5 billion tons. Essentially all of these reserves found in the Wilcox and Claiborne Groups, Wielchowsky et al (12). Typical seam thicknesses range from 2 to 7 feet. Occasional thin lignite seams are also found in the Jackson Group.

Tennessee Total estimated lignite reserves for Tennessee are 1.0 billion tons. All of these reserves are found in the Wilcox and Claiborne Formations. Typical seam thickness range from 2 to 9 feet.

Table II. Comparison of complete analyses from the Gulf Coast and North Dakota lignites.

PROXIMATE ANALYSIS (%)	WILCOX CENTRAL TEX.	CLAIBORNE MISSISSIPPI	NORTH DAKOTA*
Moisture	31.66	41.69	40.00
Ash	11.42	13.89	6.70
Volatile Matter	29.95	30.11	25.90
Fixed Carbon	26.97	14.31	27.40
Sulfur	0.93	0.54	0.48
BTU	7038	5855	5940
ULTIMATE ANALYSIS			
(Dry Basis,%)			
Carbon	59.69	54.30	59.52
Hydrogen	4.98	5.31	3.82
Nitrogen	1.03	0.64	1.07
Sulfur	1.36	0.92	0.80
Ash	16.71	23.83	11.17
Oxygen	16.24	15.00	23.62
ASH COMPOSITION (%)			
SiO_2	41.17	61.39	21.20
Al_2O_3	11.94	14.58	11.50
Fe_2O_3	7.42	6.56	5.80
CaO	19.54	7.27	21.10
MgO	2.04	1.50	3.90
K_2O	0.35	0.47	0.30
Na_2O	0.63	0.30	9.70
TiO_2	0.81	0.47	0.60
P_2O_5	0.01	0.04	0.80
SO_3	12.46	5.54	22.20
FUSIBILITY TEMPERATURES			
(reducing atmosphere, F°)			
Initial Deformation	2,156	2,318	1910
Softening	2,210	2,336	2050
Hemispherical	2,246	2,354	2070
Fluid	2,300	2,450	2140
HARDGROVE GRINDABILITY INDEX	52	101	35-70

*From Bogot and Hensel (10)

<u>Mississippi</u> Mississippi, which has an estimated 5 billion tons of lignite reserves, ranks second to Texas in total reserves in the Gulf Coast region. Most of the reserves are found in the Wilcox Formation. Seam thicknesses range from 2 to 16 feet but are typically 3 to 9 feet thick. The Claiborne Group lignites are much less extensive than the Wilcox Group lignites. Seam thicknesses range from 2 to 9 feet and are typically 3 to 7 feet thick. Williamson (13) reports lignite occurrences in the Jackson Group but states that they are mostly thin and only locally developed. In southeastern Mississippi, several lignite seams were also

penetrated in the Oak Hill Member of the Naheola Group (upper) Midway).

Alabama Total estimated lignite reserves in Alabama are 1.4 billion tons. Most of these reserves lie in the west half of the state. One relatively thick lignite seam in the upper portion of the Oak Hill Member of the Naheola Formation is the most important lignite deposit in Alabama. Self et al (14). It ranges from 1 to 14 feet in thickness but pinches out in the central portion of the state. The Tuscahoma Formation (Wilcox) also contains up to six relatively thin lenticular seams which rarely exceed 5 feet in thickness, Self et al (14). The Hatchatigbee Formation (Wilcox) and the Lisbon Formation (Claiborne) are also reported to contain occasional, very thin, discontinuous lignite seams.

SUMMARY

The total estimated strippable in-place lignite reserves for the Gulf Coast region are 22.5 billion short tons. Based on recent figures published by the National Coal Association (15), the total Gulf Coast strippable lignite reserves represent an approximate increase of 18 billion tons to the estimated total 141 billion tons of coal reserves in the United States that are potentially minable by surface methods. With the rapidly dwindling supplies of petroleum and natural gas, the Gulf Coast lignites have become an important, relatively untapped energy resource.

REFERENCES

1. Fisher, W. L., and McGowen, J. H., 1967, Depositional Systems in the Wilcox Group of Texas and their relationship to occurrence of oil and gas; Gulf Coast Assoc. Geol. Soc. Trans., v. 17, p. 105-125.

2. Kaiser, W. R., 1974, Texas lignite: near-surface and deep basin resources; Texas Bur. Econ. Geology Rept. Inv. 79, 70 pp.

3. Kaiser, W. R., 1978, Sand-body geometry and the occurrence of lignite in the Eocene of Texas; Texas Bur. Econ. Geology Geol. Circular 78-4, 19 pp.

4. Fisher, W. L.,Proctor, C. V. Galloway, W. E., and

Nagle, J. S., 1970, Depositional systems in the Jackson Group of Texas - Their relationship to oil, gas and uranium; Gulf Coast Assoc. Geol. Soc. Trans., v. 20, p. 324-261.

5. Frazier, D. E., and Osanik, A., 1969, Recent peat deposits - Louisiana coastal plain; in Dapples, E. C., and Hopkins, M. E., eds., Environments of coal deposition; Geol. Soc. America Spec. Paper 114, p. 63-85.

6. Coleman, J. M., and Smith, W. G., 1964, Late recent rise of sea level; Geol. Soc. America Bull., v. 75, p. 833-840.

7. White, W. A., 1958, Some geomorphic features of Central Peninsular Florida; Florida Geol. Surv., Geol. Bull. No. 41, 92 pp.

8. Self, D. M., and Williamson, D. R., 1977, Occurrence and Characteristics of Midway and Wilcox lignites in Mississippi and Alabama; in Campbell, M. D., ed., Geology of alternate energy resources in the South-Central United States; Houston Geological Soc. p. 161-177.

9. Sondreal, E. A., Gronhovd, G. H., and Kube, W. R., In Press, Research and development relating to lignite use in power production; in Kaiser, W. R., ed., Proc. 1976 Gulf Coast Lignite Conference; Texas Bur. Econ. Geology Rept. Inv. 90.

10. Bogot, A., and Hensel, R. P., 1976, Considerations in blending coals to meet SO_2 emission standards; Proc. NCA/BCR Coal Conf. and Expo III (Louisville, Ky, October 19-21, 1976), 12 pp.

11. Roland, H. L., Jenkins, G. M., and Pope, E. D., 1976, Lignite: Evaluation of near-surface deposits in northwest Louisiana; Louisiana Geol. Surv., Min. Res. Bull., No. 2, 39 pp.

12. Wielchowsky, C. C., Collins, G. F., Gerahian, L. K., and Calhoun, E. J., 1977, Frontier lignite exploration in the South-Central United States; in Campbell, M. D., ed., Geology of alternate energy resources in the South-Central United States; Houston Geological Soc., p. 125-160.

13. Williamson, D. R., 1976, An investigation of the Tertiary lignites of Mississippi; Miss. Geol., Econ. and Topog. Surv. Inf. Ser. MGS-74-1, 147 pp.

14. Self, D. M., Moffett, T. B., and Mettee, M. R., 1978, A study of the lignite resources in the Alabama-Tombigbee rivers region of Southwestern Alabama; Geol. Surv. Alabama Tech. Assist. Project Rept., June, 1978, 228 pp.

15. National Coal Association, 1978, Coal Facts; National Coal Assoc., Washington, D. C., 95 pp.

DISCUSSION

QUESTION: Were your results, such as maps of the sand bodies, coal quality, based on new drilling by Phillips, or existing published data?

ANSWER: As I mentioned in the introduction, maps on the sand geometry were taken from the work of Dr. Fisher, Dr. McGowen and especially Dr. Kaiser. All coal quality, except where noted, was solely based on data generated by Phillips Coal Company. In examining published lignite quality data, especially from Texas, you might find that our data may run a little lower than the published figures. This discrepancy is probably, at least in part, due to the fact that some of the published BTU data includes partially air-dried samples and also the fact that many of the published analyses were run in the early 1900's and adequate prevention of moisture loss during sample shipment was not available.

9

Evaluation of the Methane Gas Content of Coalbeds: Part of a Complete Coal Exploration Program for Health and Safety and Resource Evaluation

By William P. Diamond
Supervisory Geologist, Methane Control and Ventilation,
Pittsburgh Mining and Safety Research Center, U.S. Bureau of Mines
Pittsburgh, Pennsylvania, United States

The explosion hazard of methane-air mixtures has become an increasingly serious problem in mine planning. As mining progresses to greater depths or develops in new, previously unmined areas, an advance assessment of methane gas potential can be essential for a safe and economic mine development program. The Bureau of Mines, as part of its Coal Mine Health and Safety Program, has developed a simple, inexpensive test to accurately measure the methane content of coal samples obtained from exploration cores. The gas content of the coal per unit weight determined by this "direct method" test can be used as a basis for a preliminary estimate of mine ventilation requirements and to determine if degasification of the coalbed in advance of mining should be considered. The test results also can be used to estimate the methane resources of coalbeds in specific geographic areas. With the eventual decline in conventional domestic gas production and increased price of natural gas, commercial production of coalbed gas will become a reality.

EQUIPMENT AND PROCEDURE

The Bureau of Mines has been determining the methane content of coalbeds for approximately 7 years. In that time the "direct method" test has undergone modifications and improvements as more was learned about the occurrence of gas in coal (1,2). The direct method test requires that a small portion (1,000 to 2,000 grams) of a coal core be sealed in an airtight container soon after the sample reaches the surface. The sample container (figure 1) can be any of several types.

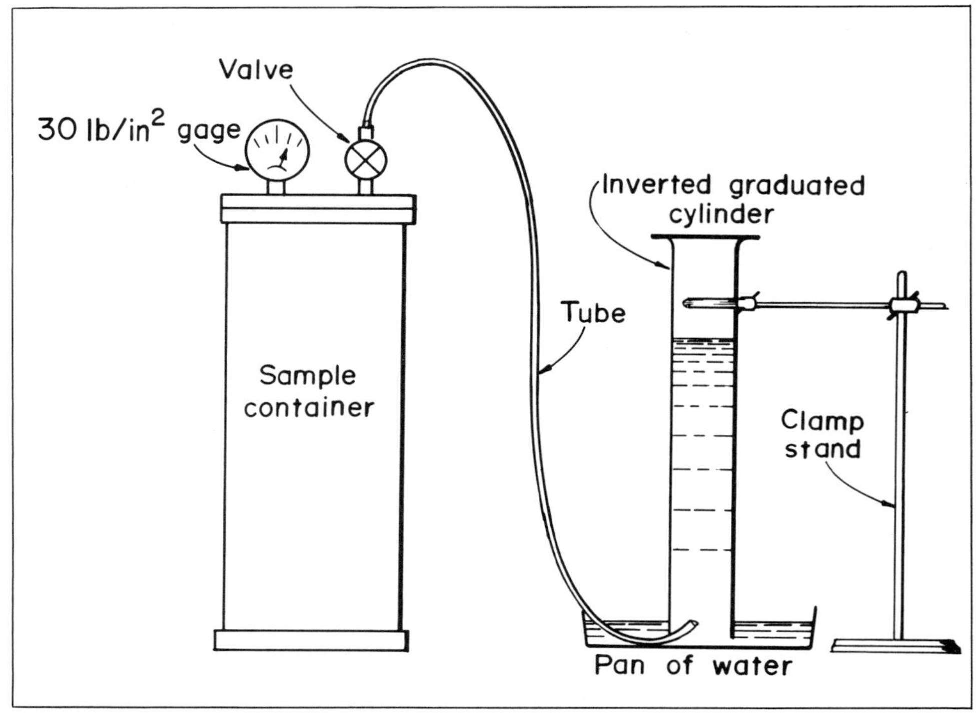

Diagram of desorption equipment. Figure 1.

The original containers were fabricated from a 12-inch piece of 4-inch-inside-diameter aluminum pipe. Less expensive alternatives now in general use are the various types of home water-filter housings. Any container that meets the following requirements can be used: It must be easily sealed airtight, it must remain airtight under an internal pressure of at least 30 pounds per square inch gage, and it must hold approximately 2,000 grams of sample. Accessory equipment includes a gage and valve for the sample container, flexible tubing, graduated cylinder, ring stand, clamp, and pan. The total cost of the complete apparatus is about 100 dollars.

Once the sample has been sealed in the container, the gas pressure builds up and must be relased periodically. The gas is bled into a water-filled, inverted graduated cylinder (figure 1), and the volume of gas released is read in milliliters. The amounts of gas released at the periodic intervals are the basic data from which the gas content of the sample will be determined. The initial gas emissions from the coal are the largest, with a gradual decrease in the rate with time. Because the temperature at which the sample is kept can affect the emission rate, it should be maintained at a uniform temperature throughout the desorption process.

The total gas content of a particular coal sample is composed of three parts: Lost gas, desorbed gas, and residual gas. Each of the three parts is determined by slightly different techniques. A core sample actually begins desorbing gas before it is sealed in the sample container. The amount of this lost gas depends on the drilling medium and the time required to retrieve the sample and seal it in the container. If air or mist is used in drilling, it is assumed that the coal begins giving off gas immediately upon penetration by the core barrel. With mud or water, desorption is estimated to begin when the core is halfway out of the hole; that is, when the gas pressure is assumed to exceed that of hydrostatic head.

The lost gas can be calculated by a graphical method. In the first few hours of emission, the volume of gas desorbed is proportional to the square root of the desorption time. In the ideal case, a plot of the cumulative emission (y-axis) after each reading versus the square root of the desorption time (x-axis) produces a straight line. A representative plot of the initial desorption results from a sample is shown in Figure 2. The intercept of the x-axis is the square root of the elapsed time (lost gas time) in minutes from start of desorption until the sample is sealed in the container. The additional data points are the square root of the sum of the lost gas time and the elapsed time between gas measurements versus the culuative gas measured at the end of each interval. The estimated value of the lost gas is the point where the constructed line fitted to the data points intercepts the negative y-axis.

The desorbed gas portion of the total gas content is the volume of gas released from the sample and measured directly in the graduated cylinder. The desorption of a sample is generally allowed to continue until a very low emission rate is obtained, generally an average of less than 10 cubic centimeters of gas per day for a period of 1 week. The length of time required to reach this point varies and is affected by many factors, including the size of the sample, the physical characteristics of the coal, and the amount of gas contained in the sample.

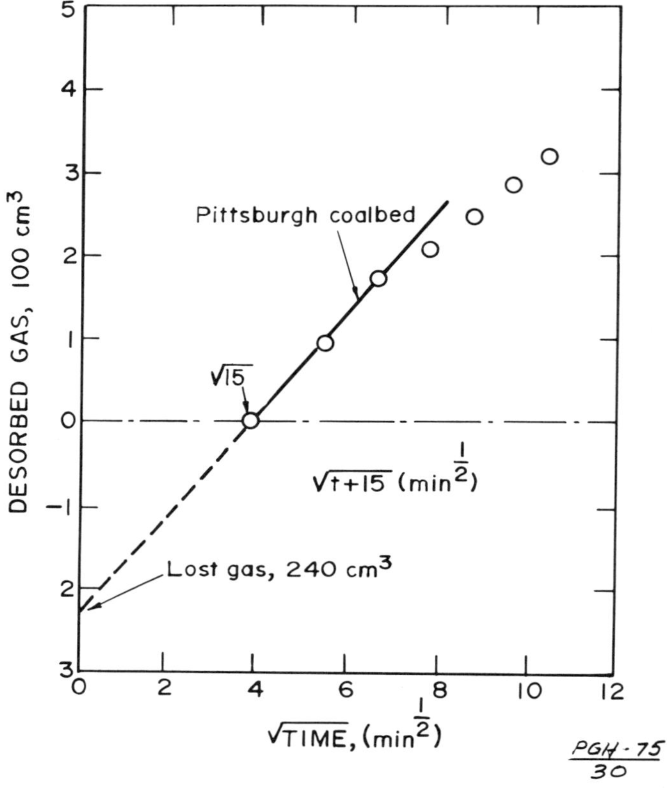

"Lost" gas graph for theoretical coalbed sample. Figure 2.

At the point when it is determined to discontinue the measurement of the desorbed gas, gas will usually still be contained within the coal sample. To complete the gas determination procedure, it is necessary to ascertain the amount of residual gas. Several methods for estimating this residua gas have been used, but the current method for determining residual gas involves crushing the sample (or part of the sample) in a sealed ball mill to approximately 200-mesh. The gas released by the crushing of the coal is measured directly by the water displacement method. The total gas content of the coal in cubic centimeters per gram is the sum of the lost gas, desorbed gas, and the residual gas, divided by the weight of the sample. The results in cubic centimeters per gram can be converted to cubic feet per ton by multiplying by 32. As part of this research, part of each crushed sample is sent to the Bureau of Mines lab for proximate, ultimate, and Btu analysis.

RESULTS

The gas content of individual coal samples has been correlated to the actual methane emissions of several mines. In Figure 3, cubic feet of methane emitted per ton of coal mined

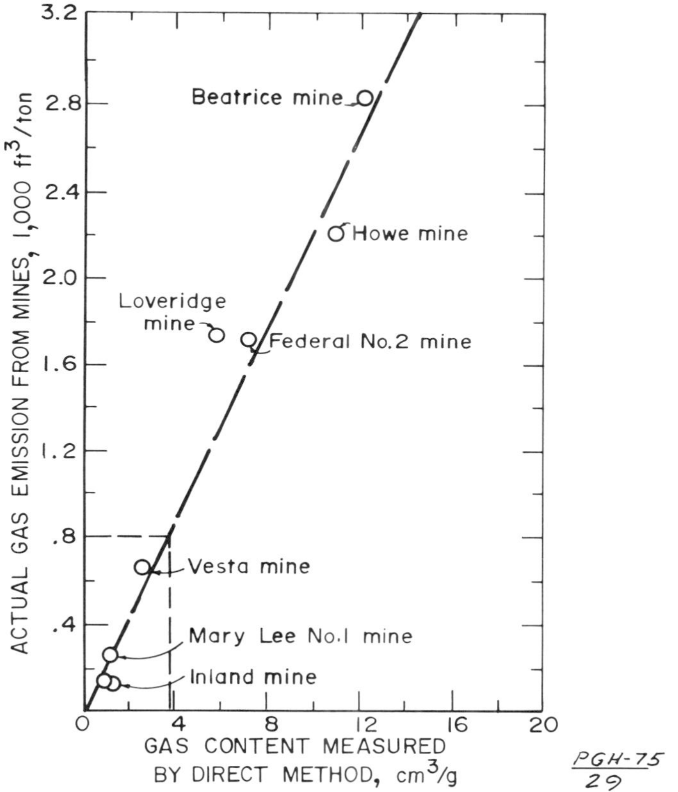

Gas content of coal versus actual mine emission. Figure 3.

is plotted versus the methane content of nearby exploration coal core samples in cubic centimeters per gram. The correlation is good for mines that are large and deep, have a sustained coal production of at least several thousand tons a day, and have been in operation for several years. This indicates that the methane emitted from a mine can be estimated using the results of the direct method test.

Total methane emissions from a mine, per ton of coal mined, is greater than the measured gas content of a ton of coal by the direct method test, because the rib and face coal, roof, floor, pillars, gob areas, and old workings all give off gas in addition to the gas contained in the actual coal mined. New mines emit less methane per ton of coal mined than older mines with extensive old workings and gob areas. Therefore, an estimated emission rate using Figure 3 may be too high for a new mine, but after the mine has been worked for some time, the emissions can be expected to approach those indicated by the graph.

It is impossible to pinpoint an exact gas content value above which all coalbeds would require degasification in advance of mining. The graph in Figure 3 can provide an estimate of mine emissions, however; other factors, not only the actual gas content, must be considered. The direct method test results can be of additional value in indicating the potential severity of gas problems when compared with previous test results in nearby active mining areas.

The Bureau of Mines has used the direct method test to determine the methane content of coalbeds for approximately 7 years. In that time, 335 coal samples from 74 coalbeds in 11 States have been collected. Gas contents of these samples ranged from near zero to 18.5 cubic centimeters per gram (595 cubic feet per ton) for an anthracite sample from Pennsylvania. Several bituminous coals, including the Lower Hartshorne in Oklahoma, Mary Lee in Alabama, and the Pocahontas No. 3 in Virginia, have gas contents as high as 17 cubic centimeters per gram (544 cubic feet per ton).

Bureau of Mines research has shown that in general the gas content of individual coalbeds increases at greater depths. However, not all deep coalbeds are gassy. Several samples of low-rank western cretaceous coals at depths of about 2,000 feet contain less than 1 cubic centimeter of gas per gram of coal (32 cubic feet per ton). In contrast, numerous samples of higher rank Pennsylvanian coals at similar depths in the east have gas contents greater than 10 cubic centimeters per gram of coal (320 cubic feet per ton).

The variation in gas content of coal samples collected at the same depth is thought to be primarily controlled by the rank of the coal. During the coalification process, which turns plant material into progressively higher rank coals, gaseous hydrocarbons, including methane are formed. Thus higher rank coals that have evolved large quantities of methane during coalification and that have remained deeply buried are usually found to have high methane contents. Low-rank coals apparently never reached the coalification stages in which large volumes of methane are produced and are expected to have generally low gas contents even when deeply buried.

Any localized geologic process that produces heat and pressures sufficient to alter the rank of the coal could produce seemingly anomalous higher gas contents than would be expected for a particular coalbed at a certain depth. Thus a coalbed that was subjected to intrusive igneous activity or abnormally high geothermal heat flow which raised the rank may have a higher gas content than the unaltered coalbed at similar depths.

ESTIMATING METHANE RESOURCES

The increasing demand for gas supplies in the United States and the probability of higher prices in the future makes gas in coalbeds an attractive commercial prospect. The results of the direct method test on coal core samples within a specific geographic area can be used in conjunction with geologic maps to estimate the volume of gas contained within a coalbed reservoir. The required geologic maps can be constructed from the data obtained from the exploratory drilling program. The maps needed include coalbed structure and thickness and overburden thickness.

At present, the preferred relationship for estimating the gas content of a coalbed in a specific area is the variation in gas content with depth. Figure 4 is a plot of the

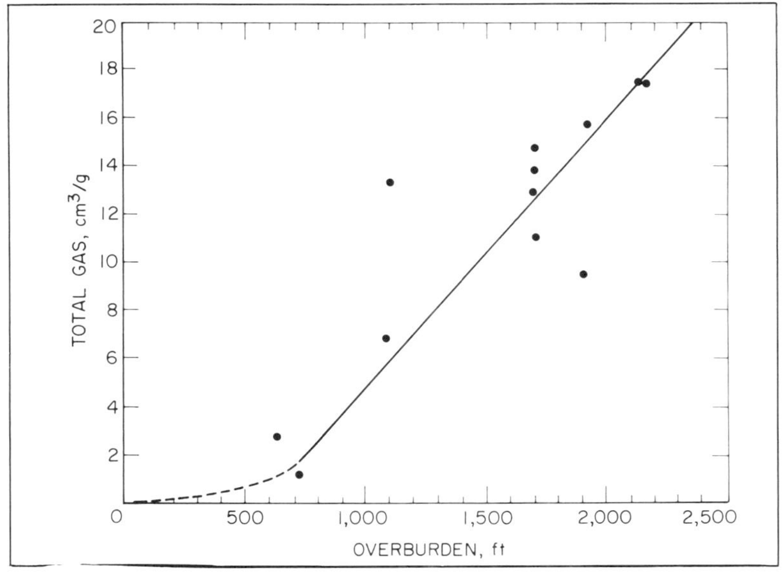

Depth of coalbed versus gas content, Mary Lee coalbed, Alabama. Figure 4.

direct method results from 12 coal samples obtained from the Mary Lee group of coalbeds in Alabama. Sample depths ranged from 600 to 2,200 feet. This graph can be used to estimate the gas content of this coal at any depth. Even though other factors, as mentioned previously, influenced the methane content of coal, the changes in overburden depth are the most accurately documented over a large area.

A structure map on the base of the coalbed is used in conjunction with topography maps to construct an overburden isopach. An overburden map for the Mary Lee coalbed in an 835-square-mile area of Alabama is shown in Figure 5. Also shown on Figure 5 is the isopach for the Mary Lee coalbed. To calculate gas resources, the gas content of the median depth of each overburden interval, as estimated from the graph (figure 4), is multipled by the coal tonnage in the interval. As an example, the 0 to 500-foot overburden interval has an estimated gas content of 0.25 cubic centimeters per gram (7.5 cubic feet per ton) at the median depth of 250 feet. Approximately 1.2 billion tons of coal are contained in the interval, which when multiplied by the gas per ton yields 9 billion cubic feet of gas. Table 1 lists the methane resources of each overburden interval in the Alabama study area. The total gas content is estimated to be in excess of 1 trillion cubic feet (3).

Table No. I
Methane Resources, Mary Lee Coalbed, Alabama

Overburden feet	Average methane content: cubic centimeters per gram	Average methane content: cubic feet per ton	Methane resources cubic feet Mary Lee Group
0-500	0.25	7.5	0.13×10^{11}
500-1,000 .	2.00	64.0	1.15×10^{11}
1,000-1,500	7.50	240.0	3.76×10^{11}
1,500-2,000.	13.10	419.0	3.54×10^{11}
2,000-2,500	18.60	595.2	1.96×10^{11}
Total ...	-	-	10.54×10^{11}

The relationship of gas content to depth as illustrated by the graph (figure 4), can be used to delineate geographic areas of high gas concentrations. The Mary Lee coal is more than 1,500 feet deep in only 12 percent of the study area (figure 6), yet more than half the methane is estimated to be in these deep coals.

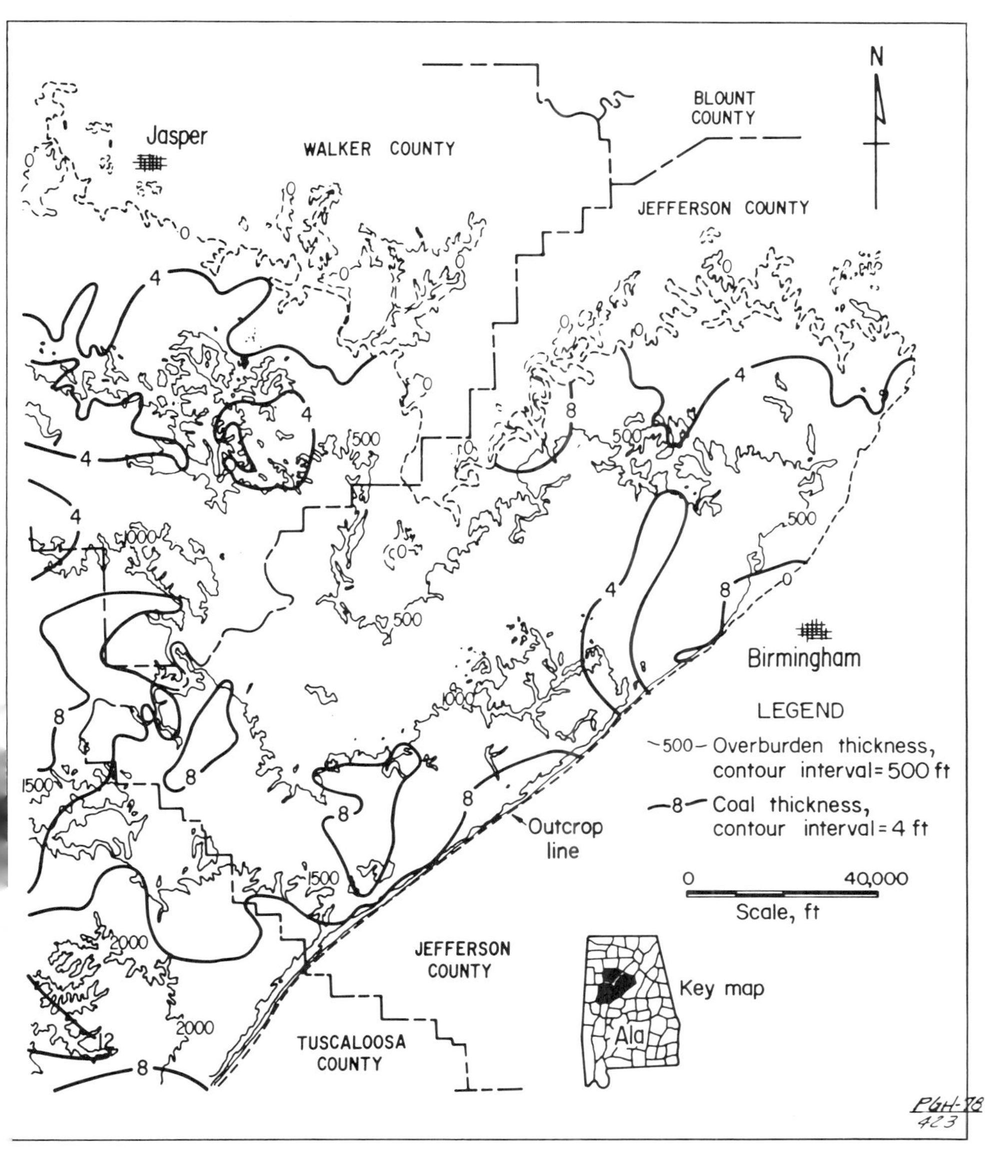

Isopach of Mary Lee coalbed, Alabama, superimposed on overburden isopach of strata overlying the coalbed. Figure 5.

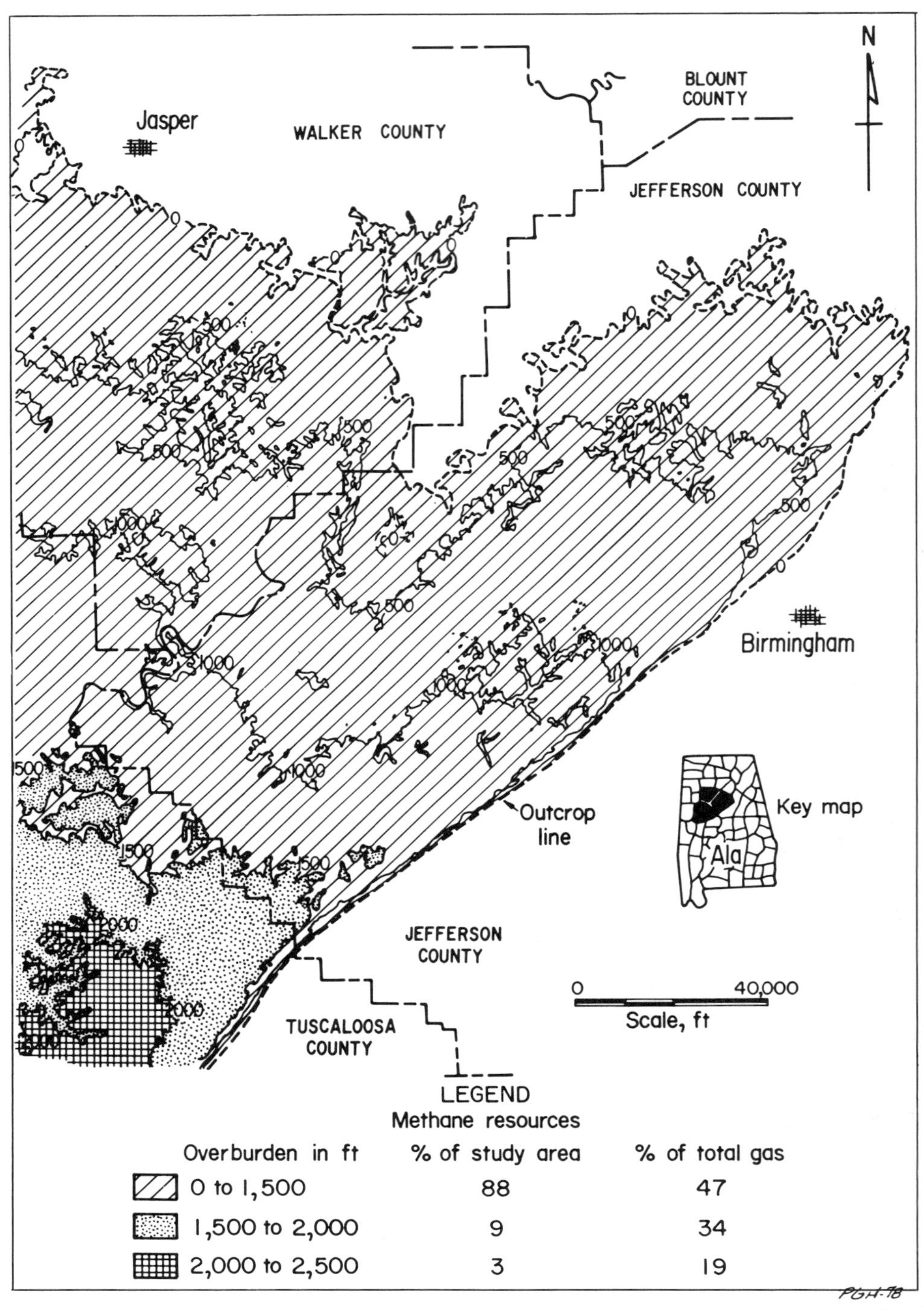

Map of gas resources related to overburden isopach of strata overlying the Mary Lee coalbed, Alabama. Figure 6.

Gas resource estimates have been made for several other coalbeds in the United States. A 1,300 square-mile-area of remaining Pittsburgh coal reserves in Washington and Greene Counties, Pennsylvania, and Monongalia and Marion Counties, West Virginia is estimated to contain 1.5 trillion cubic feet of methane (4). The Lower Hartshorne coalbed in a 600-square-mile area in LeFlore and Haskell Counties, Oklahoma, contains at least 1 trillion cubic feet of methane (5). The methane resources of a six-mine area of the Beckley coalbed in Raleigh and Wyoming Counties, West Virginia, are estimated at 108 billion cubic feet (6). Investigations of a 500-square-mile area of the Upper Freeport coalbed in Fayette County, Pennsylvania, indicate that 190 to 400 billion cubic feet of methane are contained in that coal (7).

SUMMARY

Methane has become an increasingly serious problem for the coal mining industry as operations have progressed to greater depths and entered unknown virgin areas. A simple inexpensive test is currently available to determine the methane content of coal core samples. The results of the test can provide an advance assessment of the relative gassiness of a proposed mine area, allowing the coal company management to plan ventilation or degasification systems in advance of mining to deal with the problem. The gas content data are also the basis for estimating the methane resources of coalbeds in specific areas. The large volume of methane known to be present in the coalbeds of the United States can continue to be a mining liability, or with the technology currently available can become a valuable asset. The evaluation of the methane content of coalbeds should be considered an essential part of a complete coal exploration program.

REFERENCES

1. McCulloch, C. M., and W. P. Diamond. Inexpensive Method Helps Predict Methane Content of Coalbeds. Coal Age, v. 81, No. 6, June 1976, pp. 102-106. BuMines OP 59-76.

2. McCulloch, C. M., J. R. Levine, F. N. Kissell, and M. Deul. Measuring the Methane Content of Bituminous Coalbeds. BuMines RI 8043, 1975, 22 pp.

3. Diamond, W. P., G. W. Murrie, and C. M. McCulloch. Methane Gas Content of the Mary Lee Group of Coalbeds,

Jefferson, Tuscaloosa, and Walker Counties, Ala. BuMines RI 8117, 1976, 9 pp.

4. Diamond, W. P., and G. W. Murrie. Methane Gas Content of the Pittsburgh Coalbed and Evaluation Drilling Results at a Major Degasification Installation. In Abs. with Programs, Northeastern Section Ann. Meeting Geol. Soc. of America, Binghamton, N. Y., Mar. 31-Apr. 2, 1977, v. 9, No. 3, February 1977, p. 256. BuMines OP 18-77.

5. Murrie, G. W. Coal and Gas Resources of the Lower Hartshorne Coalbed in LeFlore and Haskell Counties, Oklahoma. In Abs. with Programs, South-Central Section Meeting, Geol. Soc. of America, El Paso, Tex., Mar. 17-18, 1977, v. 9, No. 1, January 1977, pp. 65-66.

6. Popp, J. T., and C. M. McCulloch. Geological Factors Affecting Methane in the Beckley Coalbed. BuMines RI 8137, 1976, 35 pp.

7. Steidl, P. F. Geology and Methane Content of the Upper Freeport Coalbed in Fayette County, Pa. BuMines RI 8226r, 1977, 17 pp.

DISCUSSION

QUESTION: What basis do you use in selecting your samples and do you take more than one sample per seam?

ANSWER: We are essentially totally at the mercy of the coal industry for obtaining our samples. We cannot afford to go out and drill holes all over the country to obtain samples for the direct method test. In most cases coal companies will only allow us to take perhaps one foot of a six or eight foot coal bed. That does somewhat limit the confidence in the results that you can obtain on a particular coalbed. We do try to take, on a limited sample, essentially the cleanest part of the core, that part which has no obvious shale or pyrite inclusions. Since the test is based on the weight of the sample, any extraneous matter such as pyrite or shale adds weight, but will not add gas to your gas content calculation. Now we do normally try to run a proximate, ultimate, and Btu test on all the samples so that we can get an evaluation of the ash and the pyrite, which can be factored out of the calculation. However, some companies don't want you to run these tests on their samples.

In a lot of respects we are at the mercy of the companies themselves, as to the most we can do with a particular sample. We do find variations in the gas content within the same core. I've seen results vary from four to six cubic centimeters per gram on one coal bed. Some of those differences can be relegated to the ash, shale or pyrite. We have not done petrographic work on the samples, because we do not have a petrographer in our group. There still are some questions as to why multiple samples from the same core have varying gas contents. Petrographic analysis might be the clue. We don't know everything yet, but we're still trying.

QUESTION: Many coal companies do not utilize methane because the ownership of the coal gas has not been established. At the present time the methane is exhausted in the air. Do you expect a decision in the near future to determine gas ownership? Is methane found in coal seams legally considered a coal resource or a gas resource? In other words, where gas and coal are owned by different parties, who normally would own methane gas within the coal seams?

ANSWER: Now that some people realize that there is a commercial potential for gas in coal, everybody's coming out of the woodwork and wants a piece of the action. In the United States, unfortunately, in many cases one person owns the traditional gas rights; someone else owns the coal rights. These two groups are now presently squabbling and fighting, or whatever you want to call it, over who owns the gas.

Personally, I don't think it really makes that much difference who owns the gas, as long as someone decides and makes the decision that from this point on, gas in coal will be owned by the coal lease-holders or the gas lease-holders. This has been a particularly frustrating stumbling block to the various techniques that can be used to take gas out of coal beds. Some coal companies are just standing by now waiting to see what is going to happen. Several companies that feel that they must degasify or they can't mine coal have taken the opinion that they will go out and get a formal farm-out agreement from the gas lease-holders and go ahead and pay them the royalties and produce the gas. Other companies believe that by doing this you prejudice the whole situation by admitting that you do not own the gas rights, so they have either just put off doing the gasification or else they've gone ahead and done the work and waited for someone to come up and sue them. And, in fact, there currently is a suit pending in Pennsylvania over the gas rights in the coal where a group of individuals went out and bought the gas rights and started putting down vertical holes and stimulating the coal owned by a major steel corporation. The owner of the coal was somewhat perturbed at that. There are counter suits and it's really a mess. It is a big problem and I hate to suggest any sort of Federal intervention, but it is a problem that all states seem to be looking at the question slightly differently. What may be a legal opinion in one state may not necessarily be the legal opinion in another state. I am not sure that Federal regulation is the answer, but perhaps might be of some help.

QUESTION: I have one question concerning some of the techniques used for degasifying coal beds. Specifically asking about horizontal drilling in coal beds.

ANSWER: This technique is probably the most successful technique in terms of amounts of gas that have been produced from coal. This only takes place, of course, underground in coal mines, or can be done from the bottom of shafts in virgin coal. The technique involves a drilling machine, essentially positioned horizontally, which can be used underground. Part of the question asked, "How do you control your angle?" By using various drill collars and bits, and varying bit pressures the drill string can be controlled up and down to some extent.

QUESTION: How you determine where you are in the coal bed?

ANSWER: This is a problem. We do tend to go into the roof or into the floor. If we have a good geologic map we have a general idea of which way the coal bed is going and approximately every 10 or 20 feet we run a single-shot survey and keep a continuous plot of the hole as we go along.

QUESTION: How far can you drill?

ANSWER: We routinely drill up to about 1000 feet horizontally. Holes up to 2500 feet can be drilled, but the time required to survey the hole every 10 feet is excessive. In many cases we have to stop, not because we have reached the limit of our drilling, but because we are producing so much gas from the hole that the ventilation system in the mine just cannot handle it. So, a lot of times we are cut very short by the limits of the ventilation. This, of course, should indicate to you that we need some other system of getting the gas out of the mines, instead of just dumping it into the ventilation system. In fact, there are various underground piping systems combined with vertical holes into a section near where you want to degasify, to take the gas straight out of the mine and not dump it into the ventilation air.

QUESTION: Are gas pockets predictable? What about permeability in coal seams? What areas can be drained by drill holes?

ANSWER: Normally we don't think of gas being in pockets in the coal bed. Gas exists in two separate forms in coal beds. You do have some free gas in the fractures, but the great bulk of the gas is contained by an adsorption process, with the methane molecules actually attached to the coal structure. As I mentioned in the talk, the gas content seems to vary with the depth of the coal beds in most instances and, also, the rank of the coal does have some influence on distribution of gas within a certain area.

QUESTION: What about the permeability of the coal beds and what size areas can be drained by drill holes?

ANSWER: We use both vertical and horizontal holes. In terms of the horizontal holes, I am sure you drain at least as far as you can drill and, depending on how close together you put your horizontal holes, you can drain an area very fast or very slow; of course the further apart you put them, the wider an area you will potentially drain. The longer you drain gas, of course, the greater area you will affect. I mentioned in the talk that many factors will influence the production of gas. Every coal bed does behave slightly differently and in most cases we try to do site-specific small-scale drilling before we get into a very large program, just to find out what some of those formation factors are which will effect the spacing of holes. The same thing is generally true on the vertical holes.

QUESTION: What do you consider a commercial gas well from a coal seam in terms of volumes of gas per day at today's pricing?

ANSWER: I've not really done that much economic analysis on the production of gas from coal. Unfortunately, perhaps. From horizontal holes in the Pittsburgh coal bed with which we've had a lot of experience, we find that for every foot of horizontal distance drilled in the coal bed we can produce about 120 cubic feet of gas per day. So if you have a 3,000-foot hole multiplied by 120 cubic feet, you do have quite a lot of gas. One installation in the Pittsburgh coal bed at the bottom of the shaft had seven horizontal holes drilled into it, and for about five years now it has been producing an average of about 600,000 to 700,000 cubic feet of gas per day. And if you go back and calculate the value of the gas produced in that five years, that is a fairly substantial sum of money which could be used to off-set the cost of putting in a shaft perhaps five or 10 years in advance of mining. Vertical wells are still undergoing active research to improve their effectivness. The government has a project in Alabama right now in the Mary Lee Coal Bed which is now just coming into production, and several holes are producing in excess of 100,000 cubic feet of gas per day. So the potential is there and the more holes drilled, of course, the more gas you'll get.

QUESTION: I have a question about gas in open-pit mining versus underground mining.

ANSWER: I don't really think gas is that much of a problem in strip mining. We've not done anything with it.

QUESTION: What is the percentage of recovery of methane in various cases?

ANSWER: That, again, is something that takes a fair bit of production time and analysis to determine. Again, all coal beds behave slightly differently. Friable coal beds tend to give up their gas much faster because of the greater surface area in the fracture system. Coal beds that are blocky and which tend to give up their gas much slower, potentially would not give up as much of their gas as a friable coal. I hate to even mention the percentage--perhaps maybe 50 percent. Again, the longer you degasify, the more gas you will eventually get out. Fifty percent may be somewhere in the ball park. It may get better if you wait five or 10 years before you mine through your degasification pattern.

QUESTION: Do your gas reserve calculations take into account the amount of gas contained in the overburden and immediate floor of the coal seam? If not, by what factor might reserves be increased if these sources were taken into account?

ANSWER: No; in the calculation of the reserves for a coal bed in a particular area, we do not take into account the gas in the rocks surrounding the coal bed--roof and floor. Substantial amounts of methane, as I am sure you all know, can come into the mine atmosphere from the surrounding strata. The direct-method test is not a valid test to use on rock types other than coal. We have actually done it on oil shale, but it's not the proper way to evaluate a sandstone reservoir. We have not really tried to assess the amounts of gas which will come into a mine from the surrounding strata.

QUESTION: Are methane drainage holes uncased for some distance above the coal and drilled below the bed to allow escape of gas from these areas?

ANSWER: I am assuming this is for a vertical hole. It can be done in several ways. In most cases we will drill our holes down to a point directly above the coal bed and cement casing. Then we will drill down through the coal bed and usually leave a sump perhaps 50 feet below the coal bed to allow the materials slumping off the coal to fall to the bottom. We then install a water pump to pump out water, since you must first remove the water from the coal before you get any substantial gas production.

Computers are finding new uses in the coal industry. A growing field of application is in analytical laboratories. Computerization in small coal laboratories is now used for the standard major test methods.

Seismics and Geochemistry

10

Surface Reflection Seismic – Looking Underground from the Surface

By David G. Peace
Western Geophysical Company of America
Isleworth, Middlesex, United Kingdom

INTRODUCTION

The past five years have seen the gradual introduction of the words 'Surface Reflection Seismic' into the vocabulary of coal explorationists worldwide. The ability to 'look underground from the surface' in a real two dimensional way represents a major advance for the coal industry. Conventional low resolution seismic is already going into a 3d mode so the promise for coal looks even better in the future.

However, there are as yet very few geophysicists active within the coal industry. Just the geophysical terminology alone can be daunting enough to the poor coal geologist, and learning practical geophysics from a 'do it yourself' seismic survey book is just not possible.

To try and help the coal geologist out of this predicament, I have tried to answer in the following pages the 3 main questions that often seem to crop up. These firstly concern the logistics of a seismic survey what happens where and when and why? Secondly, which is everyones worry, the 'will it work in my area?' type of question. This question can and should be answered very early on in the proceedings. By quite straightforward types of test it is possible to evaluate the seismic characteristics of an area. Ignore these tests and it may make or break your entire survey.

Finally, wheneverything is fine just how accurate is it possible to be?

THE LOGISTICS OF A SEISMIC SURVEY - (OR WHAT HAPPENS WHERE AND WHEN)

A seismic exploration survey is invariably conducted in 3 major phases. The data acquisition then the data processing and finally the data interpretation. Each phase may be undertaken individually and carried out by one or different geophysical companies.

For present purposes I will assume that the initial decision to perform a seismic survey has been taken; the area to be surveyed is well established and all preliminary parameter experimentation is complete.

Data Acquisition - the first Phase

Figure 1 gives an idealised flow for the field data acquisition. Most field crews will operate from a base camp. If near a convenient town, this will be used. In remote locations some form of portable accomodation such as trailers is used.

The average field crew will have about 6 key personnel, observers, party chief, seismologist, drillers etc. and a few labourers (usually local hires) to assist with tasks like laying out receivers, connecting cables and helping with drilling operations.

Firstly, the surveying party will accurately position the line and mark in the locations for receivers and shot points. Figure 2 shows a survey party using theodolites and marker poles to measure elevations and bearings.

After the surveyor will come the drilling rigs. Figure 3 shows a HYDREQ drilling rig and a water bowser. This rig uses a separate compressor to drive the drill and is capable of penetrating to shot hole depths down to 100' very quickly. The water bowser can be used for dual purposes, for the drilling mud when wet drilling is used or later to fill the receiver holes with water if hydrophones are being used as receivers.

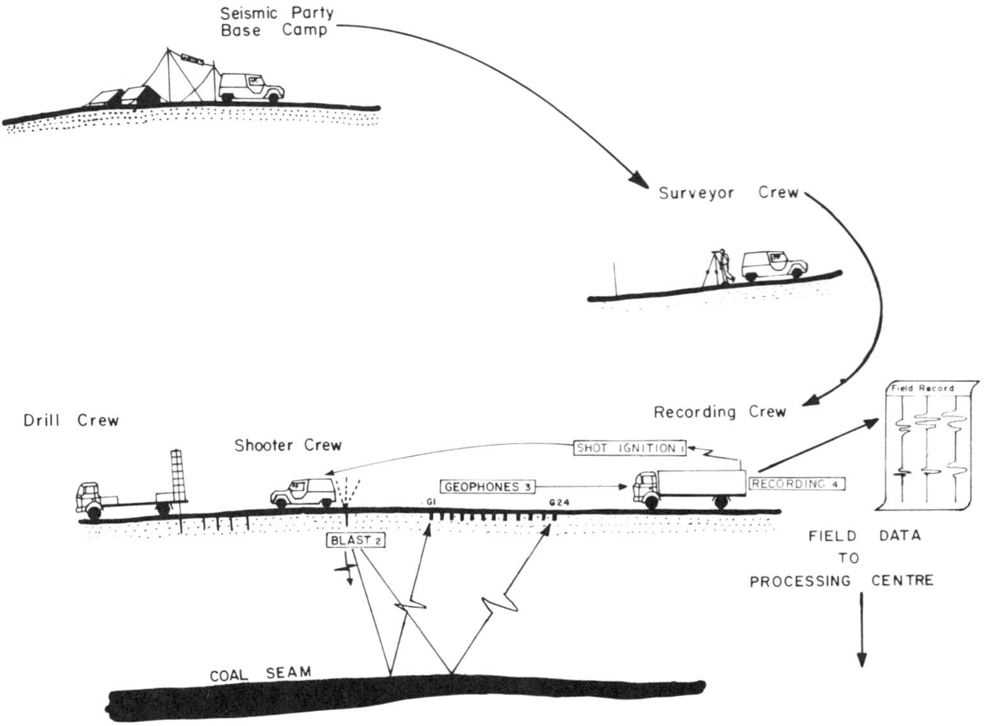

Seismic Party
Base Camp

Surveyor Crew

Drill Crew

Shooter Crew

Recording Crew

Field Record

SHOT IGNITION 1

GEOPHONES 3

RECORDING 4

G1

G24

BLAST 2

FIELD DATA
TO
PROCESSING CENTRE

COAL SEAM

Figure 1

Figure 2

This particular rig was drilling 32' shot holes.

In Figure 4 the hole has been completed, the drilling mast folded down and the shooter in the foreground is placing a cap and booster into the explosive charge used for the shot. 1lb charges were being used here. Charge size is one variable that needs to be tested at the initial experimentation stage. Charges from up to 10lb per hole down to 1 firing cap only per hole have been used. The shooter then places the primed charge at the bottom of the hole. The charge will be covered over and the hole carefully tamped down to prevent a blow out. If a blow out occurs, most of the shot energy goes up the hole and is wasted in the air blowing backfilling out of the hole. This would necessitate a reshoot or give a lost hole.

Figure 5 shows a similar drilling operation in obviously much more hilly terrain. Here the hole is complete and the driller is tripping out the pipe. Air drilling is being used instead of the bit and water drilling shown in the other figures.

While the shot holes are being drilled, the receivers are being laid out along the line. Figure 6 shows a collection of different types of receiver. The receivers in the lower centre which have caps and planting spikes are geophones, while most of the others are hydrophones for use in water retaining holes. The long slim silver detector in the centre is a hydrophone which has changeable electronic components which may be varied to suit areas of different attenuation characteristics.

The firing line and receivers are connected to the recording truck. This truck contains the recording system, magnetic tape transport and a monitor record camera. This particular crew, Figure 7 was using a small portable 24 trace recording unit which conveniently fitted in the back of a small landrover. Admittedly, there was not much space left inside for more than one observer!

When all is ready, the recorder is switched on, the shot fired and the resulting seismic energy recorded. This process of drilling/loading/shooting/recording is repeated until the traverse is complete.

Figure 3

Figure 4

Figure 5

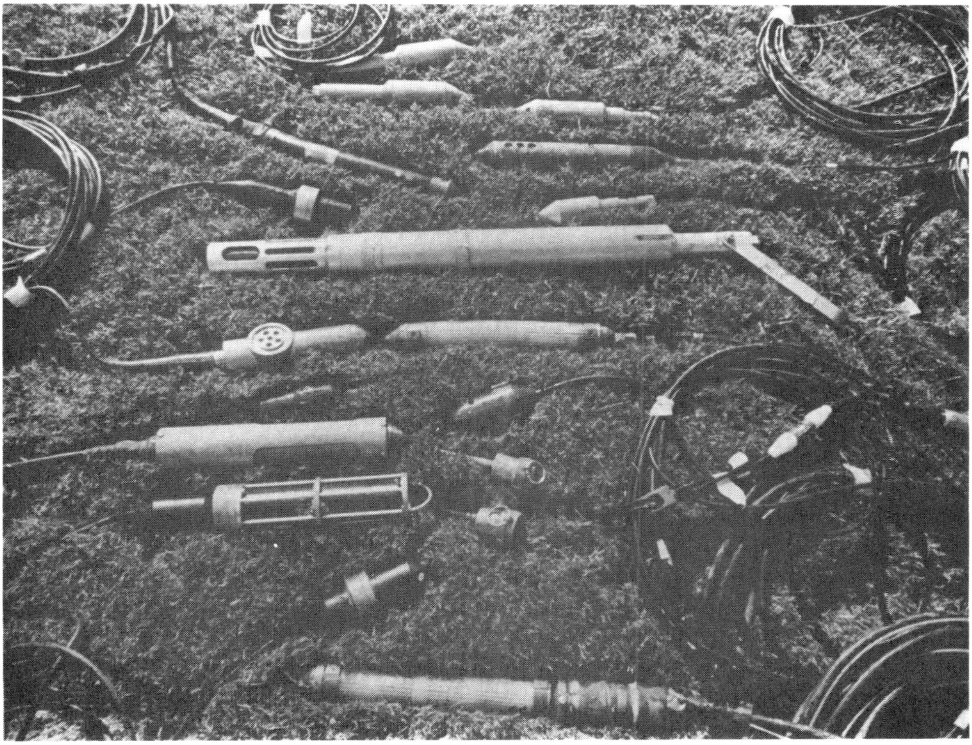

Figure 6 - Different detectors used in high resolution seismic profiling. Hydrophones are used where holes retain water; otherwise moving coil detectors are employed.

The final output of the data acquisition phase is Magnetic tapes, recording logs of what is on tape and closely noted details of all the shooting and recording parameters used. This field data is then sent to the processing center.

THE SECOND PHASE

The data processing

Figure 8 gives the next part of the seismic flow.

This part of seismic is for most geologists the part that resembles a black box. That is: The field data goes in and sometime later a final section comes out - but what exactly happens in the middle they are not too sure about.

Seismic data processing, like most other types of paper data processing, is merely the computer application of various software packages to the data. These packages are designed to change, update, correct or refine the data in various ways in connection with:

* The limitations imposed by field acquisition methods.
* Making the necessary corrections for near surface low velocity layering.
* The alteration and attenuation of the seismic signal that occurs on its travel through the earth from the shot to the receiver.

Figure 9 shows a mini computer that would be suitable for use in the base camp area of the field crew. This is a Western Preseis, many of which are working in remote areas of the world. Having a mini computer on site gives many advantages, especially in giving results as fast as possible so that the field crew downtime waiting on processing is kept to a minimum. Mini computers are today quite sophisticated and powerful enough to cope with most data processing situations.

I don't propose here to get into the finer details of processing, but would like to just briefly illustrate some of the potential.

Figure 7

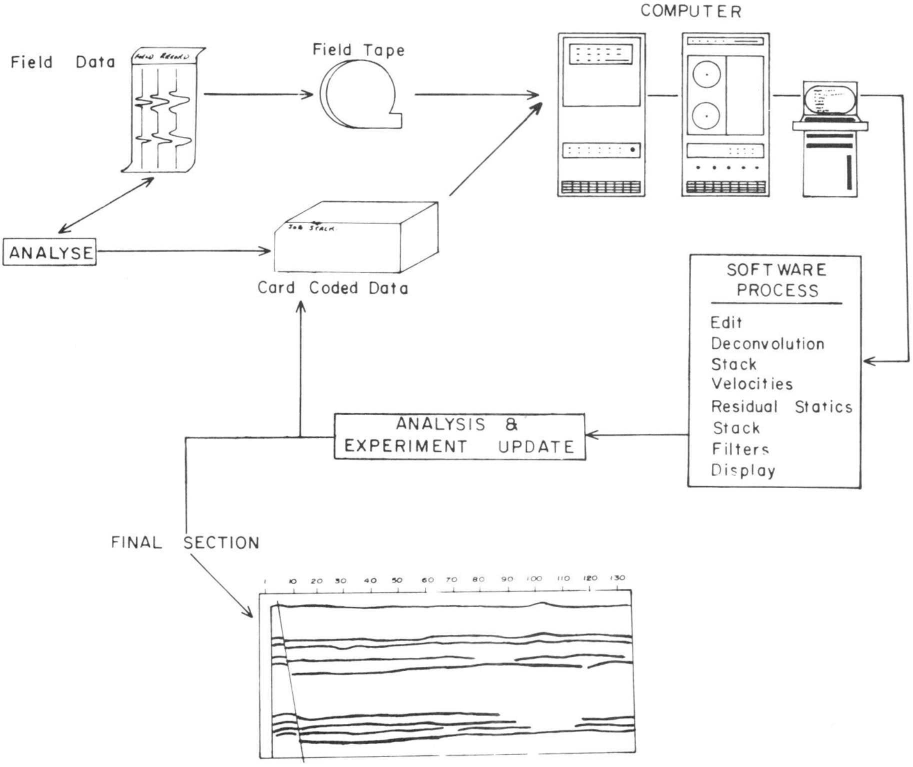

COMPUTER

Field Data

Field Tape

ANALYSE

Card Coded Data

SOFTWARE PROCESS

Edit
Deconvolution
Stack
Velocities
Residual Statics
Stack
Filters
Display

ANALYSIS & EXPERIMENT UPDATE

FINAL SECTION

1 10 20 30 40 50 60 70 80 90 100 110 120 130

Figure 8

Figure 9

Data processing techniques can:-

* Compensate for the natural decay in signal amplitude for increased signal travel time.
* Edit out bad or noisy traces on a record by record basis.
* Adjust shot to geophone relationships in areas of limited access or unfavourable terrain.
* Remove or attenuate various coherent or non coherent noise patterns.
* Deconvolve data.
* Compute average, R.M.S. and interval velocities and provide time/velocity functions.
* Filter out undesirable interference frequencies such as 60 Hz power line pick-up.
* Adjust data timing to correct data for:-
  1. Reduction to a common elevation datum.
  2. Variations in near surface weathering corrections.
* Stack data originating from common subsurface locations.
* Apply various 'cosmetic' processes post stack to try and enhance primary signal by attenuating various background noises.
* Display data in many forms and sizes to suit individual requirements.

For example, the records shown in Figures 19, 20 and 21 are unstacked and represent the data recorded from single shot points. Whereas, the section shown in Figure represents stacked common depth point data from many shot points.

One of the biggest problems for land seismic is the calculation of the accurate near surface statics. These are needed to correct the seismic data to a reference datum. These static corrections vary from shot to shot and their calculation is dependent upon a very accurate knowledge of the velocities in the near surface rocks. Also, the higher the frequencies we are trying to record, then the more accurate must our knowledge of the velocities be. Invariably, we do not have as accurate a knowledge of these velocities as we would like, so we are left with a residual static problem.

There are many residual static correction programs available, which are specifically designed to overcome this problem, but care needs to be taken in their use.

Figure 10 shows a seismic cross section both before and after the application of Westerns residual static program the MISER (2). The effect is quite remarkable, changing an apparently very broken section into one with very good continuity. This is very nice, but there are problems to be aware of when using these programs:-

* The program must be of a 'surface consistent' approach, that is the program searches for similar static anomalies for all the various combinations of shot/receiver travel paths that reflect energy through a given CDP. If it is not surface consistent, then it may apply apparent residual statics that have no suface consistent basis, this is highly undesirable and wrong!

* When recording very high frequencies the size of the model wavelet used in the residual static analysis becomes smaller. When the model and the data being analysed are of high frequency, then the possibility of cycle skipping occuring increases. This effect mis-correlates apparently similar waveforms and then applies residual statics as if to smooth the events together. Unfortunately, this may even happen on a surface consistent basis when the frequencies analysed are high enough.

  When frequencies are lower however, this is less likely to occur - for instance on the Figure, where the frequency content is fairly low, in this case approximately 30 - 40 Hz, which is very low for coal exploration.

  This slide illustrates another important point though. Many people critisise surface reflection seismic for not being accurate enough at greater depths. The deeper you look, the less high frequencies come back so obviously our bed thickness resolving power does indeed decline. However, this section shows quite clearly a probable washout on the right hand side at approximately 1.75 secs. two way time. Depth computation shows the feature to be approximately 5250' deep and 200' thick, very useful information even using 'low resolution' techniques.

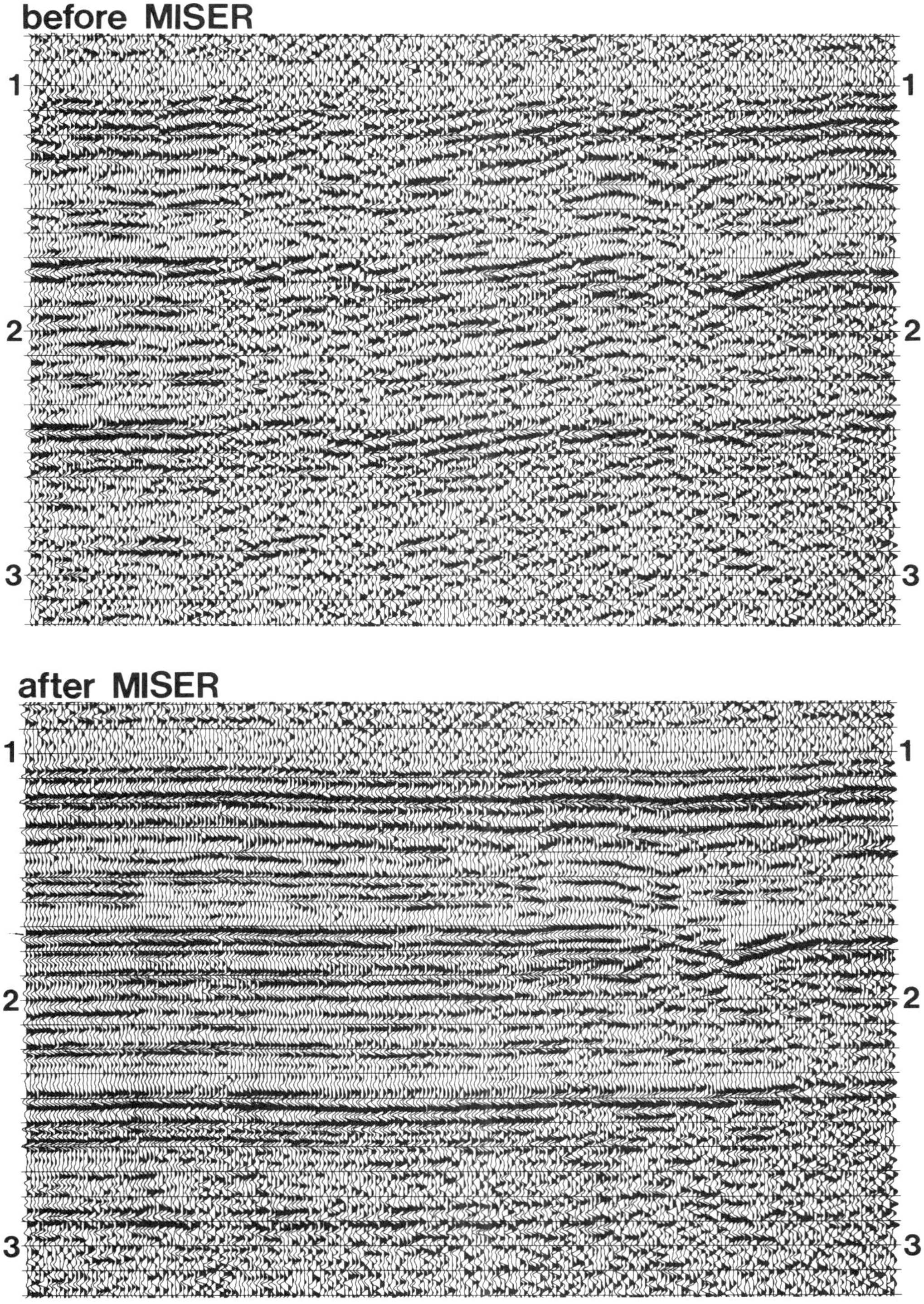

before MISER

1 1

2 2

3 3

after MISER

1 1

2 2

3 3

Figure 10

THE THIRD PHASE
Data Interpretation

Figure 11 gives the data flow. This shows that seismic is really only one part of an Integrated Exploration Program.

We now have the seismic cross sections. Borehole data will invariably be available together with whatever local knowledge there is of the prevailing geologic conditions.

These 3 sets of data must now be carefully integrated together to form an accurate 3d interpretation of the mining prospect.

Each data set on its own is unlikely to give sufficient information or accuracy to make a reliable interpretation.

For instance, in the case of a working face encountering an unexpected fault, as in Figure 12. The fault has a throw larger than seam width. Which way the coal is, is not ascertainable at this time. No control exists immediately in front of the fault. One way to find where the coal is would be to drill a borehole close to where the fault was encountered. Then it would be known exactly where the coal is. Unfortunately, as Figure 13 indicates, one borehole may not show everything we would like it to. There may be more faulting, channel sand washouts or pinchouts of various forms. By utilising a seismic line in conjunction with the borehole, a more reliable picture would be apparent.

While on the subject of channel sand washouts, we know these are notoriously difficult to locate accurately using borehole control alone. Such a channel sand may only be found in 1 or 2 holes out of many in a prospect, but rarely will these give enough data to predict their direction or number.

Figure 14 shows a typical sand channel. The circles are boreholes. With only borehole control we would only locate the sand at borehole J at top centre. If, however we linked the boreholes together with seismic lines then a clearer interpretation could be made, similar to Figure 15.

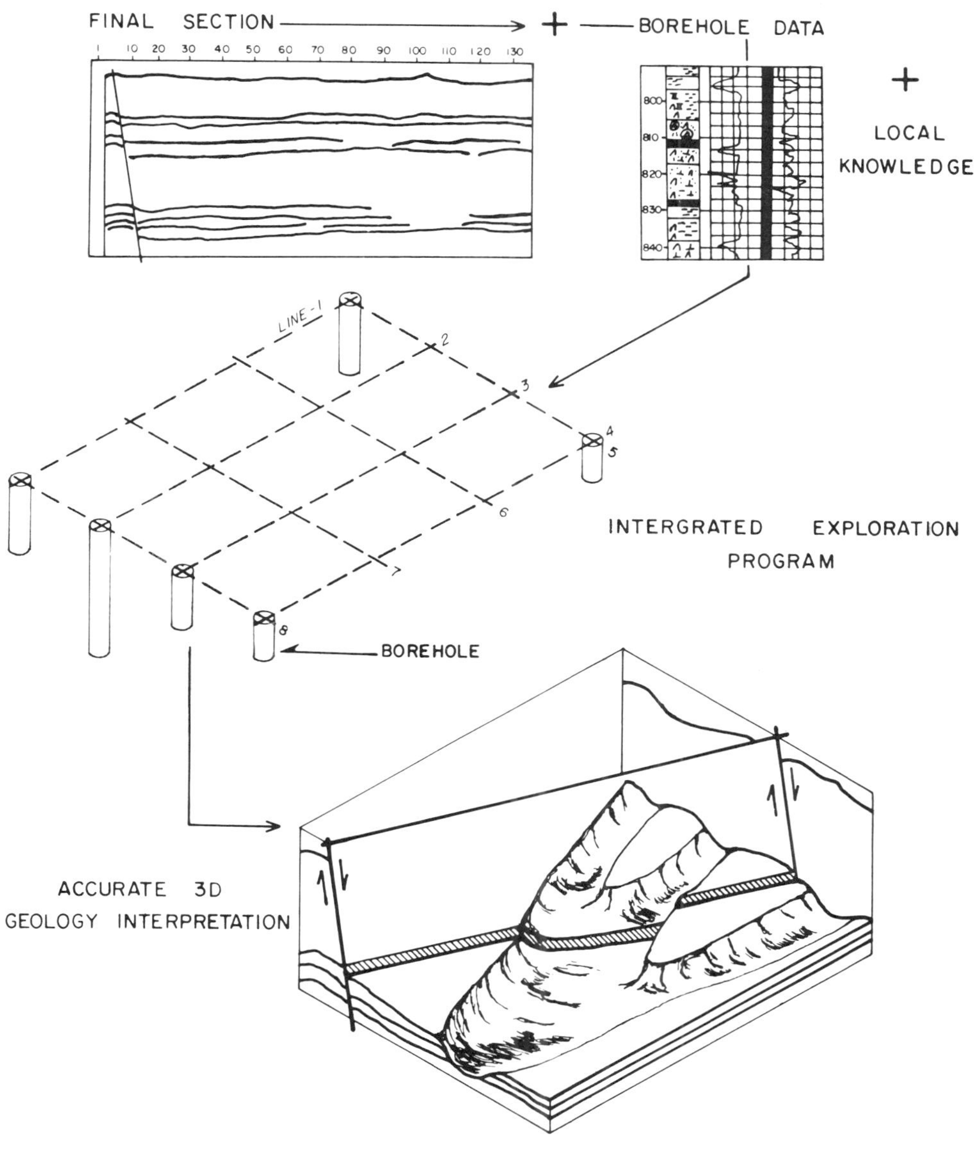
FINAL SECTION
1
10
20
30
40
50
60
70
80
90
100
110
120
130
BOREHOLE DATA
800
810
820
830
840
LOCAL
KNOWLEDGE
LINE-1
2
3
4
5
6
7
8
INTERGRATED EXPLORATION
PROGRAM
BOREHOLE
ACCURATE 3D
GEOLOGY INTERPRETATION

Figure 11

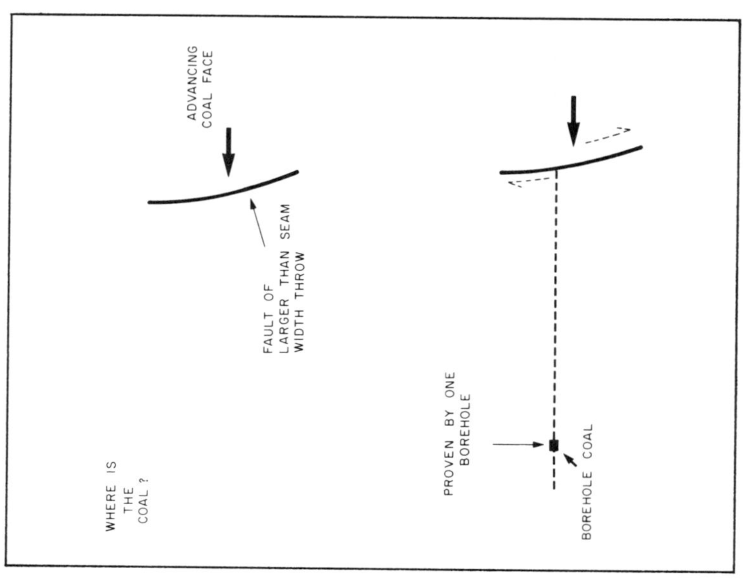

Figure 12

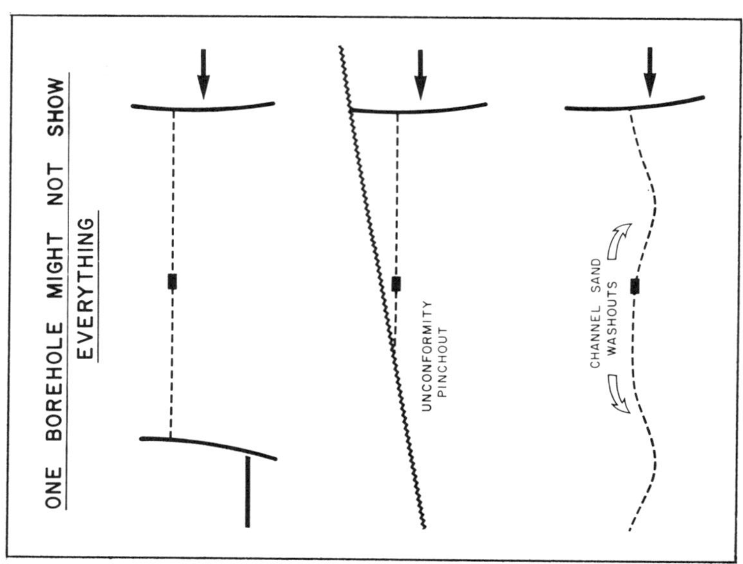

Figure 13

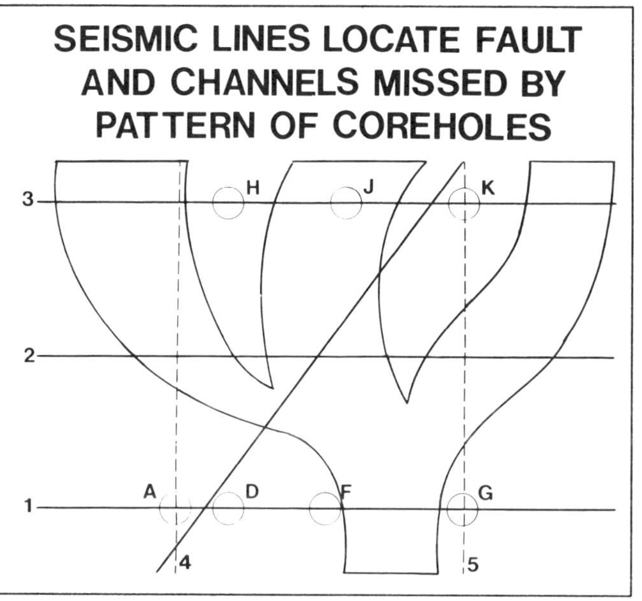

Figure 14 - Map of corehole locations, sand channels which dissect coal and fault which displaces seam.

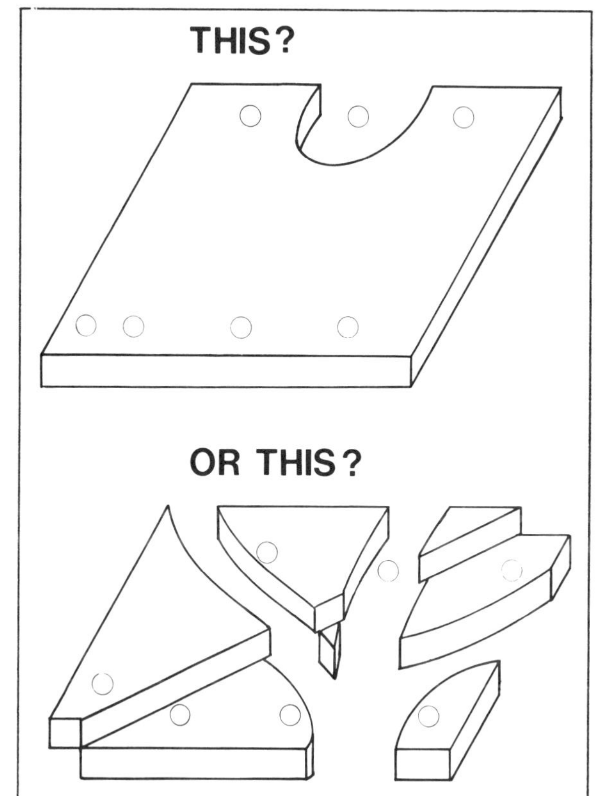

Figure 15 - Map of corehole locations relative to high resolution seismic lines which reveal the fault and channel locations to within 5 meters.

THE BIGGEST PROBLEM

WILL IT WORK IN MY AREA?

One of the peculiarities of seismic from the Coal Explorationists point of view, is the apparent ease with which a 'good' seismic section appears to deteriorate into 'bad' seismic section, or worse into unusable noise only. How to evaluate the likelihood of this happening is a fundemental question, which needs to be answered BEFORE the prospect is shot.

The function of quality on a section, goes back directly to the ratio of Primary Seismic Energy.

to

Various coherent or non coherent noises recorded in the field. Unfortunately, within a given field configuration it is only possible to record the two together, they are inseperable at this point, the recording equipment cannot distinguish between primary reflected signal that we want and noise, that we don't want. The field layout can however be biased towards retaining the most primary signal and rejecting as much noise as possible. Early experimentation with field parameters is essential if this is to b achieved.

BEFORE ANY FIELD OPERATIONS COMMENCE, WE SHOULD:

* Evaluate as accurately as possible the near surface geology to a depth of about 100'.
* Divide the seismic prospect into areas of quite simila geology/rock type.
* If geology is consistently the same - single test site are probably adequate.
* If geology is significantly varied, ie. Surface cover changes from sand to limestone then we can reasonably expect the seismic response to change.

In such cases, multiple test sites are desirable, one for each different rock type.

In each test location sufficient experiments should be carried out to ensure that:-

1. The shot is placed at the optimum depth to minimise noise generation.

2. The respective merits of geophone versus hydrophone can be evaluated.

3. The receiver is placed at the optimum depth to minimise noise reception.

4. The shot to receiver distances and inter-receiver distances are at the best compromise positions, ie. to maximise the surface dimensions without reducing the sampling interval to such an extent that it will not meet the objectives of the survey.

With respect to the respective merits of geophone versus hydrophone, the current geophysical opinion is somewhat divided about which to use. I believe there is a place for both however. Figure 16 shows some typical response curves for a geophone and a hydrophone. Frequency increases towards the right on the horizontal axis.

As may be seen quite clearly, after an initial peak value of about 10-12 Hz for the geophone and 15 Hz for the hydrophone, both receivers have settled down to an essentially flat response to any increase in signal frequency. For the range of frequencies we are likely to encounter at present (ie. a range from approximately 50+ to 500+Hz) both receivers have a flat response. This suggests that any increased resolution is due to the superior coupling of one receiver against another, not particularly in their individual design. It is a well known geophysical fact that the near surface low velocity layer is responsible for generating low frequency seismic noise while at the same time severely attenuating the higher frequencies we are trying to record. A typically bad near surface geological condition is caused by the presence of unconsolidated gravels. These tend to 'shake' badly and distort any seismic signal travelling through them.

How important are these receiver differences though? Are they worth worrying about? In a word - YES, they can be vital to the success of a survey. In order to find out more about some of these differences, Western shot a series of tests which were recorded simultaneously by some 24 different geophones and hydrophones. The various receivers were placed in close proximity to each other. Figure 17 shows these positions. The 5 x 30' holes were drilled 1 mtr. apart across the line of shooting, both hydrophones and geophones were placed in these deep holes. Geophones were also planted on the surface around hole

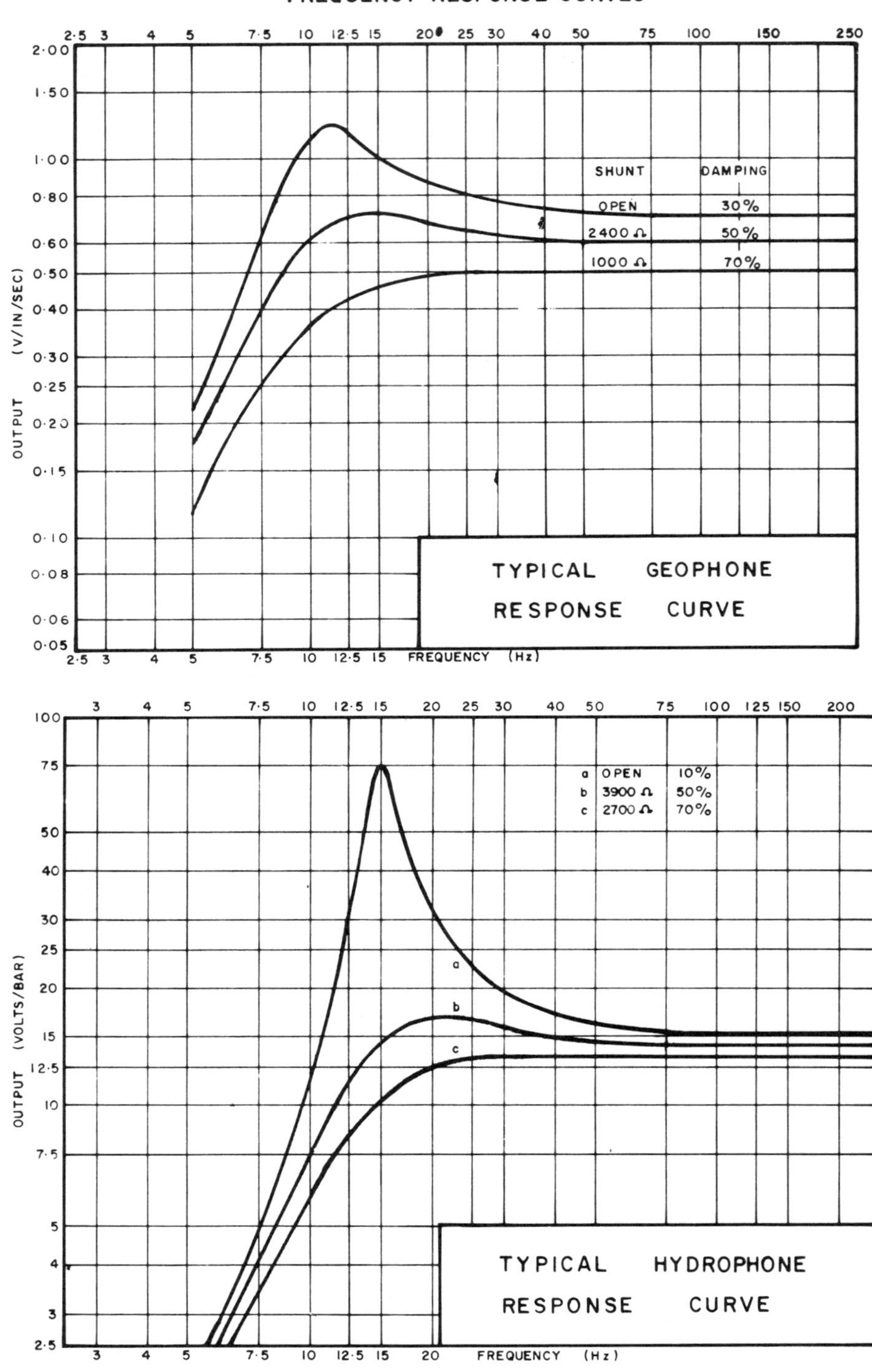
FREQUENCY RESPONSE CURVES
OUTPUT (V/IN/SEC)
SHUNT
DAMPING
OPEN 30%
2400 Ω 50%
1000 Ω 70%
TYPICAL GEOPHONE RESPONSE CURVE
FREQUENCY (Hz)
OUTPUT (VOLTS/BAR)
a OPEN 10%
b 3900 Ω 50%
c 2700 Ω 70%
TYPICAL HYDROPHONE RESPONSE CURVE
FREQUENCY (Hz)

Figure 16

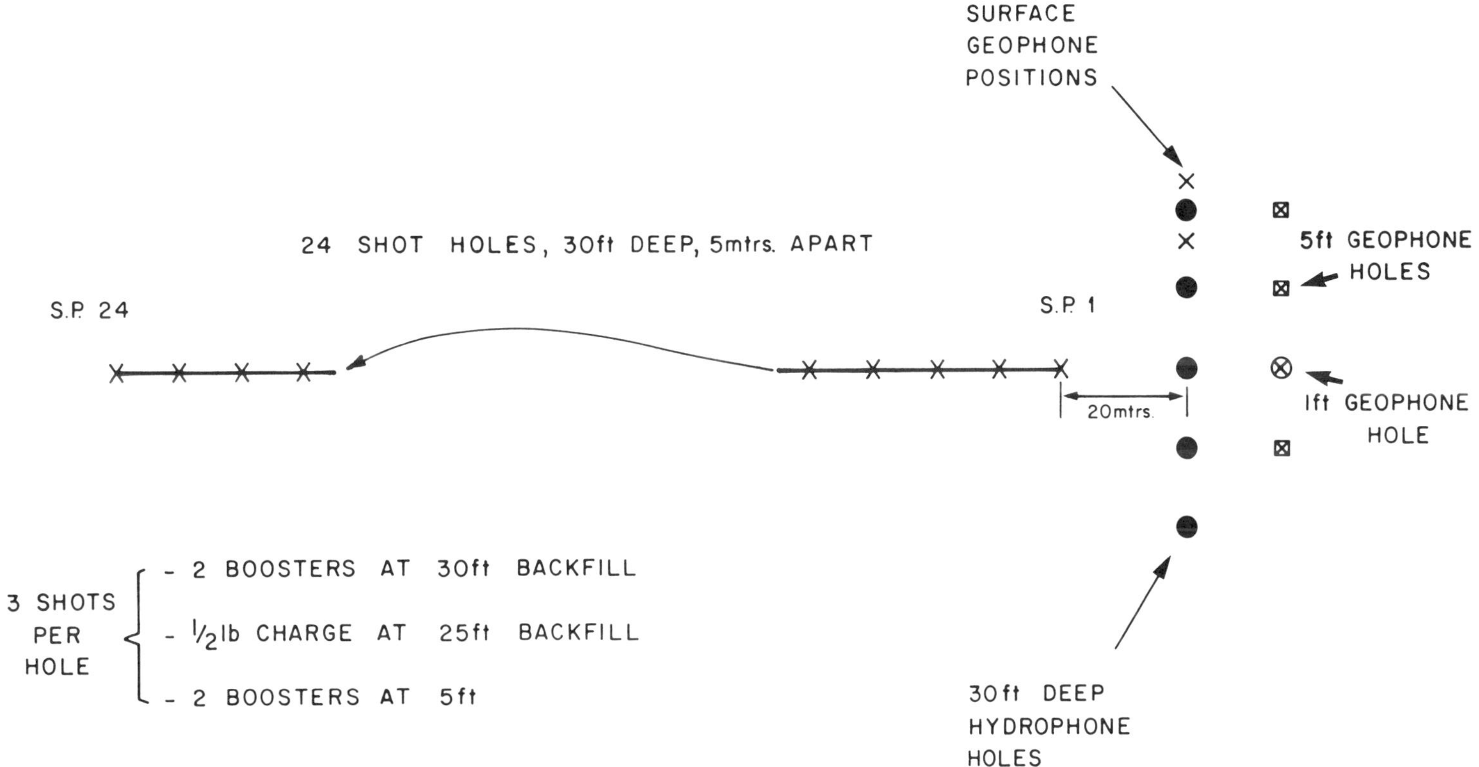
RECEIVER COUPLING TEST
SURFACE GEOPHONE POSITIONS
24 SHOT HOLES, 30ft DEEP, 5mtrs. APART
5ft GEOPHONE HOLES
S.P 24
S.P 1
20mtrs.
1ft GEOPHONE HOLE
3 SHOTS PER HOLE
- 2 BOOSTERS AT 30ft BACKFILL
- 1/2 lb CHARGE AT 25ft BACKFILL
- 2 BOOSTERS AT 5ft
30ft DEEP HYDROPHONE HOLES

Figure 17

one and in 5' holes behind the 30'holes. The deep 30' hydrophone holes were saturated with water immediately before each shot was fired.

24 shot holes were drilled to a depth of 30' each hole 5 mtrs. apart, the first hole 20 mtrs. away from the nearest receiver. This gave offsets of between 20 and 135 mtrs. or 66 and 443 ft. respectively.

The near surface geology conditions were typically sandy and flinty for the first 2 - 5 ft. and then straight into heavy London clay beds. Holding water in the holes was not a problem. The shots were fired, from SP1, firstly using 2 boosters only at 30' depth. The shot was recorded using Western's MANPAQ® portable field recording system, onto magnetic tape in SEGC tape format. 3 seconds of data were recorded at ½ millisecond sample rate.

After the deep 30' shot fired, the hole was gravel back-filled and loaded with a ½lb charge at 25', this was then fired. The backfilling process was repeated and 2 boosters were shot at 5' depth.

This cycle was carried out for each of the 24 shot point locations, giving the input data required for a 'regular' 24 trace single shot record. The data was sent to the processing center where it was demultiplexed from the field format, and rearranged to combine similar traces for each hydrophone and geophone. The records were subsequently processed with various frequency filters.

These records contain much information about the problems we can expect and how we might avoid them. Figure 18 shows a selection of the different receiver records. It becomes immediately apparent that there is a vast difference in response between nearly every receiver. These differences vary between very good on some records which exhibit good clean continuous primary reflections, and not too high a level of background noise,and one particular hydrophone which has suffered an almost total failure. These records have had <u>no</u> processing filters applied.

COMPARISON OF VARIOUS GEOPHONES/HYDROPHONES RESPONSES TO THE SAME INPUT SIGNAL

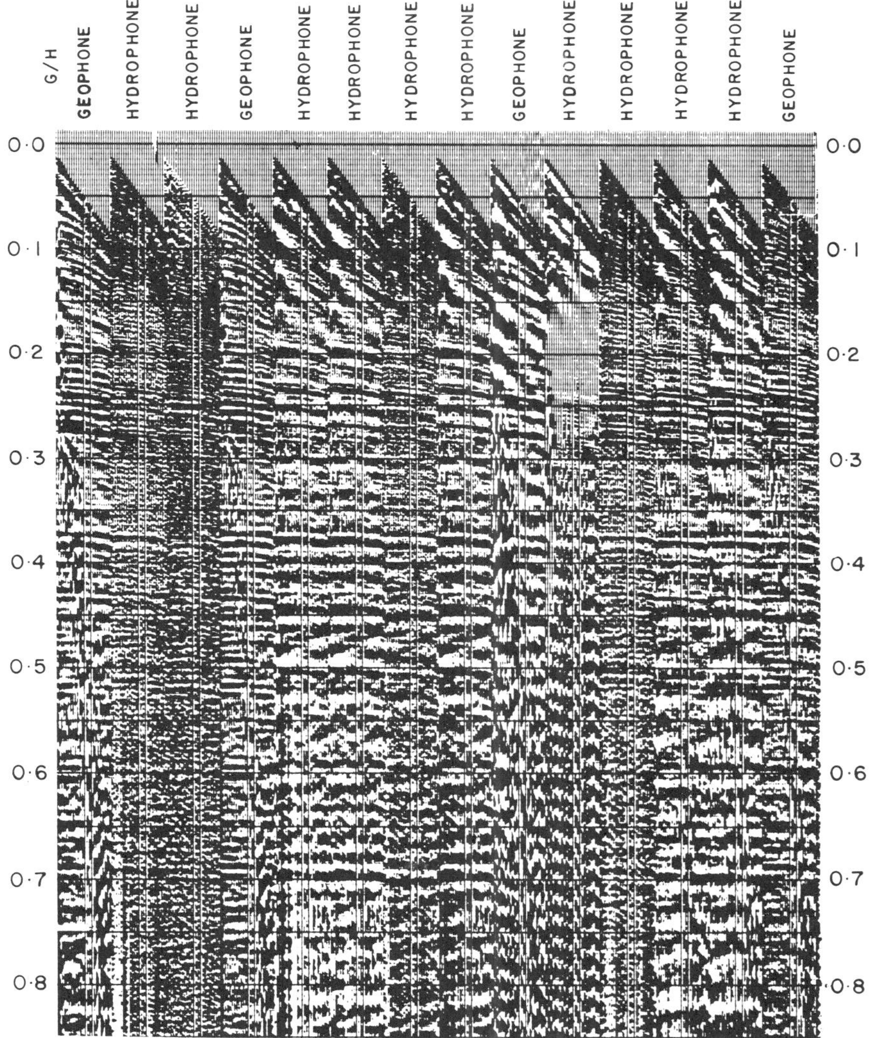

Figure 18

Taking a closer look at some records. Figure 19 shows a typical land geophone, L25E-40. This type geophone would be considered a good standard land seismic geophone found on many higher resolution crews. Its response is similar to that shown on the geophone response curve earlier. The record shows several good, fairly continuous reflectors. Its also shows the presence of a strong surface wave noise train which dominates the entire record. In addition, the first 150 to 200 ms. are heavily overlain with first arrival P wave noise which makes it difficult to identify any primary reflections. This geophone was planted on the surface by Hole 1. No filtering has been applied. The highest frequency apparent on this record, is about 60 Hz at just over 200 milliseconds, which will give us a minimum resolvable bed thickness of only some 8 - 10' at best. This is clearly not as good as we would like for coal exploration purposes.

Now, examining one of the hydrophone records (Fig. 20, the MP24B, we see that this receiver was at a depth of 30 ft. in one of the water saturated holes.

Comparison between this and the previous geophone record show a number of improvements:-

* Much better signal/noise ratio (ie. much less noise interference from both P waves and ground roll sources

* Excellent continuity on most refelctors.

* Much higher frequency content.

This higher frequency content is quite marked when you look closely at the time interval between 200 and 300 milliseconds. On the geophone record we can see a maximum number of 7 reflectors. On this hydrophone record a maximum of 11 reflectors may be seen. That's 4 extra resolvable geologic horizons just for the price of a hole and a hydrophone! The maximum observed continuous frequency is approximately 125 Hz - or double that seen on the geophone output.

Note also the improvement at times of between 500 - 600 milliseconds. With the better ground coupling of the hydrophone, the severe noise train is less apparent, and with most of the noise removed, several good primary reflections can be seen.

Even on this record though, there is still some ground

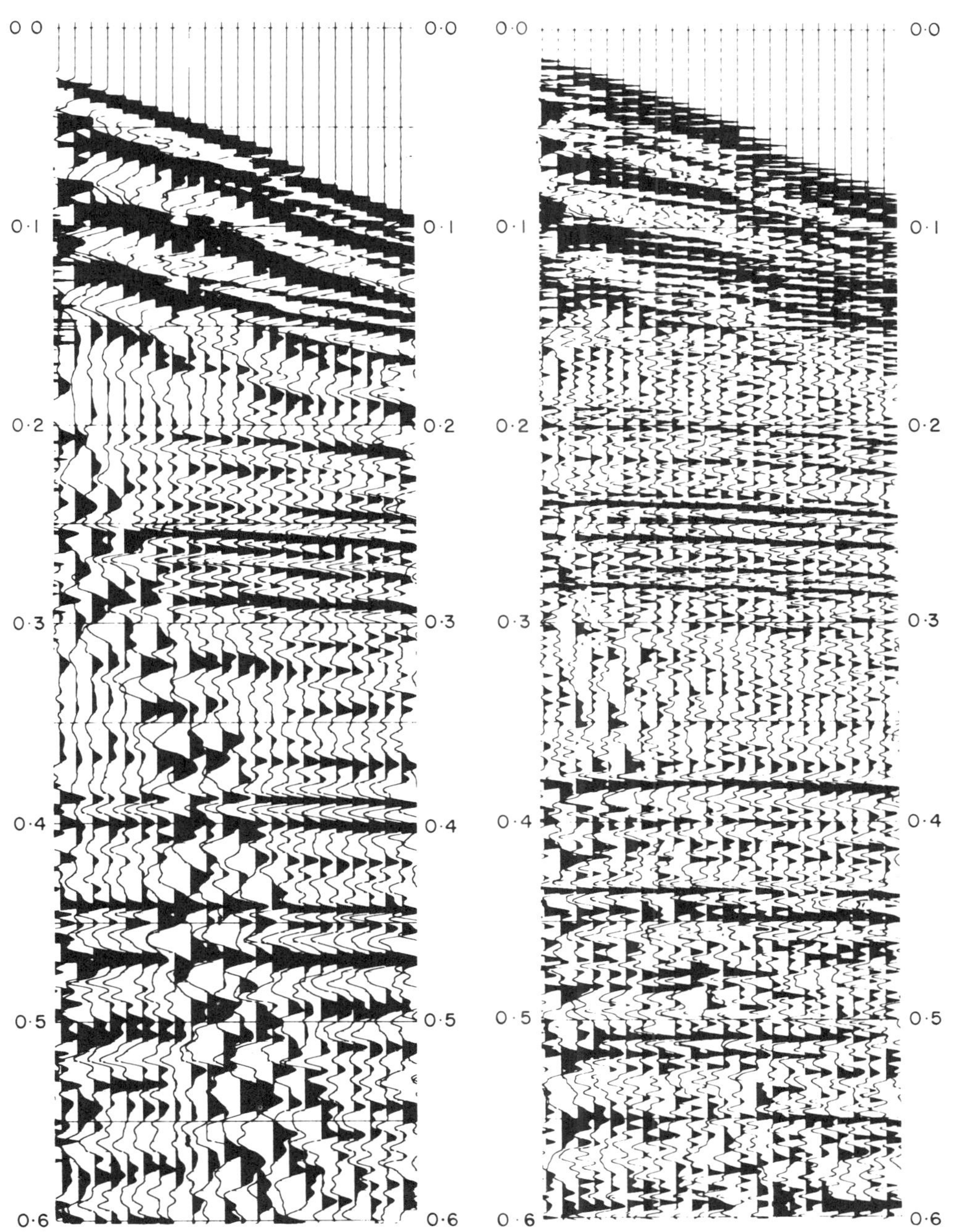

Figure 19

Figure 20

roll present, from about 300 ms. onwards down the record. This may be quite easily removed by the application of a lo cut processing filter. Figure 21 shows the same record where such a filter has been applied. On this filtered record we can see improved resolution both in the very shallow section and the deeper area below 400 milliseconds as well.

Can we do any better, should we need to filter the data to remove noise trains? Yes, we can do better and yes, it may be necessary to use field filters. Not all hydrophone are the same and we should naturally expect some different results.

Figure 22 shows a Western B6 hydrophone. This was buried at 30 ft. like the Mp 24D. Don't forget all these receivers were recording the energy from the same shots. Again, we can see differences. The event at about 160 ms. is much better defined. The zone between 200 - 300 milliseconds now shows some 14 different reflectors that have been resolved. At velocities of 6500 7000'/sec typical at this depth, we see for the highest frequencies, a total be thickness resolving power of about 18'. The event representing that thickness is about 5 - 6 ms. long in time. Present display techniques allow us to diplay such an even at a scale of some 5 ms. = ½ cm. on a regular time section It is a relatively straightforward task for the interprete to pick out discontinuities, like faults along such reflections. An accuracy of about 1 mm should be attainable under good conditions, giving a resolving power of approximately 3 - 4'.

The B6 record also shows some differences at depths where increased resolution is apparent. Note the event at 440 to 450 ms., this event is broken about 2/3rds the way to the right hand side. While there are certain dangers in inferring too much from raw records, there is obviously a deep discontinuity of some kind producing the time delay we observe. This is unlikely to be caused by near surfac variations because of the good continuity of the overlyin reflectors.

Finally a word about geophones, these measure particle displacement and are generally either placed on or buried close to the free surface. This particle displacement is reduced for near vertical reflected energy as the phone is placed deeper in the ground; but it is not reduced for

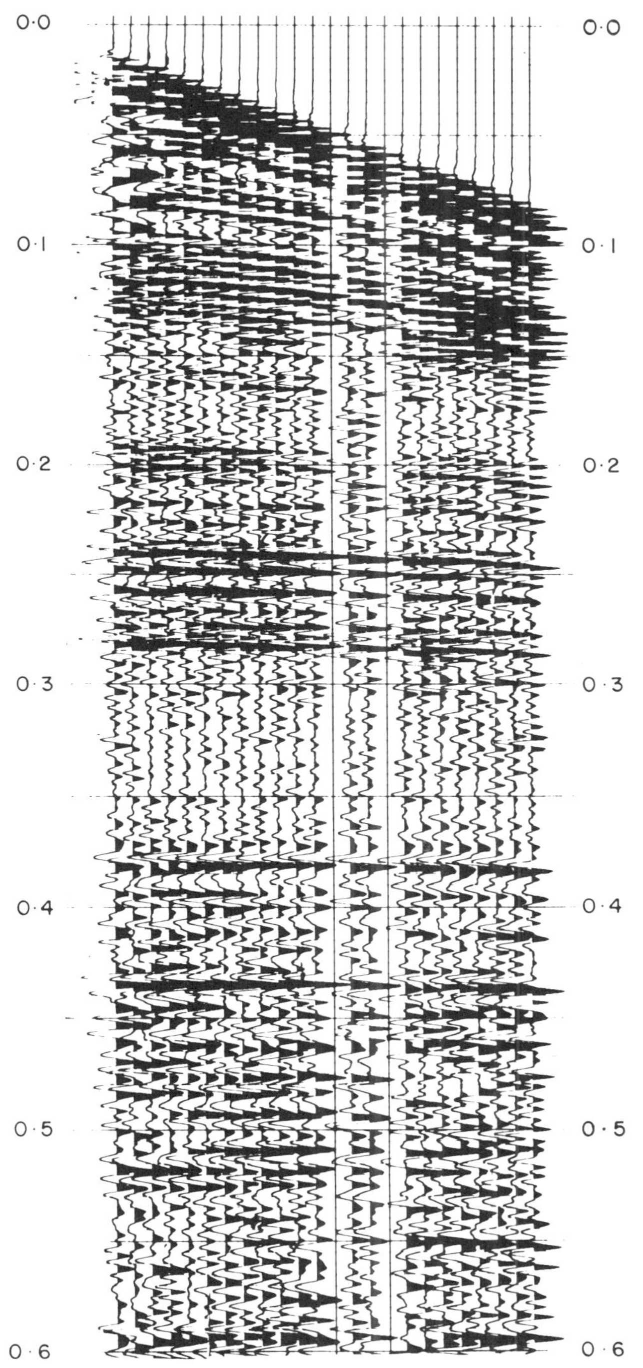
MP 24B HYDROPHONE FILTERED
0·0
0·1
0·2
0·3
0·4
0·5
0·6
0·0
0·1
0·2
0·3
0·4
0·5
0·6

Figure 21

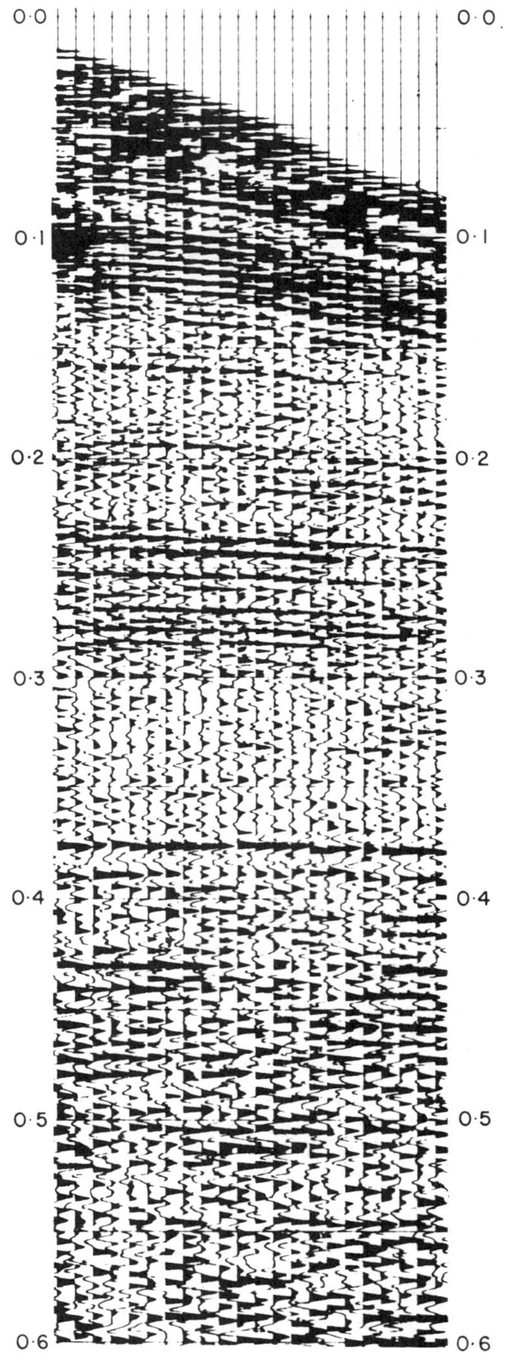

Figure 22

horizontal noise wave movement. Thus we might expect an increase in noise and a reduction in primary energy when the geophone is placed deeper in the ground away from the surface.

In our test we placed two identical geophones in different depth holes. One at 5 ft. and the other at 30' depth. The results were very interesting. Slide 23 shows the 2 records, the geophone at 30' depth is on the left, the geophone at 5' on the right.

As we had expected, the horizontal travelling waves have had an increased effect on the deeper geophone, note the increased noise train. Also there is a reduction in reflection energy at the deeper geophone. Several reflections on the shallow RHS record i.e. at 220, 400, 470 ms. are much stronger than those on the deeper record. Unexpectedly though, the reflected energy that has been recorded by the deeper geophone is of higher frequency and thus greater resolving power than that of the shallow geophone - which is what we are looking for. Also comparison of the deeper data with other records shows that the deep buried geophone is working as well as many of the hydrophones - if not better, at times greater than 500 ms.

By running this type of analysis as a pre curser to a full seismic survey, the optimum possible recording parameters may be selected. This ensures that at least if the earth decides to reflect us back high frequencies, we are in the best position to record them.

HOW ACCURATE IS IT POSSIBLE TO BE?

This is quite a tough question and not one to which an exact answer can be given, especially in an unknown area. In essence, you really have to try it to find out. Accuracy in terms of measurable depths to a target seam or distance across a washout that we can see on a seismic cross section, are a function of:-

* Adjusted one way travel times to the feature of interest.
* The velocity which the energy travelled with along the particular travel path. (In practice, this seismic rock velocity is highly variable and a variable velocity function of T versus VRMS is used.)

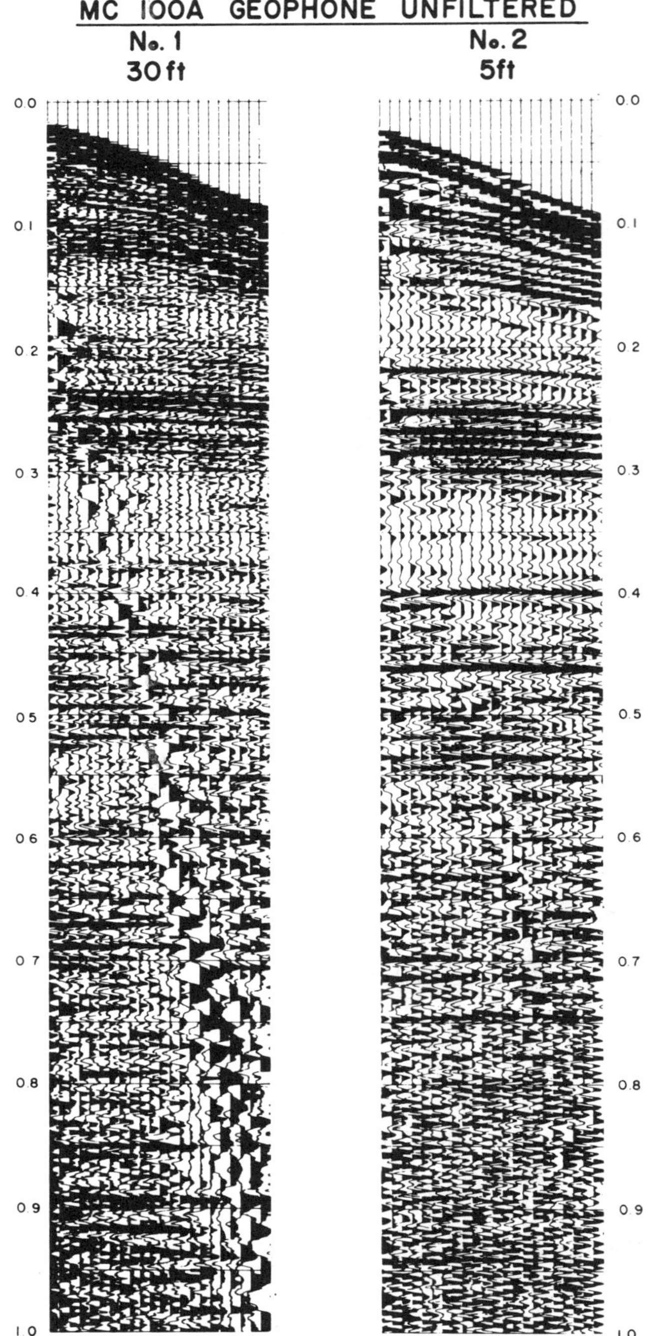
MC 100A GEOPHONE UNFILTERED
No. 1
30 ft
No. 2
5 ft
0.0
0.1
0.2
0.3
0.4
0.5
0.6
0.7
0.8
0.9
1.0

Figure 23

Consider the experimental line in Figure 24. A growth fault and a channel washout are clearly visible. Fig. 25 shows an enlargement of the washout, the edges of the strata surrounding the washout are quite distinct and from careful measurements we may estimate the size of the eroded channel. The depths on the figure have been approximated, using seismically derived RMS velocity functions. (Approximating depths from RMS velocities can be problematic when using short shot to detector distances as in coal exploration work. Velocity surveys run in nearby boreholes are likely to be much more reliable). The depths calculated place the top of the washout as approximately 465' deep, and the center of the washout at approximately 580' deep, giving a 115' total washout thickness. Although this section has not been velocity migrated to place the dipping washout slopes in their correct position,it is a simple matter to estimate the apparent width of the feature. I use apparent width, because on a single line we do not know the angle at which the seismic line is intersecting the washout.

In this case, each seismic trace represents 5 metres. About 90 traces cover the feature so we have about a 450 metre apparent width.

To demonstrate the accuracy possible with seismic, Western carried out a further test some 2 years ago, primarily for this purpose.

On the LHS of the line in Fig.26, a growth fault is visible. In the field, two trenches were dug across the fault to 'prove' its surface existence. The fault throw observed in the trenches was about 15 cms. The 4 vertical lines represent boreholes that were drilled across the fault close to the seismic line. These were drilled to a depth of approximately 300 metres, for the purpose of locating the fault at depth and establishing its throw. Electric logs were run in each borehole and then correlated. As the holes are close together (approx. 200') and the sand/shale sequence quite distinctive, the log correlation from hole to hole is very good.

RMS velocities were derived from the seismic data, and depths to various horizons on both sides of the fault established. From these, the fault throw was estimated seismically, these are the 'squared' figures on the LHS of Figure 26. The 'actual' fault throws taken from the

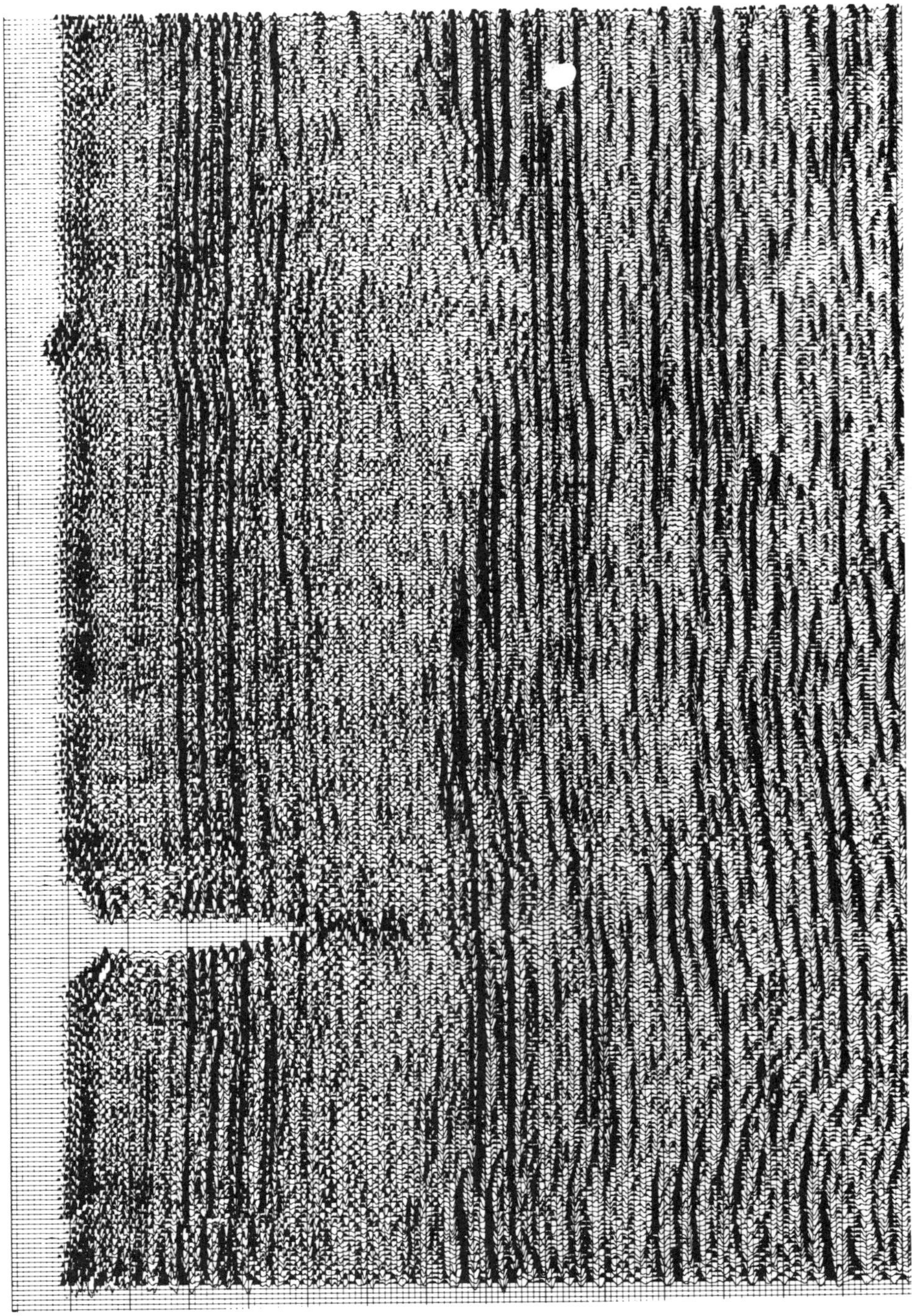

Figure 24 - An uninterpreted seismic cross section display after computer processing has been completed.

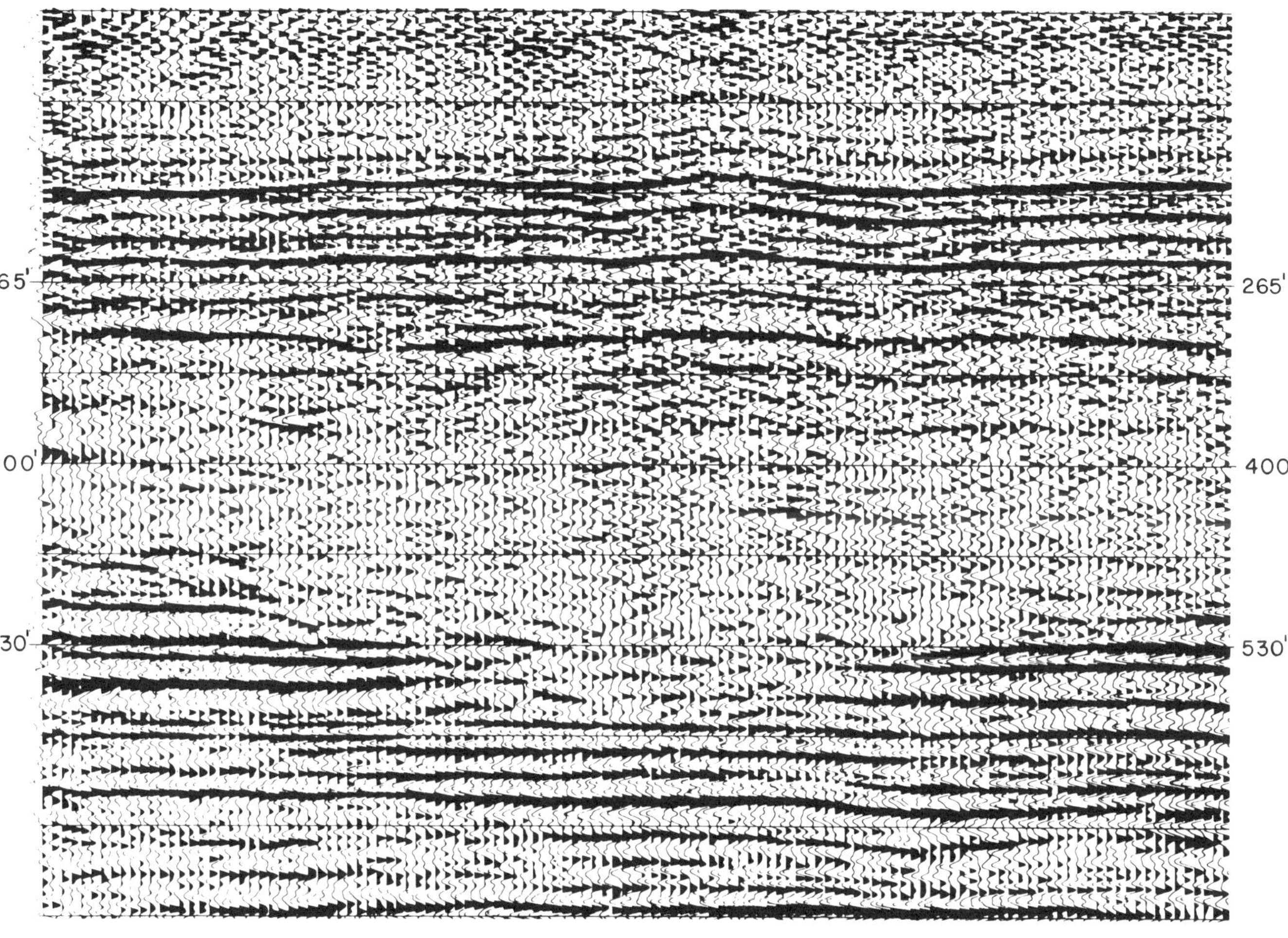

Figure 25. *The fine details of a buried channel at a depth of about 500 feet can easily be seen in this example from the Texas Gulf Coast.*

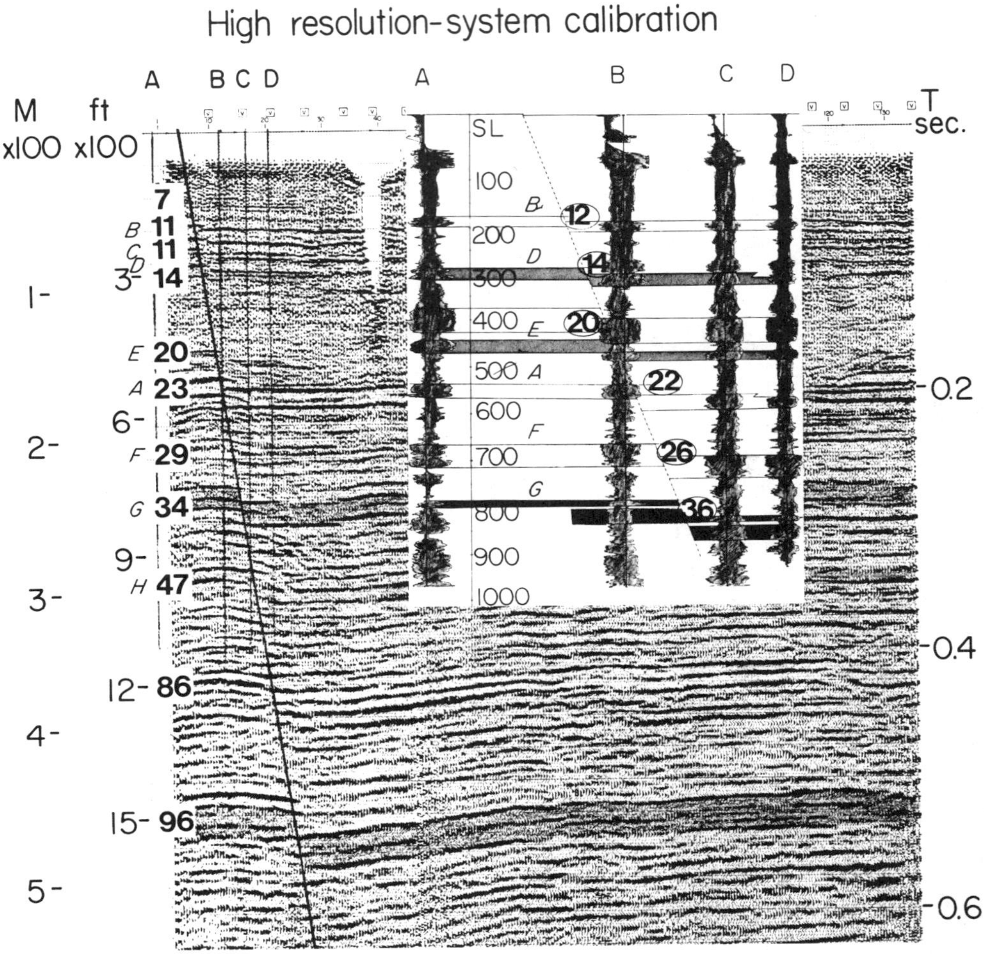

Figure 26 - Comparison of subsurface borehole information with seismic section showing geologically established fault throws in circles on log section and seismic throws in squares.

borehole control are the circled figures on the logs. As may be seen, the seismic throws agree very well with the actual throws, to within 1 foot down to about 700' and then within 3 ft. down to maximum correlatable depth. The seismic throws continue to greater depth with an approximate 96' throw at 1500' deep.

CONCLUDING COMMENTS.

Although this final test may be considered a little shallow for some mining purposes, it shows that surface seismic profiling can reasonably attain the accuracy required by the coal exploration industry. Our job now is to avoid the 'no reflections' problem, improve the accuracy of the method to greater depth and ensure that the coal exploration geologist is aware of the potential of seismic, it is there to make his life a whole lot easier.

I would like to take this opportunity to thank my Company for allowing me to present this paper, and add that the views are necessarily those of the author.

DISCUSSION

QUESTION: Comment upon the effect of mountainous or deeply dissected topography on obtaining good high resolution seismic data.

ANSWER: At first impression, one might think that mountainous or deeply dissected topography would probably always have an adverse effect on obtaining good high resolution seismic data. This is not the rule however. It is frequently the case that good data may be obtained in quite rough terrain, and equally so that very poor data be obtained in areas that are flat.

Data quality is more likely to be affected by:

1. The near surface geology conditions where the shot and receiver are positioned. For instance, unconsolidated gravels and sands will usually give a poorer response than a well cemented sandstone.
2. The underlying rock types and sequences of rock types down to target depth.

3. The depth and thickness of the target seam itself.

For instance: if the following three conditions are present,

1. Well consolidated medium on surface.
2. Nice homogeneous sandstone/shale down to target seam.
3. Relatively flat thick (>20 feet) coal at shallow (<1500 feet) depths.

then the effect of very rugged or mountainous terrain is likely to be minimal.

If, however, the conditions are,

1. Thick unconsolidated gravels, conglomerates on sand on the surface.
2. Highly variable rocks - sands - shales - limestones sands etc. in complex sequences.
3. Thin (<6 feet) coals at greater depths (>1500 feet).

then the effect of rugged or mountainous terrain could void the survey completely.

It is very important to realize, though, that one does not always get a bad survey if the survey conditions are bad; likewise with the converse. We are presently unable to quantify 'seismic conditions' so that we can say such and such a condition will give bad results or that another condition will give good results. But we do know that invariably the only sure way to find out is to try it!

QUESTION: Is it possible to identify coal seams in reflection/refraction records or is drilling information required for interpretation?

ANSWER: It is difficult to uniquely identify coal seam reflection from seismic data where the presence of coal has not been confirmed by drilling. Depending upon the quality of the seismic data, it is possible to identify coal seams given suitable borehole data. I would always recommend a combined seismic and borehole program. The boreholes to confirm the coal, its thickness and quality and the seismic to interpolate continuity or lack of continuity between boreholes. In this way, it is possible to vastly reduce the number of boreholes yet still achieve a greater detail of the subsurfaces.

QUESTION: How precise is the seismic reflection method to determine seam thickness? What is the minimum seam thickness detectable? At what distance from the receiver can a small (2- to 3-foot fault) be recorded?

ANSWER: The precision with which we can predict small faults, determine seam thickness and obtain a discreet reflection from an individual coal is a function of several variables:

* Predominant wavelet frequency in Hz.
* Radius of survey investigation.
* Velocities of rocks overlying and surrounding the coal.
* Rate of earth attenuation of signal.
* Level of recording noise.

These vary quite significantly from area to area and, therefore, it is impossible to give accurate answers to the question without knowing far more about the above variables. As a guide, though, under ideal conditions it is possible to detect fault throws of about 3-5 feet and seam thicknesses of about 3-5 feet; under worst conditions you should not rely on seismic!

QUESTION: What is cost per mile in flat and mountainous terrain?

ANSWER: Again, difficult to answer without knowing more about specific projects. On high resolution data acquisition the cost would probably be on a monthly or other period basis. For this the client receives the services of the field crew. Now how much the crew is able to record within a period will obviously be affected by the terrain (whether it is flat or mountainous), climate, geology of prospect, accessibility and several other variables dependent upon the objectives and precision necessary to achieve them. The worse the conditions, the less the crew is likely to record and therefore the higher the cost will be. Data processing is charged separately. The cost of doing seismic will usually be far less than the cost of gaining data by saturation borehole programs. That is, in a 10 by 10 kilometer prospect it is likely that borehole control would be placed at 1/2 to 1 kilometer spacing, i. e. at 1 kilometer some 121 holes or at

0.5 kilometer some 443 holes. By utilizing seismic inbetween control boreholes it would be quite feasible to increase the borehole spacing to maybe 2½ kilometers, which would give a saving of the cost between 96 and 418 boreholes!

For the price of that many boreholes, you could run several seismic surveys and know that your resulting exploration plan would have both the necessary borehole control to uniquely identify the coal, its quality and thickness and the seismic control to determine the coal continuity (or lack of continuity, i. e. faults etc.) between control boreholes.

11

In-Seam Seismic Methods for the Detection of Discontinuities Applied to West German Coal Deposits

By Horst Rüter
and Reinhard Schepers
Institut für Geophysik der
Westfälischen Berggewerkschaftskasse
Bochum, Federal Republic of Germany

INTRODUCTION

In the Ruhr-district, West Germany, as well as in many other coalfields the majority of coal is mined by highly mechanized longwall faces. Using this technique the face moves between two premined roadways on a day to day basis. A sudden stop of the working face caused by a geologic fault not previously detected can cease the production perhaps for a long time. Even a throw of 1 to 2 meters can lead to such a situation depending on the geologic conditions. Since the maximum depth of mined seams is about 1000 m in the Ruhr-district, it is not possible to detect minor faults by surface seismic methods.

The in-seam seismic exploration technique using channel waves is the only developed geophysical method to detect such faults a few hundred meters ahead of the face. The physical

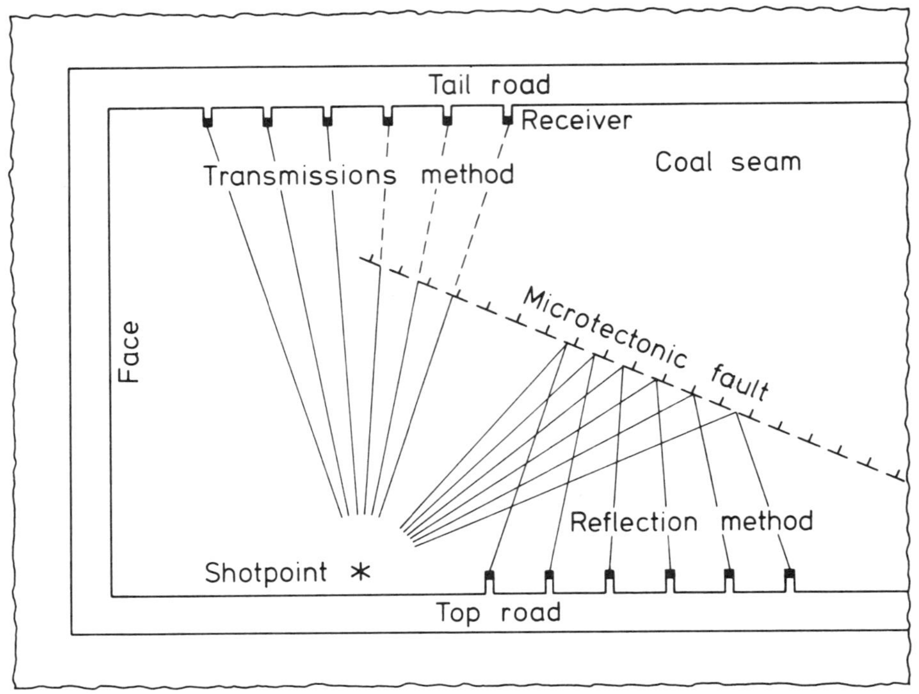

Fig. 1. Principle of the in-seam transmission survey and reflection survey

basis of this technique, introduced 1963 by Krey (1), is the existence of guided waves in low velocity channels. Because the seismic velocity in coal is only about one half of the velocity in the surrounding rock, a coal seam is a very good channel for seismic waves. The advantage of channel waves as compared to body waves is given by the fact that body waves spread into three dimensions whereas channel waves spread only into two dimensions. Hence, the attenuation due to geometrical spreading is smaller for channel waves.

In routine surveys the in-seam seismic method can be used in two different ways as shown in Fig. 1. The so-called "transmission survey" is the simpler one of both. Shot and receivers are located in different roadways. Given a fault with a throw greater then the seam thickness, the reception of channel waves will be disturbed by such a fault, if the fan (triangle of shot, first and last receiver in the spread) intersects the fault line. A localization of the fault is not possible by this method. Using a great number of shot and receiver positions it is possible in most cases to separate faulted and not faulted areas within the seam.

The localization of discontinuities is possible by the second method, the "reflection survey". Shot and receivers are

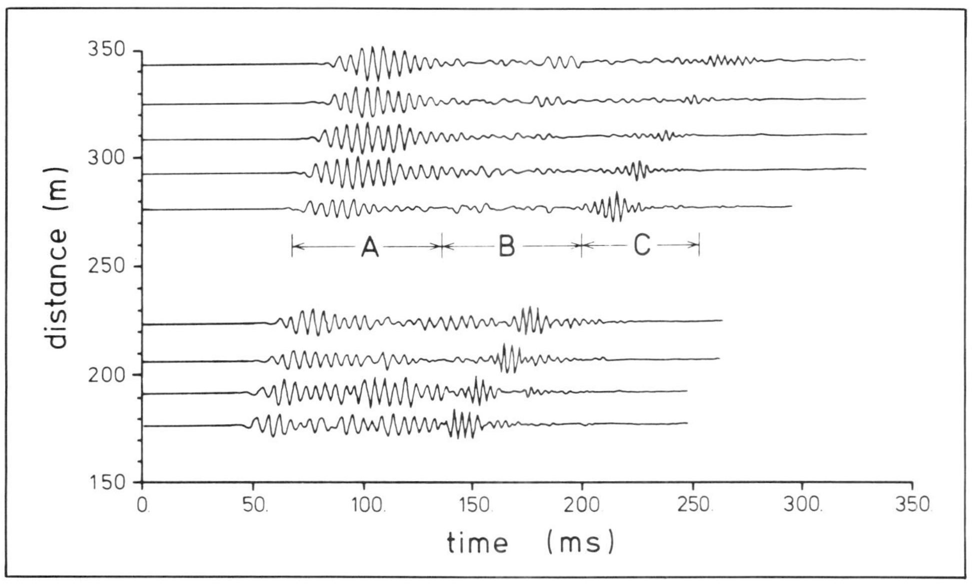

Fig. 2. Transmission seismograms for a fixed receiver location and shots at different distances. Displayed is the component which is parallel to the seam and parallel to the face. The seam is 2.4 m thick but the bottom layer of a thickness of 1.1 m includes several dirt bands

located in the same roadway. By means of the recorded travel-times of the channel waves reflected by a fault, the location of the fault can be determined supposing the propagation velocity of the waves is known.

Using channel waves in a transmission survey or in a reflection survey it is essential to have some basic information on the physical properties of channel waves. The channel wave is not <u>one</u> special wave but the sum of all seismic waves guided by the seam.These waves can be divided into SH (Love) and P-SV (Rayleigh) types according to their polarisation, or into symmetrical and antisymmetrical types according to their amplitude distribution relative to the centre plain of the seam. Each type can theoretically be divided into an infinite number of modes. Fortunately, only the lowest mode (fundamental mode) of these wave types is relevant in practical applications. The result of a transmission survey, shown in Fig.2, may demonstrate the complexity of channel wave propagation.

The important differentiation into Love and Rayleigh types is illustrated by Fig. 3. The symmetrical Love wave has particle motion only in one direction parallel to the seam with maximum amplitudes in the middle of the seam. The symmetrical Rayleigh wave has particle motion parallel and vertical to the seam, but its component parallel to the seam is perpendicular to the particle motion of the Love wave. For the symmetrical Rayleigh wave the vertical component is zero in the middle of the seam.

Refering once again to Fig. 2 it should be pointed out that the waveform of the different groups (A,B,C) spreads out from trace to trace due to the fact that channel waves have

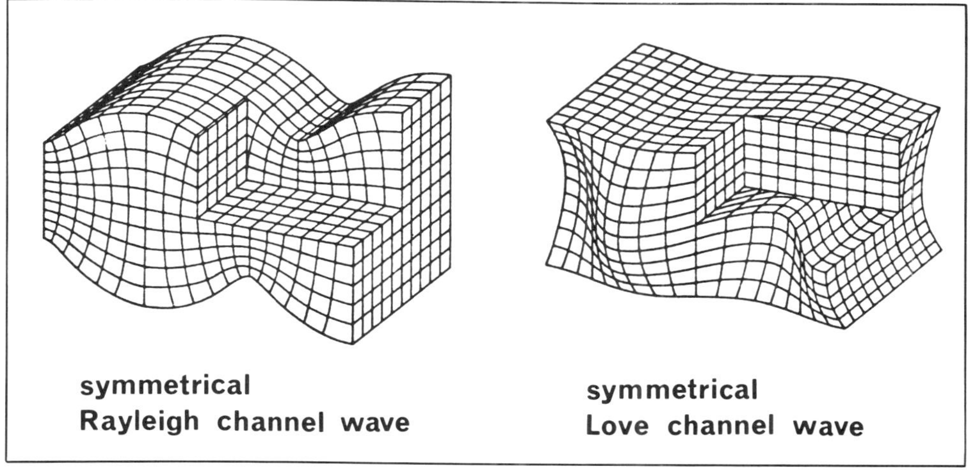

Fig. 3. Particle motion of channel waves (after Arnetzel (2)

the important property of being dispersive. This means, that the propagation velocity of a signal depends on its frequency. This phenomenon is described by the dispersion curve. A dispersion curve for the fundamental mode of the symmetrical Love channel wave is shown in Fig. 4. In the lower part of Fig. 4 the ratio E_R of the kinetic energy of the channel wave inside the seam to the total kinetic energy of the channel wave is shown in its dependence on frequency. We see that in the case of a 1 m thick seam the total energie is confined to the seam for frequencies above 700 Hz. The characteristic minimum of the group velocity curve U belongs to the well known Airy phase. Signal frequencies of this phase should preferably be used, because the geometrical attenuation is smallest for the Airy phase. Additionally, extensive series of model experiments of Dresen and Freystätter (4) have shown that a good detection of faults should be possible, if the value E_R exceeds 0.5 (s. Fig. 4). Taking into account both arguments the frequency for a survey of a 1 m thick seam must be at least 500 Hz. This is in conflict with considerations

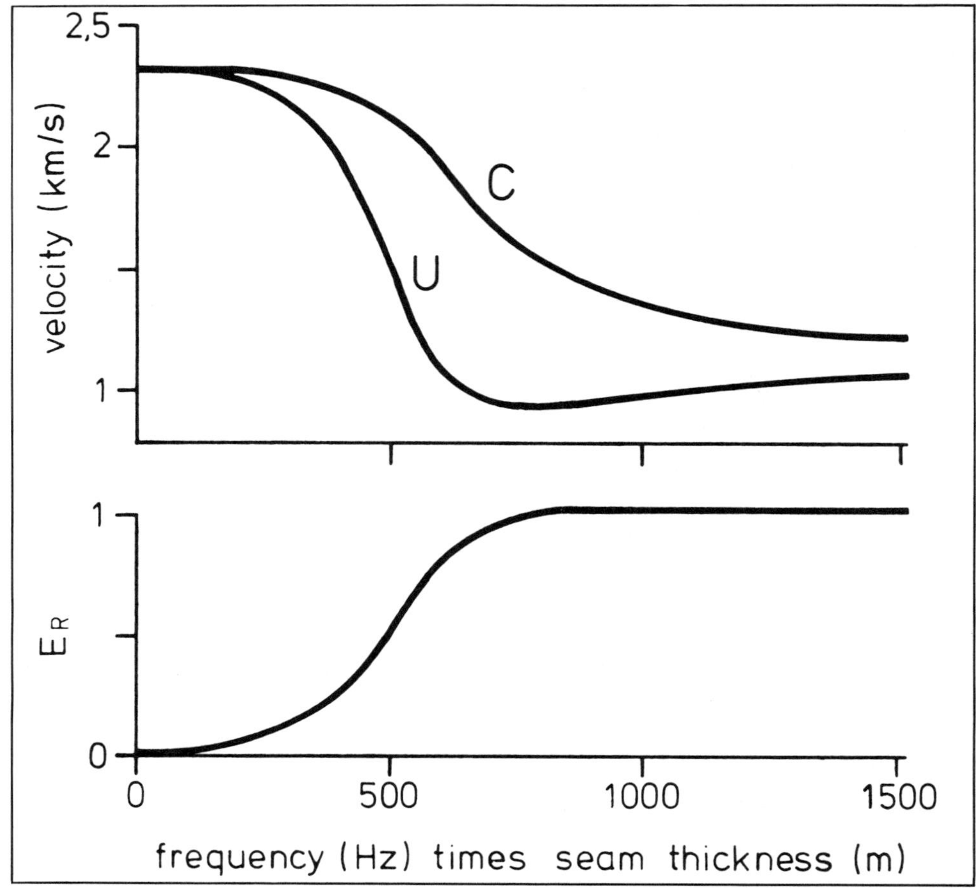

Fig. 4. Dispersion curves for the fundamental mode of the symmetrical Love channel wave (C = phase velocity, U = group velocity), and the ratio E_R of the kinetic energy of the channel wave inside the seam to the total kinetic energy of the channel wave

concerning attenuation due to absorption (5).

The proper choice of the optimum frequency band should be supported by field experiments. As result of a field trial the frequency content of transmitted channel waves in its dependence of shot-to-receiver distance is shown in Fig. 5. It is worth noticing that the maximum of the spectrum does

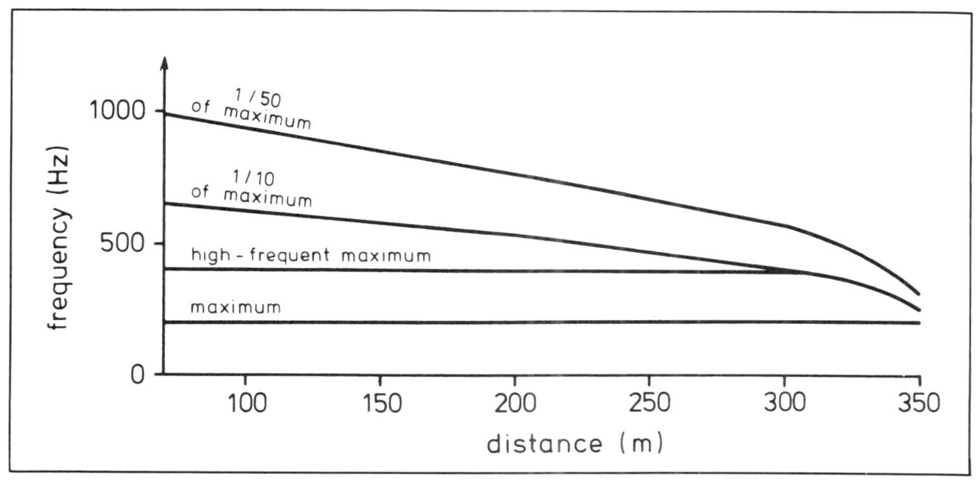

Fig. 5. The frequency content of transmission seismograms as function of the distance between shot and receiver.

not depend on distance. The upper limit of the recorded spectrum, defined as 1/10 of the maximum, slowly decreases with increasing frequency. This leads to less resolution in fault detection (and perhaps less reliability) above a certain range.

SURVEY METHOD

Until now underground seismic measurements have been carried out with analog recording systems. As transmission surveys could be interpreted by visual inspection of the field seismograms analog recording gave satisfactory results in these cases. But on reflection seismograms clear reflection events could be identified only in a few cases. Therefore the development of a digital recording system for underground seismic measurements suggests itself to gain multiple coveraged data and to improve the signal-to-noise ratio by suitable data processing methods.

Since the end of the last year (1977) the BERGBAUFOR-

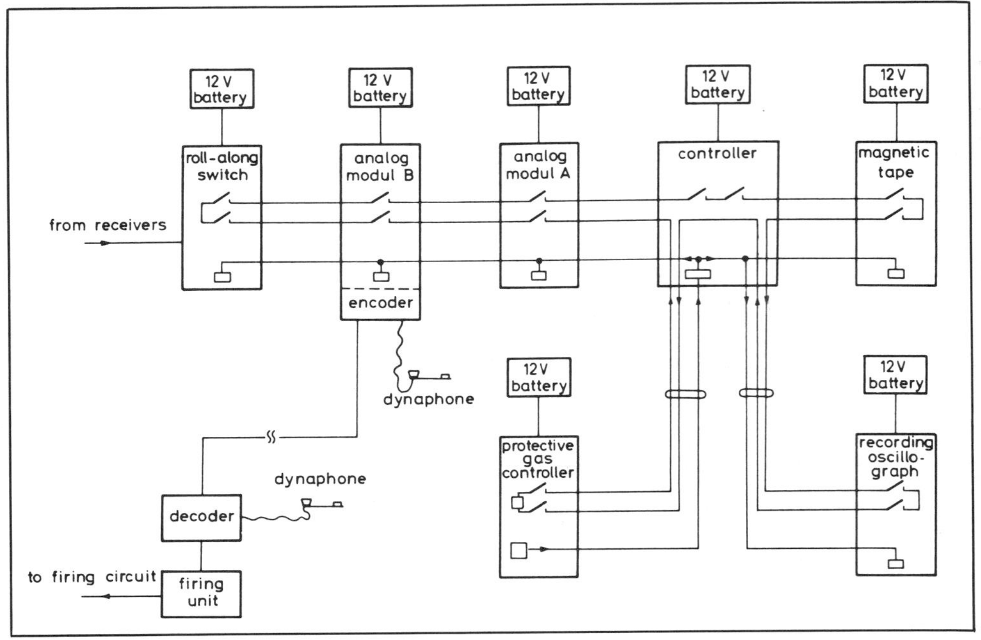

Fig. 6. Block diagram of the firedamp-proof digital recording system

SCHUNG GMBH at Essen, West Germany, has a firedamp-proof digital seismic recording system (6) which can be used in coal mines. The recording system MDH, as it is called, consists in the main part of a DFS V from TEXAS INSTRUMENTS. To meet the requirements of firedamp-proof and the additional difficulties of underground work PRAKLA-SEISMOS GMBH adapted the system and developed some new peripheral units.

A block diagram of the seismic recording system MDH is shown in Fig. 6. The DFS V consists of the two analog moduls A and B, the controller and the magnetic tape. According to the measurements (see Fig. 5) now all surveys are done with a sampling interval of 0.5 ms giving a flat response of the system in the range from 20 Hz up to 512 Hz. Recording with a sampling interval of 0.5 ms the number of input channels is 12 for each analog module. The roll-along switch has 120 input channels. A whole spread of 120 geophone groups can be layed out and tested before shooting. Then, a reflection survey can be done, applying the principle of multiple coverage, without the necessity of rearranging the spread.

The seam waves are generated by shots. The shot ignition is triggered by the recording system. Therefor a digital word is transmitted from the encoder via a telephone cable to the decoder (see Fig. 6). The digital word is checked by the decoder. If it is correct the decoder gives a trigger signal to the firing unit which has to be manually charged before. As the magnetic tape is triggered simultaneously with the shot a maximum of 100 records can be stored on the tape. This number is sufficient to store the data of one survey on one tape. A change of the tape during the survey is not possible because of the bad underground environment.

The oscillograph can directly display the input signals. To controll the operation of the magnetic tape play-back from the tape is possible.

The units of the recording system, as shown in Fig. 6, are built in compression-proof boxes. All units are connected to the protective gas controller by means of flexible tubes. To keep up the firedamp protection, all of the apparatus are rinsed with a protective gas before measuring, and then the gas is compressed. Each unit can only be switched on when the pressure of the gas within the housing is kept above a certain level. If the pressure drops below this level the gas controller automatically shuts down the whole apparatus.

To generate seismic channel waves explosive charges were detonated in small boreholes which are drilled in the middle of the seam and parallel to it. The depth of the boreholes may vary between 2m and 3m. The charge normally is 0.125 kg to preferably generate high frequencies.

The seismic receivers are placed in boreholes which are drilled in the same way as for the shots. The depth of the boreholes is 2 m in most cases. To attache the receivers firmly within the borehole, a borehole probe was constructed (7), which has at one side an expandaple rubber jacket. When air is injected into the probe, it expands the rubber jacket, pressing the probe tightly against the wall of the borehole. Now, always two geophones of a natural frequency of 28 Hz are mounted within each borehole probe. The directions of maximum sensitivity of the two geophones are perpendicular to each other. The borehole probe is orientated such that two components of the particle velocity parallel to the seam are recorded. Thereby it is possible to discriminate between channel waves of Rayleigh or Love type.

For a reflection survey boreholes are drilled every 3 meters along the spread line. To suppress those waves, which travel along the face of the coal seam, three geophones are connected to produce one input trace (seg. F. g. 7). As indicated in Fig. 7 the group seperation is 9 m in this case. Surveys have been run, where the distance between boreholes was 2 m, resulting in a spacing of 6 meters for the groups. If ever possible the same spread is used for a transmission survey.

As the strike of the faults we are searching for can have

any angle with the direction of the spread line, it is important to plan the lay-out of the spread and of the shotholes such that reflections from all parts of the fault can be received. We will elucidate this problem by means of the Fig.8. In the lower part of Fig. 8 one possible arrangement of shot and receiver locations is depicted. The blacked-in areas at locations in the coal seam give the range of angles between fault strike and spread, for which at least one receiver will pick up a reflection event from that specified location. Fig. 8 shows that at distances from the geophone array greater than the length of the spread the possible range of strikes becomes small and its orientation depends on the location relative to the middle of the spread. The length of the spread can not simply be made longer by increasing the space between the geophon groups; otherwise a dip aliasing (8) will result. This means, if the angle between fault strike and spread exceeds a certain angle, depending on the group spacing, reflection events from the fault can not correctly be migrated.

Another point is, that the degree of coverage, too, depends on the angle between fault strike and spread. This is shown in Fig. 8 for one location nearest to the geophone spread and right in the middle of it. The shape of the blacked-in area results from drawing for every angle one vector,the length of which is proportional to the number of traces that can be stacked for this location. In the case of Fig. 8 this number is 6 for a fault striking parallel to the spread line end the number increases to 15 for moderate angles.

The planning of an appropriate arrangement of shots and

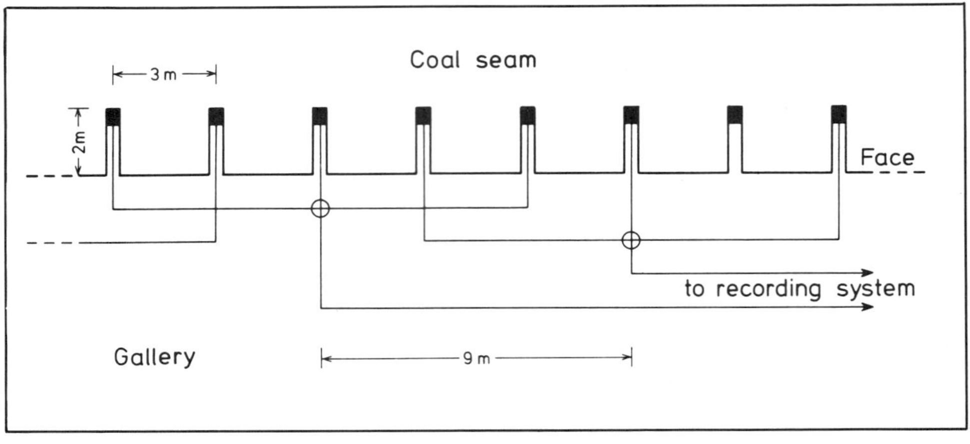

Fig. 7. Part of receiver spread showing geophone group connection

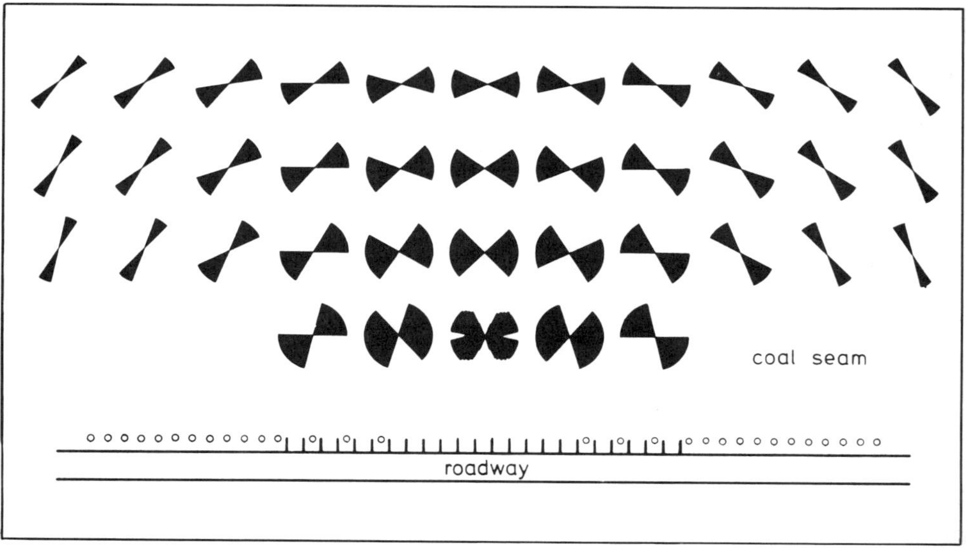

Fig. 8. The dependence of fault detection on shot and receiver arrangement. Circles indicate shots

receivers will be easier, if there is some knowledge of the probable fault strike.

THE NEED FOR NEW DATA PROCESSING METHODS

The use of digital recording systems in in-seam seismic measurements is essential for a successful application of the reflection method. The propagation of channel waves differs very much, however, from the propagation of body waves in layered media, which are observed in surface reflection surveys. Very sophisticated processing routines have been developed for surface reflection data, but we cannot expect to apply these computer programs with the same success to in-seam reflection data. Adapting programs or developing new programs the following differences have to be taken account of.

i) Channel waves are dispersive. That means, the velocity by which a signal is propagated is a function of its frequency. An input pulse becomes spread out and thus interpretation of arrival times gives less meaningful results.

ii) Simultaneously channel waves of the Rayleigh and Love type were recorded, which obey different dispersion curves. The complexity of the dispersion curves leads to a separation of the channel waves into several wave groups. The reflectivity for the wave groups is different according to their different frequency content.

iii) In surface reflection work the horizontal layering prevails, but regarding in-seam seismic every angle between fault strike and spread line has the same probability.

iv) The seam is more or less homogeneous and isotropic.

v) At each receiver location two components of the particle velocity are recorded. The two components should be processed simultaneously.

vi) Some data processing methods are based on a statistical distribution of reflectors, but there are only a few reflectors within the surveyed seam area.

The first three points lead to more complex problems. The latter three points offer some chance to overcome these problems by developing new data processing concepts. Computer programs to be applied to in-seam seismic data can be classified as analysis programs and processing programs. Analysis programs should deliver those parameters (e.g. the dispersion curve) which are necessary to successfully apply processing programs. Furthermore, analysis programs can provide a better insight into channel wave propagation and thus improve the mathematical model.

Developing processing programs assumptions have to be made which are based on the mathematical model. The usefullnes of the model will be appreciated by the degree of success in interpretating field data.

Analysis programs

Since at each receiver location the two components parallel

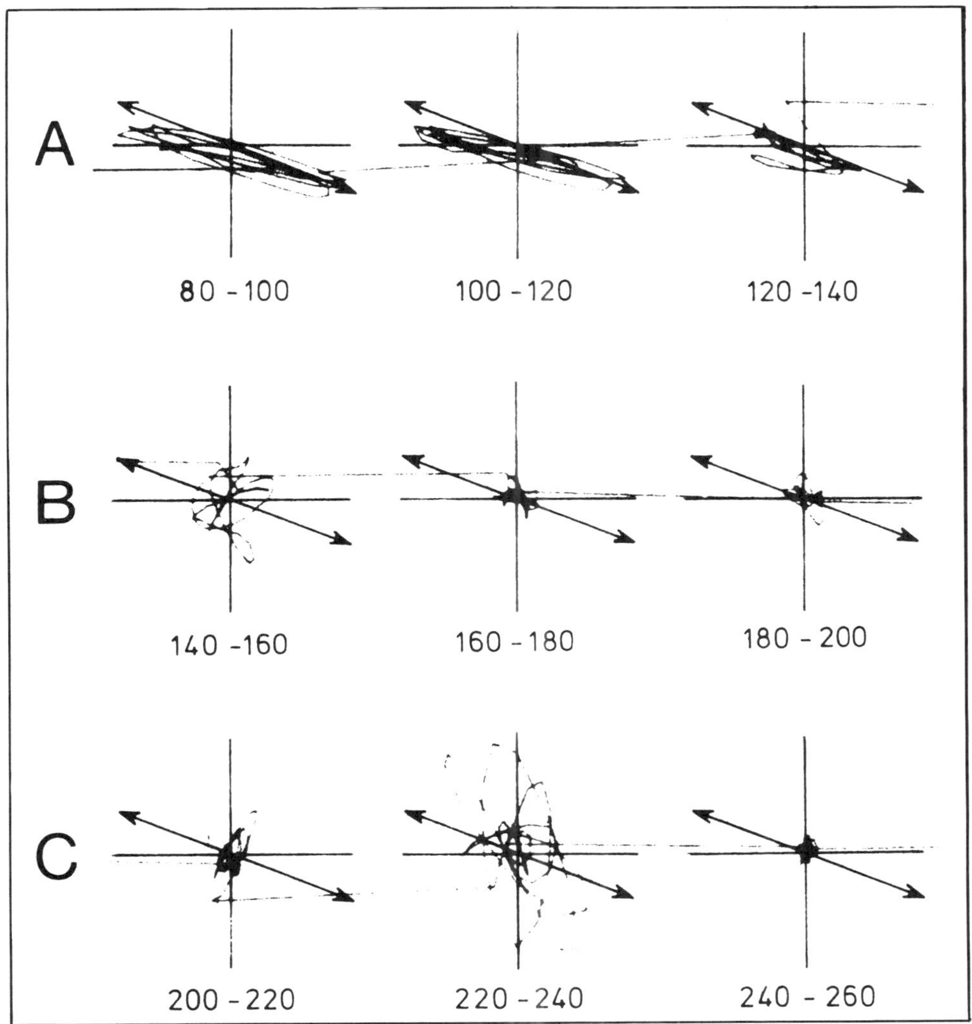

Fig. 9. Hodographs of particle velocity parallel to the seam. A, B, and C refer to Fig. 2

to the seam are measured the hodograph of the particle velocity parallel to the seam can be constructed. The hodograph indicates the direction from which the incident seismic wave arrives at the receiver location. Furthermore the polarisation of the particle motion (linear or elliptical) can help to identify different types of waves.

Examples of hodographs are shown in Fig. 9. The nine hodographs were computed from the transmission seismogram shown in Fig. 2 at a distance of 275 m. The corresponding seismogram of the component vertical to the face is not depicted in this paper. Each hodograph describes the particle motion for a time interval of 20 ms. The arrow in each hodograph indicates the incident angle which is computed from the coordinates of the shot and the receiver. The upper three hodographs belong to the wave group A as marked in Fig. 2. In the whole time intervall the particle motion shows linear polarisation pointing to an arrival of a compressional wave or a Rayleigh wave. The first hodograph for the wave group C reveals a linear polarisation vertical to the direction of incidence. This means, that in this time intervall the Love wave is predominant. Once knowing the incident angle a separation of Rayleigh and Love waves is possible.

After the separation of the different wave types, further analysis programs can be applied to determine the group and phase velocities as function of frequency.

Processing programs

Stacking and migration. As the channel waves are dispersive the stacking velocity is frequency dependent. Therefore it is not surprising that the stacking of broad-banded field data yielded poor results.

It is more promising to filter the data by a set of narrow bandpasses and then to produce a stack for each bandpass. The correct stacking velocity should be delivered by analysis programs applied to transmission seismograms. Otherwise the stacking velocities can be taken from theoretical dispersion curves. An application of such a stacking process to field data is demonstrated in the next chapter.

If the reflected signal is high-frequency, difficulties arise from the fact that high-frequency signals travel with a velocity of about 1000 m/s and even less depending on the seam thickness. Such low velocities can yield move-out corrections of more than 50 ms. Given a signal with a dominant frequency of 500 Hz the move-out has to be accurate to less than 1 ms.

The stacked sections gained in a way as described above are narrow banded and hence the resolution is poor. A stack of all narrow banded sections was tried but no positive result could be achieved. To overcome these difficulties it may be advantageous to leave stacking and instead try migration before stack. As for a given frequency the signal velocity is constant along the travelpath a migration in the frequency-wave number-domain is most promising (9). In the case of great angles between spread line and fault strike dip migration (8) should yield improved results. Especially this is true, if there is some a priori knowledge about the fault strike.

Detection and deconvolution. As in surface reflection work it is the aim to produce an image of the subsurface as close as possible to the real situation, deconvolution processes play an important role. Regarding the in-seam reflection seismic we can assume that the block of coal to be investigated will contain only a small number of discontinuities and it is the distance to these discontinuities we are interested in. From our field experiments we learned that the signal-to-noise ratio of reflected events is poor in most cases. Therefore the nearest problem in in-seam seismic is the detection of reflection events. This detection must be reliable. Reliable means: we have to say the miners that within a well defined range there will be no fault or there will be a fault which will stop the working face.

As we are eager to stretch out the range of detectibility of faults the detection problem will be still more severe in future. A good knowledge about the behaviour of channel waves, that means a better definition of the signal's organisation, and the implementation of array processing methods (10) can help to solve this problem.

The aim of deconvolution filters is to improve the resolution but at the same time deconvolution can make detection easier. One such deconvolution process known as crosscorrelation filter, has been proposed by Krey (11). An application of the crosscorrelation filter is given in the paper which Krey presented at the first Coal Exploration Symposium (11). The crosscorrelation filter assumes that transmission seismograms have been recorded for travelpaths which are nearly equal to the travelpaths of the reflected signals.

Another deconvolution process has been proposed by Buchanan (12). Given the dispersion curve he computed the inverse operator which compresses the dispersive wave train into a spike. Until now, the process has only ben applied to transmission seismograms. Our field surveys proved that high-frequency signals are better reflected at discontinuities than low-frequency signals. Corrections allowing for this effect have to be taken account of in the deconvolution process.

Assuming the propagation of channel waves to be deterministic and the additive noise due to nearby underground activities to be random prediction filters should be suitable for separating signals from noise. There are two different starting points to construct prediction filters, that is Wiener's autocorrelation principle (13) and Burke's maximum entropie principle (14). We propose to use the latter one, because it can be better adapted to a time varying problem as it is the propagation of channel waves.

FIELD CASES

A new research project on in-seam seismic has been started this year (1978) financially supported by the Government of the Federal Republic of Germany. Within the scope of this project many field trials will be done to develop the field technique and to compile field data recorded under different geologic conditions. The field data will be used to test new processing programs. We present first results in this chapter.

Fig. 10 shows the map of a seam area in a West German coal mine. The homogeneous top layer of the seam is about 1.4 m thick. The bottom layer has several dirt bands and it is about 1.2 m thick. The fault which is depicted in Fig. 10 exhibits a throw of 3 - 4 m at those points, where it crosses the roadways.

Three transmission surveys have been carried out (see Fig. 10): from shotpoint 2 to spread B (D1), from shotpoint 1 to

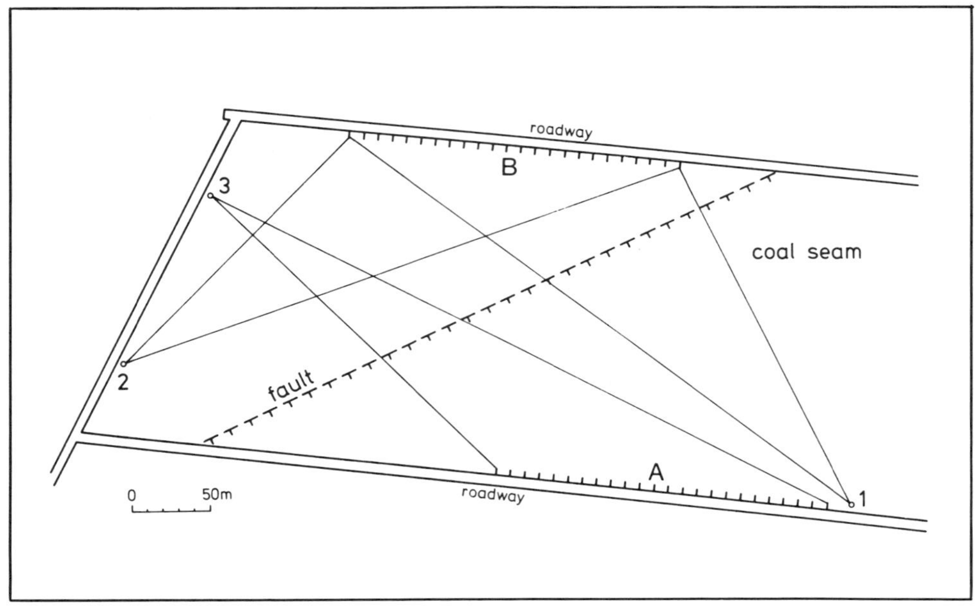

Fig. 10. Location of shots and receiver spreads for three transmission measurements. The spread B was additionally used for a reflection survey

spread B (D2), and from shotpoint 3 to spread A (D3). The results gained by the three transmission surveys D1, D2 and D3 are shown in Fig. 11. In each case both components parallel to the seam are displayed side by side without any data processing applied. As can be derived from Fig. 10 the transmission D1 should be undisturbed whereas the transmissions D2 and D3 should be disturbed by the intersecting fault. Inspection of Fig. 11 tells us that these predictions are true according to the transmission seismograms D1 and D3, but the transmission seismogram shows a distinct arrival of a seismic wave where there should be none. As compared to the seismograms D1 the signals on the seismograms D2 reveal no dispersion and the second event on the seismograms D2 clearly is more low-frequency. We may infer from this, that at the region of the fault swept by the transmission D2 the throw

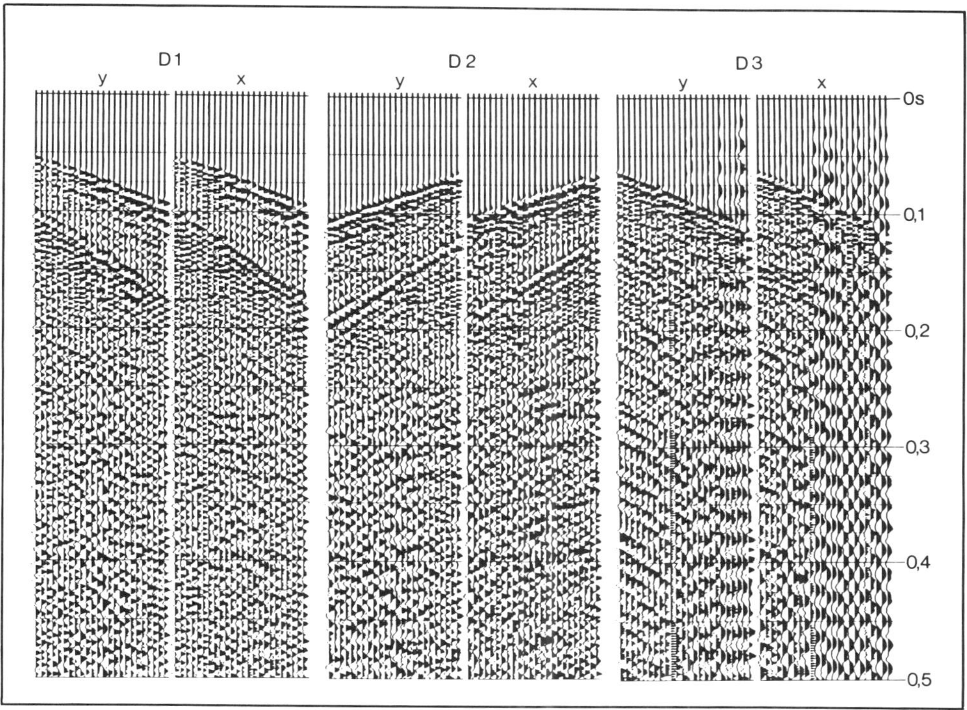

Fig. 11. Results of the three transmission measurements as depicted in Fig. 10 (D1 = 2 to B, D2 = 1 to B, D3 = 3 to A; x = component parallel to the seam and parallel to the face)

is a little more or less than the seam thickness. As for low frequencies a considerable portion of the seam wave travels outside the seam, low-frequency signal can be transmitted across such a fault perhaps with some decrease in amplitude. One should try to display real amplitudes to make visual record interpretation easier. Furthermore, a detailed frequency analysis combined with a dispersion analysis should be done.

The spread B (see Fig. 10) was used to run a reflection survey. The obtained data was filtered by a set of bandpasses.

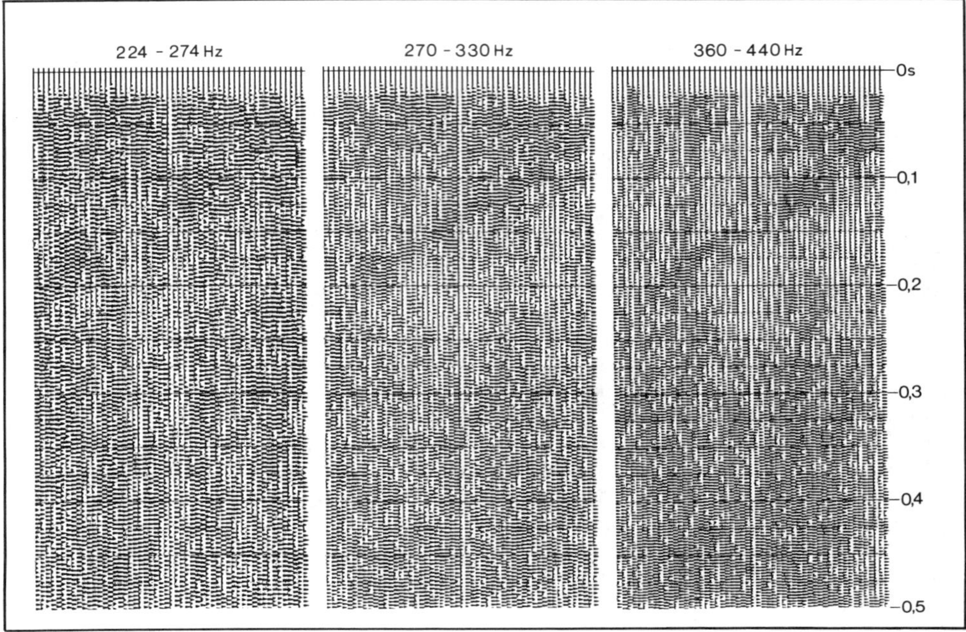

Fig. 12. Stacked sections for three different frequency ranges as result of a reflection survey

Then, move-out corrections were applied to the frequency limited data and stacks were produced. Three stacked sections of different frequency range are shown in Fig. 12. On the two most high-frequency sections a reflection event cearly can be identified at traveltimes decreasing from left to right from 200 ms to 50 ms.

These traveltimes are in good agreement with traveltimes computed from the fault location shown in Fig. 10. The same reflection is not clearly visible in the most left section of Fig. 12. For frequencies below 225 Hz the results became even less meaningful. The high-frequency reflection corresponds to the fact that low frequencies were transmitted past the fault (see Fig. 11, D2).

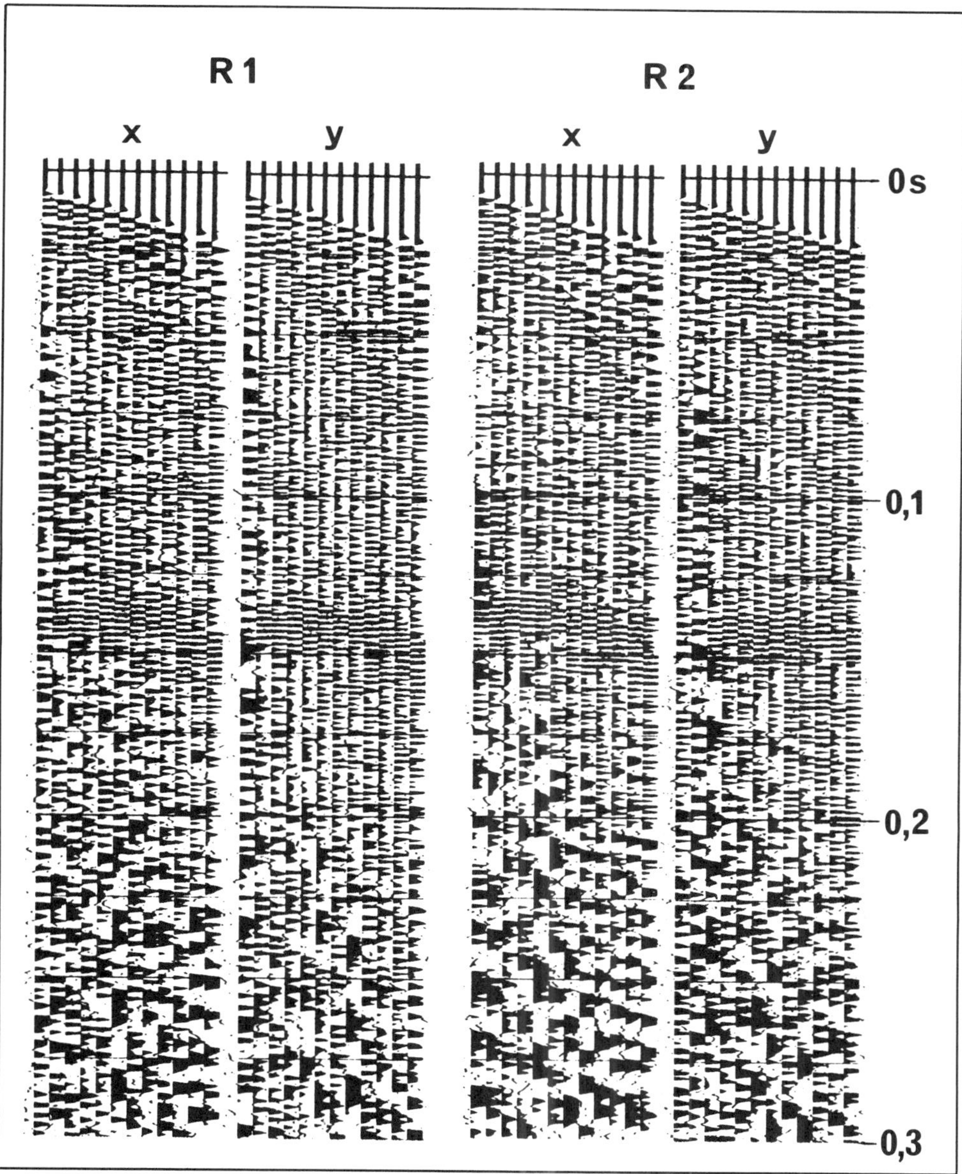

Fig. 13. Unprocessed field records displaying an excellent channel wave reflection. (x = component parallel to the seam and parallel to the face)

The high-frequency reflection event is particularly noteworthy, because recordings with 1 ms sampling intervall, which were usual in the past, would have given no results.

An even more high-frequency reflection from a fault could be recorded during a field trial carried out to survey a seam at a depth of about 1050 m. The seam is 1.4 m to 1.2 m thick. It has only one thin dirt band in the top layer. In Fig. 13 two unprocessed field records are presented enclosing both components parallel to the seam. In the first part of the records, up to about 100 ms, the direct channel wave can be observed increasingly spreading out with low frequencies at its beginning and higher frequencies toward its end. At a traveltime of about 130 ms a highfrequency reflection event is clearly visible. The reflection mainly includes frequencies in the range from 400 Hz to 500 Hz. The distance to the fault is about 100 m.

The excellent example of a reflected channel wave shown in Fig. 13 encourages us to reinforce our efforts to establish the in-seam seismic as a method, that can help the miner to look deep into the seam ahead of the face.

ACKNOWLEDGEMENT

We thank the RUHRKOHLE AG for their permission to present some field data. We are gratefully indebted to Dr. S. Freystätter for stimulating discussions.

REFERENCES

1. Krey, Th., "Channel Waves as a Tool of Applied Geophysics in Coal mining", Geophysics, Vol. 28, 1963, pp. 701-714

2. Arnetzl, H., "Grundsätzliches über seismische Untertagemessungen mit Flözwellen im Steinkohlenbergbau", Bericht der PRAKLA-SEISMOS GMBH, January 1978

3. Freystätter, St., "Modellseismische Untersuchungen zur Anwendung von Flözwellen für die untertägige Vorfelderkundung im Steinkohlenbergbau", Bericht Nr. 3 des Instituts für Geophysik der Ruhr-Universität Bochum, 1974

4. Dresen, L. and Freystätter, S., "The Influence of Obliquely Dipping Discontinuities on the Use of Rayleigh Channel Waves for the In-Seam Seismic Reflection Method, GEOPHYSICAL PROSPECTING, Vol. 26, pp. 1-25, 1978

5. Buchanan, D. J., "The Propagation of Attenuated SH Channel Waves", Geophysical Prospecting, Vol. 26, pp. 16 - 28, 1978

6. Brentrup, F.K., "Entwicklung einer schlagwettergeschützten Digitalapparatur für die Flözwellenseismik", Glückauf, 1978, in press

7. Rüter, H., "Entwicklung und Erprobung eines pneumatisch angekoppelten Mehrkomponentengeophons für die Flözwellenseismik", Bericht der WBK, Bochum, January 1978

8. Robinson, J.C. and Robbins, Th. R., "Dip Domain Migration of Two-dimensional Seismic Profiles", Geophysics, Vol. 43, pp. 77 - 93, 1978

9. Stolt, R.H., "Migration by Fourier Transform", Geophysics, Vol. 43, pp. 23 - 48, 1978

10. Robinson, E.A., "Statistical communication and detection", Griffin & Company LTD, London, 1967

11. Krey, Th., "In-Seam Seismic Exploration Techniques", Proceeding of the first International Coal Exploration Symposion, London, pp. 227 - 254, 1976

12. Buchanan, D.J., "Fault Location Using Seismic Technique", ECSC Progress Report, 1976

13. Robinson, E.A., "Multichannel time series Analysis with digital computer programs, Holden-Day",San Francisco, 1967

14. Claerbout, J.F. "Fundamentals of Geophysical Data Processing", Mc Graw Hill, New York, 1976

DISCUSSION

QUESTION: Can the "in-seam seismic method" be applied to boreholes during a preliminary exploration stage?

ANSWER: The "in-seam seismic method" has yielded satisfactory results, only, if we have had a sufficient redundancy of our data. In the reflection work we normally shoot 6- to 10-fold coverage data, and for a transmission survey at least 24 channels at different geophon positions were recorded. Applying the "in-seam seismic method" to boreholes you can use one source position and one receiver position in each borehole. Therefore, the results of such a survey will not be very reliable.

QUESTION: Have you investigated using beam stearing techniques to improve your data?

ANSWER: As far as the field technique is concerned we have not yet applied any beam stearing technique. But we will soon accomplish beam stearing within the computer, by applying the slant-stack technique, as published by Claerbout, or some similar methods.

QUESTION: Is the VLF a new development which might revolutionize the seismic exploration as we now know it?

ANSWER: If the meaning of VLF is "very low frequency" you should better speak of VHF (very high frequency), because frequencies of about 500 Hz are very high with respect to normal seismic exploration. We have the advantage that we can attach our receivers directly to the rock whereas in normal seismic exploration the receivers are placed into the soil. Another reason for higher frequencies is guidance of the waves within the coal seam.

12

Exploration in the East Midlands, United Kingdom: Some Procedural and Interpretation Principles

By R. E. Elliott
Regional Geologist
National Coal Board
East Midlands Region
Eastwood, Nottingham, United Kingdom

INTRODUCTION:

The East Midlands embraces the southern half of the East Pennine Coalfield, covering an area of some 1,700 square miles, of which a major part is concealed beneath Permian and Mesozoic rocks. The present output of some 28 million tons per annum from the Upper Carboniferous Coal Measures is produced by longwall faces at 39 collieries.

Exploration for economic coal reserves for existing and new collieries at depths of 2,000 to 4,000 feet has been taking place since 1950 and has considerably increased in pace since early 1974. This occurred with the introduction of additional drilling contractors creating competition and significant improvements in technique. Seismic surveying was introduced in 1973 and adapted by degrees to the coal mining situation, bringing significant improvements, particularly in the last three years (1).

Some 1,200 square miles have now been explored to varying degrees of detail east of the active East Pennine Coalfield in the East Midlands and Yorkshire and several large collieries are proposed as a result, with a potential for more. The planned output from new collieries at the time of writing is 17 million tons.

EXPLORATION STAGES:

The style of exploration referred to naturally falls into about five stages, after each of which a 'go' or 'no-go' situation arises. These are a modification and extension of those discussed by A.M. Clarke (2) at the first international coal exploration symposium:-

1) Location of the area to be explored;

2) Determination of the approximate boundaries of adequate potential for at least one colliery;

3) Investigation of detail regarding the structure and limits of the coal bearing formation and also the sedimentary environment and quality of each potentially economic seam; the objective being to establish to what extent and where is continuous economic mining feasible;

4) Acquisition of more precise and geotechnical data where the overall mining plan requires: for example, at the site of proposed shafts and adits, at subcrops and along the line of main development roadways;

5) After a producing colliery is established, underground observations, the addition of boreholes and perhaps seismic traverses to help elucidate further problems as they arise.

At each of these stages the objective is to reduce the risk attached to the mining venture.

Stage One:

The first stage in the East Midlands has usually been based on general geological extrapolations, aided by broad geophysical surveys. Old seismic surveys carried out in the search for oil and, to a lesser degree, country-wide gravity surveys, have been of assistance. However, such evidence has always been regarded as tentative and it has been the policy to drill one or two holes for confirmation of the existence of the coal-bearing formation.

Stage Two:

Following the 'scout' drilling of stage one, a decision to continue necessitates the planning of a few seismic survey lines to span the area of interest, controlled by a few additional boreholes so as to achieve stage two. These seismic surveys must be 'high-resolution' surveys and the boreholes should be cored from near the base of the cover rocks down to the lowest potentially economic seam. A hole density of one per five square miles has been found to be adequate. Surface permitting must be arranged so as to endeavour to site the holes on or very near the seismic lines, but unfortunately this has not always been possible under the conditions appertaining in the United Kingdom. After drilling, 'target seams' requiring detailed attention at the next stage are identified; at least one, and preferably two or more are required if full investigation of the prospect is to be justified. Various factors must be taken into account before a decision to continue to stage three can be made. These include 'target seam' thickness, the degree of concentration of the seams, their depth, their quality in relation to markets, their degree of disturbance however imperfectly known, and the remoteness of the prospect from existing markets and manpower.

Stage Three:

This stage is not necessarily planned as a single operation. The infilling of both boreholes and seismic traverses to achieve detailed plots of various iso-lines, depicting structural and sedimentological features, has an element of trial-and-error and is best kept continually under review, making several tentative geological studies during the course of operations. In several East Midlands prospects, infilling has continued until the holes have reached a density of about one per square mile and the seismic traverses have an average spacing of one mile. These traverses are always run in two directions but unfortunately they tend to form an irregular network due to permitting problems.

The density of data required to achieve stage three varies according to the complexity of both tectonic and sedimentary structures. The pattern of faulting, and particularly the lengths of detectable faults, dictate the spacing of seismic traverses. The lateral variability of the 'target seams' and their associated measures is the principal feature governing the density of boreholes. The knowledgeable geologist should have some notion of the likely complexity of these controlling factors. Often the final mileage of seismic profile and the final number of boreholes to complete stage three is arrived at by a process of 'topping-up' until the three-dimensional 'jigsaw' of data clicks together to give a structural pattern that 'rings true', a stratigraphy correlatable in detail and key sedimentological features that are mappable with conviction. Down-hole logs are run at all holes so that checks on coal seam thicknesses and depths are made and a skeleton log of the uncored parts of holes can be drawn-up complete with any faults that may be evident after correlation with neighbouring holes.

Stage Four:

The mining plan evolved out of stage three will inevitably point to locations at which more precise data is required. These may be on the site of shafts or access drifts, in the vicinity of subcrops to a water-bearing formation, or along main access roadways leading to the various mine districts. In the latter case at least, additional seismic profiles may be required and in all such locations extra boreholes may be involved. Excepting holes drilled purely to control seismic profiles, these stage four boreholes have an additional geotechnical value. On-site core observations, water flow tests, down-hole logs and laboratory tests on samples are designed to answer questions regarding rock strength, porosity, transmissability, rock temperature, etc.

Stage Five:

This last stage is spread over the life of the mine, even over the life of a geologist. This is still exploration in the dictionary sense of 'search, investigate and examine thoroughly'. Previously unknown elements in both the structural and sedimentary fields are discovered as underground workings progress, especially faulting below the limit of resolution of the seismic surveys and small sedimentological features falling between the scatter of boreholes. Both can set problems relating to the maintenance of steady outputs from large collieries with longwall faces.

In the East Pennine coalfield, a mine geologist is appointed to every two or three collieries; it is his duty to assist the mine management to reduce any element of risk built into the colliery plan that has a geological aspect. The mine geologist studies relationships between current mining problems and the geological environment of the workings; he relates the latter to data collated from previous exploration and by considering the current mining proposals produces hazard assessments which forecast where and when problems are likely to arise from variations in the geology encountered by advancing coal faces. He must design his own work programme so as to produce accurate hazard assessments and also reports making recommendations for the acquisition of critical data by means of surface or underground boreholes or by the careful positioning of underground development roadways or even by seismic means. The outcome of his efforts should be the provision of an updated geological picture of the ground ahead of his collieries' current workings so that improved colliery proposals can be formulated, either by avoiding potentially serious geological features or by laying contingency plans to alleviate problems as they arise. This process is repeated periodically, usually annually, as the workings advance and more geological information comes to hand.

STRUCTURAL EXPLORATION:

It has to be acknowledged that present day seismic techniques cannot locate all those faults of interest to the deep-mines engineer. At around 2,500 feet deep the smallest faults detected in the East Midlands have throws of about 23 feet. This low figure has only been achieved where the surface formation and 'shooting' medium is claystone and where there is no disturbance due to previous opencast operations, mining or natural causes such as solution by groundwaters leading to loosening or partial collapse of strata. Thus, the aim must be limited to establishing the pattern of faults that will control the general layout of access roadways and production faces, not including those lesser faults that could stop production. The gap between fault throws that can be carried on a coal face and the minimum detectable during exploration may be narrowed slightly in the near future, but this might incur a very large increase in survey costs and is very unlikely to be achieved by greater flexibility in longwall mining methods. In the author's opinion, only a radical change in technique in either field could completely resolve this problem. However, the cost at the time of writing of one mile of high resolution seismic survey in the United Kingdom is equivalent to the selling price of about 350 tons of power-station coal, worked by a 600 foot long face in a five foot seam in say one shift. By these standards, expenditure on seismic surveying is small in comparison with the value of satisfactory results if at least the structural boundaries of blocks of reserves are located.

In addition to faults, gradients are often crucial, especially where locomotive haulage is envisaged. There is a need for accurate relative times read from seismic profiles. Thus, lateral variations of low velocities in the near-surface or so-called 'weathering zone' are crucial. A low-velocity layer refraction crew operating over the same traverse lines as the main reflection crew is now frequently included in contracts for the National Coal Board. Thus, data is collected that enables more accurate static-corrections to be completed. Such a procedure also helps to optimise the shooting depths which are usually just below the low velocity layer.

As many as possible of the proposed boreholes are sited on or very close to the traverses so that borehole depth and seismic time can be correlated to produce contoured velocity maps for key horizons. In the East Midlands, the most effective correlation is at the Permo-Carboniferous unconformity. Using the unconformity as a base, extrapolations downwards to the highest 'target seam', using approximate Coal Measure velocities, suffice to identify the usually strong reflections at or close to that seam. The occasional down-hole synthetic seismogram has served to check this procedure.

It is not known how reflections in the coal-bearing formations relate to coal seams, however, some relevant statements may be made. Since coal has very different density and velocity characteristics to the rocks normally adjacent to a seam, the two boundaries to each bed of coal must play an important role in the integration of primary reflections and indeed of complex multiples. In spite of the complexity of this situation, it has been found that, in practice, relatively strong reflectors are found on seismic profiles at or close to 'target seams' that are of the order of 5 to 10 feet thick in the East Pennine coalfield. Moreover, character is imparted to the profiles which enables horizons at or near the 'target seams' to be identified, traced around closed loops and time-contoured. Thus, after the corresponding velocity maps are produced, the near-correlated horizons can be depth contoured with values automatically keying-in to the borehole depths. The maps should then be moulded, as it were, to the style of structure met in nearby collieries or in accordance with surface or general knowledge of the structure, so that in their final form they take account of all lines of evidence.

SEDIMENTOLOGICAL EXPLORATION:

Occasionally, a broad sedimentary feature has been detected on the best quality seismic profiles. In one prospect, boreholes show that a ten foot coal seam splits gradually into three component seams and the associated measures double in thickness over a distance of one mile. This change shows up

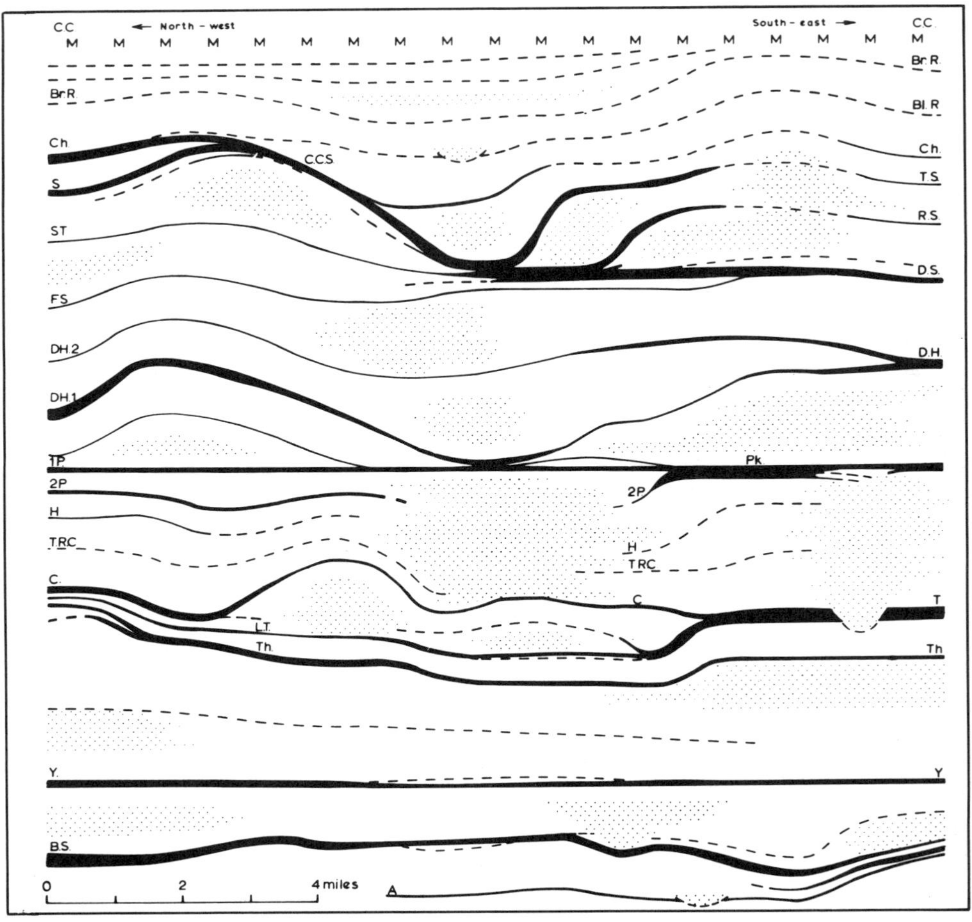

Figure No. 1: A north-west to south-east profile showing variable stratigraphy across 13 miles and five colliery takes, from near Sheffield across Sherwood Forest to near Southwell. The sequence is 600 feet thick and the thickest coal seam shown is six feet thick. Details are shown semi-schematically with the three most consistent horizons horizontal. Coals are shown black; Sandstones are stippled; M. : marine band; C.C. : Clay Cross, D.S. : Deep Soft, D.H. : Deep Hard, Pk. : Parkgate, T. : Tupton, Y. : Yard, B.S. : Blackshale.

well on several seismic profiles as gradually diverging reflectors with the development of two or three extra reflectors. However, such evidence is difficult or impossible to interpret by itself and cored boreholes are essential for sedimentological investigations.

Many coal-bearing formations with economic seams contain from 2 to 12 per cent of coal by thickness and occasionally more. As originally deposited, this coal was represented by peat forming some 20 to 60 per cent of the sequence. The formation of this peat was periodically interrupted by sedimentation in both widespread and local environments; in general, the former occur where marine environments were dominant and the latter where fluviatile environments were dominant. In deltaic areas, situations within this range interplay and sedimentation is further complicated by vagaries in the strengths of distributaries and by delta-switching.

Because peat formation was so variously interrupted, the lateral continuity of coal seams with their roof and floor rocks can be relatively great in the more marine sequences, or especially unpredictable as in largely fluviatile successions. An example of the latter is illustrated by Conolly and Ferm (3). Their Figure 3 illustrates a section through over 800 feet of Upper Permian coal measures across nearly 14 miles of the north-western part of the Sydney Basin in Australia. It shows much sandstone and generally impersistent, sometimes splitting coal seams, especially in the higher part of the section above the Montrose seam. They describe these seams as being "typical of those formed in fluvial backswamps". A relatively marine sequence with productive coals occurs in Scotland in the Limestone Coal group of the Lower Carboniferous; there, seam splits are rare and do not give rise to recurrent mining problems.

The East Pennine Carboniferous Coal Measures, being paralic and largely of deltaic origin, are intermediate between the Australian and Scottish examples; Figure no. 1 illustrates lateral variations in a 600 foot sequence over a distance of 13 miles across the takes of five active collieries. It shows that there are stratigraphical zones deposited down-delta with relatively continuous but thinner seams accompanied by less sandstone and other zones deposited up-delta with thicker seams liable to splitting, accompanied by relatively more sandstone. This kind of variation is to be found over the whole of the East Pennine and nearby coalfields and is attributable to the varying development and character of the deltas responsible for depositing the sediments and thereby interrupting peat formation. Judging from the description by B.A. Collins (4), the Mesaverde Group of the Eastern Piceance basin, Colorado, was deposited on a delta with a particularly strong fluvial influence, since marine bands are rare or absent and sandy beds are commonly associated with coals which are variable in development. Some idea of the lateral variability of coal seams in a prospect may often be obtained from outcrop, nearby worked or previously explored areas, and hence the spacing of boreholes for stage three may be planned with this in mind, making some allowance for 'topping-up' the density where and if necessary. This density must be sufficient to 'net' the key sedimentary features and achieve bed-by-bed correlation. The density and hence expenditure is minimised if cores are taken throughout the sequence likely to contain economic seams; this is because the sooner correlation is achieved the sooner maps can be produced outlining those sedimentary features that delimit the life of longwall coal faces, the working heights for which their equipment is designed and the occurrence of difficult roof and floor conditions that interrupt or slow down production.

Bed-by-bed correlation refers to the identification of the individual components into which an economic coal seam splits; this is essential for the compilation of satisfactory maps on which forecasts can be based. To achieve this, all possible correlation aids have to be invoked but not necessarily including those that merely zone the sequence. In general, the generic field identification of fossils has proved to be adequate for this purpose in the East Midlands. Those aids utilised in the detailed correlation of the East Pennine Coalfield are listed in Appendix I. If, as in most coalfields without a very high fluviatile content, it is possible to achieve a bed-by-bed correlation, the cored sequences will 'click' together like a three-dimensional jig-saw as stage three drilling draws to a close.

The cost of drilling deep holes at a density of say one per square mile is considerable, but exploration at the correct degree of detail to suit the geological circumstances is the keynote of efficiency. At the time of writing, the cost of a borehole in the East Midlands, drilled to 3,000 feet and taking 800 feet of cores, is equivalent to the selling price of about 3,500 tons of power-station coal, worked by a 600 foot long face in a five foot seam in say 10 shifts. A good understanding of the hazards confronting each coal-face can readily save far more capital than the cost of an exploration programme, but to achieve this it is necessary to build on to sound stage three exploration during stage five. The description of two map compilation procedures follow and emphasise the value of sound correlation as a basis for this work.

Coal-seam splits:

Certain seam-split isopachytes are critical and form limits to economic working or boundaries between longwall faces equipped to work different thicknesses. To locate these precisely is impossible using scattered point data. A best-fit position is desirable so that modifications during stage five are minimised and the value of forecasts is maximised. It may be necessary to map such lines, for example, between some 20 boreholes across 10 miles. A plot of thickness against horizontal distance where information is dense produces a hyperbola or similar curve and shows that linear interpolation between more widely spaced data points is not adequate. The gradient of a typical seam-split increases significantly and directly with thickness, especially around one or two feet. The relationship between horizontal distance and thickness, from a few inches towards but not up to the maximum interval, produces a scatter on log-normal graph paper to which a straight line can be fitted (Figure no. 2). A graphical method based on this is recommended, see Appendix II. The true seam-split isopachyte is always more irregular than the forecast line, but the aim is to balance the errors either side.

East Midlands' experience shows that seam-splits vary considerably in character, some are quite regular and readily contoured but a few are particularly irregular. These latter need special study at stage five using data collected on coal faces at a high density, which in one current case is one measurement per 1,600 square feet.

Floor-coal splits in the East Midlands are more gradual than roof-coal splits (5, Figure 6). The sediment forming the latter caused considerable compaction of the thick underlying peats, whereas that forming the floor-coal splits compacted a much smaller thickness of peat. This is the likely reason for them differing in thickness gradients, although many other factors play their part.

Channel-fills within seam-splits:

It is advisable to pay considerable attention to the form and occurrence of sandy rocks between seams. This is

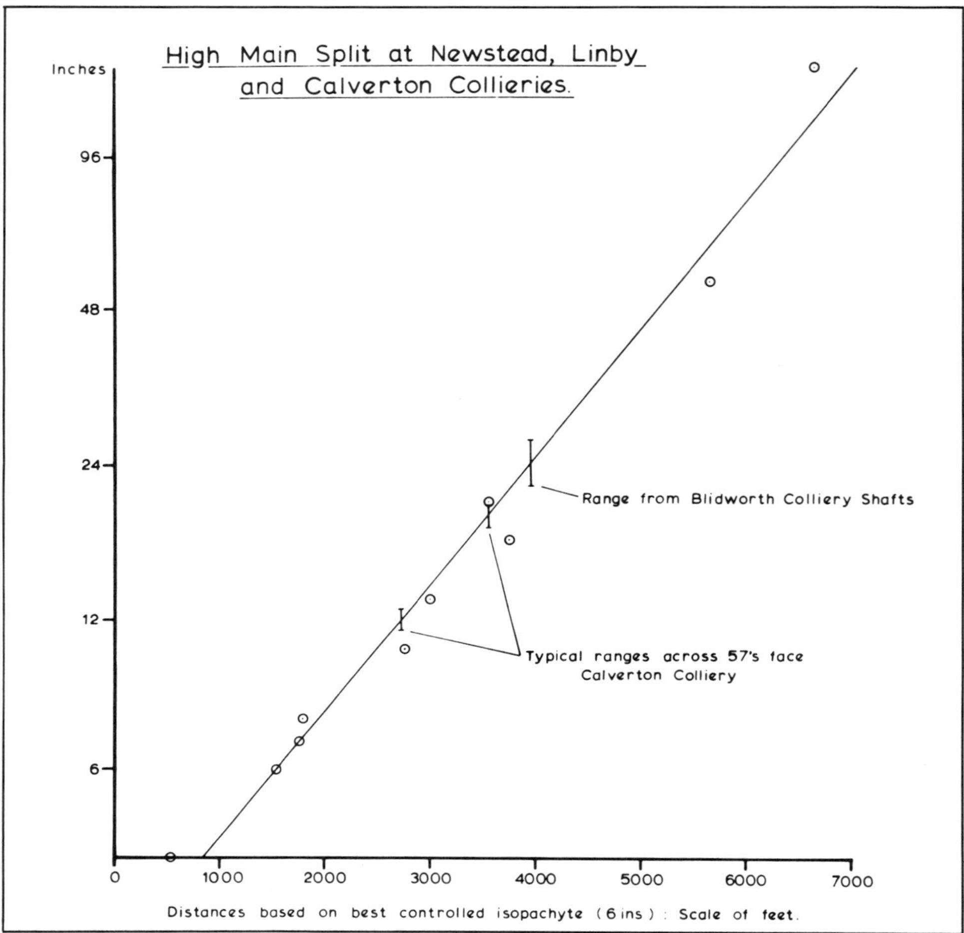

Figure No. 2: A log-normal plot of the thickness of a seam split which limits workings at three collieries. Vertical bars represent local ranges measured underground. Details of construction are given in Appendix II.

necessary because they are associated with at least eight mining problems (Appendix III). A brief statement of their origin will emphasise the complexity of this matter. Deltas of various types and in some cases braided rivers, switched about across the site of coal-formation, a swampy plain, producing a kaleidoscope of geographies through periods of the order of a few million years. During the life time of any one water course, and in particular during flood stages, sediment was carried overbank and deposited as a large lobe or ribbon in an area that might otherwise be the site of peat accumulation. Deposition from all of these water courses was the main cause of the interruption of peat forming conditions. Thus, seam-splits developed lateral to these lobes and ribbons and along their cores were narrower ribbons of the coarser sands and silts or of additional coal (Figure no. 3) filling and finally choking the water courses. It is these channels that form the second-most important sedimentological hazard to coal mining.

At any one time there must have been a full complement of channels. Their location was semi-random and more random down-delta, being tied to relatively fixed parent rivers in the hinterland and yet switching left and right across the plain when flood conditions gave rise to breaches and the establishment of new courses. Thus, some prospects contain more sand-silt ribbons than others and their date and hence horizon with respect to any one target seam is variable. Naturally, the water courses were either primary, secondary, tertiary or even higher order branches and accordingly the dimensions of the lobes and ribbons of sediment for which each is responsible are variable. These generalisations are strongly supported by experience in the active East Pennine coalfield.

Figure no. 3 shows a variety of somewhat idealised examples of channel-fills and their associated sediments. Towards the top of this figure there are sand-filled channels (a, b and c) that occur in large lobes of sediments so that the seam-splits that are lateral to them are well outside, perhaps 5 or even 10 miles outside the figure; towards the bottom of the figure are peat-filled channels (f and h) with more restricted overbank sediments. The 23 foot deep channel (f) is taken from a well-documented example (6) having levees as shown, extending only 2,000 feet away from the channel centre. Each structure shown on Figure no. 3 creates its particular mining problems, ranging from losses of production associated with local steep gradients (f) and roof compaction structures (a, b, c & f) to more serious incendive temperature sparking

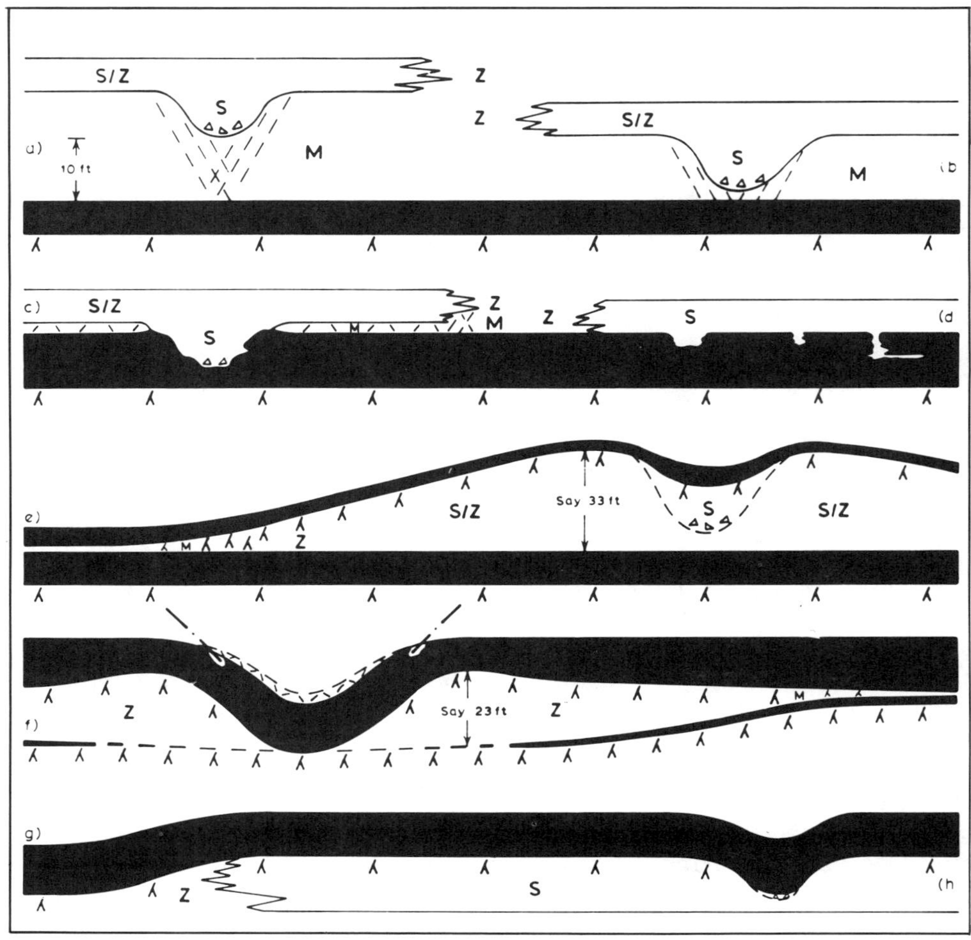

Figure No. 3: Channels at various levels with respect to coal seams: (a) and (b) within the seam roof, (c) cutting into seam, (d) sandstone sheet on seam with sedimentary dykes and sills in seam, (e) roof coal split over levees and channel part-filled with coal, (f) floor coal split under levees and channel completely filled with peat now compressed to coal, (g) compaction monocline over edge of sandstone sheet, (h) channel cut into sandstone sheet and filled with coal. S : Sandstone; Z : siltstone; M : mudstone; ⋌ : seat earth; broken lines represent compaction induced fractures.

problems (eg. c and d) and the unexpected loss of an economic section (e). Clearly, it is necessary to map these features at as early a stage as possible; this is largely within stage five but some general guides to planning, particularly concerning the larger channels, can be obtained at stage three.

The vertical section from a borehole provides a sample of the make-up of the coal measure formation thereabouts; the frequency of the channel-fills in the cores gives some notion of the frequency to be met horizontally in mining operations. Records from a group of holes give improved statistics for the incidence of channels; a large channel may be located by 5 to 10 per cent of all holes in the prospect. The distribution of these records, perhaps coupled with a general study of sediments at the one horizon, often suggests that just one ribbon of sandstone is mappable. With a borehole density of one per square mile, it is usually feasible to determine the likelihood of that ribbon being any one of a range of average widths and then select and plot the most likely width. A simple exercise using overlays placed semi-randomly over the borehole location plot suffices and is detailed in Appendix IV.

CONCLUDING REMARKS:

Throughout the detailed correlation, mapping and interpretation of the measures associated with the target seams, it is essential to inter-relate studies of roof, seam and floor by considering both the lateral and time-equivalence of coal and sediment, and also the sympathetic relationship between the palaeotopography of floors and seam thicknesses. Examples of both of these are to be found on Figure no. 3. Moreover, the varied expression of tectonics in different stratagraphical settings brings together the whole of the exploration field of studies during stage five.

Through experience of both prospects and existing mines, the geologist assembles and periodically updates a series of 'models' that have special value in reducing mining risk in the coalfield concerned; the more important ones are:-

(a) the three-dimensional geometry of facies relationships;
(b) compaction stages; distortion and fracture at critical situations in (a);
(c) the tectonic fracture and fault lattice;
(d) mining induced rock-behaviour around a coal-face;
(e) the source and movement of water around a coal-face.

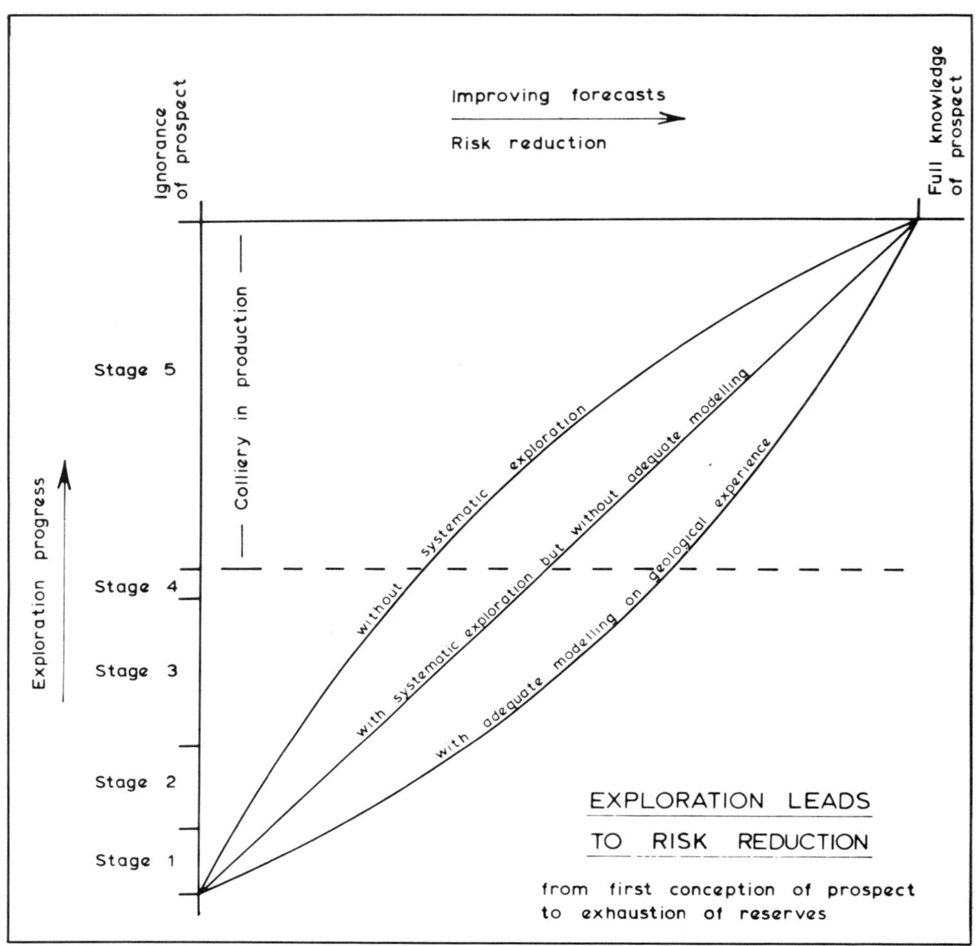

Figure No. 4: Exploration leads to risk reduction. Illustrating the effect of exploration planning and the application of continuously updated geological models on the progress of risk reduction.

A gradual approach to the truth regarding the geology of the seams and their environment is achieved as progress is made through exploration stages one to five. The full truth is not on paper until the coal is worked out, but it is the geologist's job to input experience and know-how into mine planning procedures by successive applications of his 'models'. It is by this procedure that the risk attached to a mining venture is progressively reduced and the truth is approached and approximated relatively early as represented on Figure no. 4.

ACKNOWLEDGEMENTS:

The author wishes to express his thanks to his superiors and colleagues within the National Coal Board who made it possible for this paper to be written and presented in person. The opinions expressed are those of the author and not necessarily those of the National Coal Board.

REFERENCES:

(1) Ziolkowski, A., in the press, "Seismic profiling for coal on land", in "Developments in geophysical exploration methods", Ed. A.A. Fitch., Applied science publishers limited, London.

(2) Clarke, A.M., 1976, "Seismic surveying and mine planning : their relationships and application", Proc. 1st International Coal Exploration Symposium, London, May 1976, p.158-191.

(3) Conolly, J.R. and Ferm J.C., 1971, "Permo-Triassic Sedimentation Patterns, Sydney Basin, Australia", Amer. Assoc. Petroleum Geologists Bulletin, vol. 55, p.2018-2032.

(4) Collins, B.A., 1976, "Coal deposits of the Carbondale, Grand Hogback and Southern Danforth Hills coalfields, Eastern Piceance basin, Colorado". Quarterly of the Colorado school of mines, vol. 71, 138 and xiii p.

(5) Elliott, R.E., 1969, "Deltaic Processes and Episodes : The interpretation of productive coal measures occurring in the East Midlands, Great Britain". Mercian Geologist, vol. 3, p.111-135.

(6) Elliott, R.E., 1965, "Swilleys in the Coal Measures of Nottinghamshire interpreted as palaeo-river courses". Mercian Geologist, vol. 1, p.133-142.

APPENDIX I

Aids to correlation utilised in the East Pennine Coalfield.

(a) Sequential patterns of lithofacies.
(b) Marine and Euestheria bands.
(c) Non-marine lamellibranch bands.
(d) Trace fossil mudstone.
(e) Brown seat-earths with sphaerosiderite.
(f) Grey seat-earths with nodular siderite.
(g) Dull-coal layers.
(h) Sulphur rich coal layers.
(i) Tonsteins, identified by their kaolin or high titanium oxide content.
(j) Carbonate rich bands.
(k) Very carbonaceous mudstones, canneloid mudstone or cannel coal between coal beds.
(l) Kaolin oolith bands.
(m) Quartz rich coal.
(n) Very fusainous coal layers.

The longer range or more useful features occur towards the top of the list. 'Layer' is here equivalent to 'ply' as used in Australia.

APPENDIX II

A graphical trial-and-error method of constructing seam-split isopachytes.

(1) Draw a series of isopachytes by the 'eye-ball' method choosing an interval that is closer than is normally required for finished plans.
(2) Select the best controlled isopachyte, or two en-echelon covering different parts of the map if the control is such that no one line is well established.
(3) Measure distances from each data point to the control isopachyte(s).
(4) Plot a graph of the thickness of the split against distance, the thickness being plotted on a log-scale and the distances being either side of the control isopachyte(s) (Figure no. 2).

(5) Draw, by eye, a best-fit straight line.
(6) Scale distances from the plot of each data point to the best-fit line and use these to reposition the control isopachyte(s) on the map. This process will usually render the isopachyte more sinuous and will be such that a revised plot of the graph would place all points on the straight line.
(7) Reposition other isopachytes to conform with the control isopachyte(s) using distances read off the graph.

If points are plotted near the thin-end of the split or near the maximum thickness a wide scatter will result. It is probable that the log-normal relationship is tenable only where over-bank conditions of sediment deposition were dominant. Where the split is known in workings adjacent to the exploration area it is convenient to add vertical bars to the graph representing local variation about the best-fit line (Figure no. 2). This gives some idea of the extent to which any one borehole might vary from the local mean.

APPENDIX III

The mining problems associated with sand-siltstone ribbons in the East Pennine coalfield.

(a) Excessive weighting due to sandstone with widely spaced joints or joints at obtuse angles to the face.
(b) Coal getting machine impeded by sandstone or siltstone-filled washout or sedimentary dyke or sill within the seam (Figure nos. 3c and 3d).
(c) Machine cutting into excessive floor thickness due to thinning of seam over a linear palaeotopographic 'high' such as a levee (Figure no. 3f).
(d) As (c) but 'high' due to differential compaction over edge of underlying sandstone ribbon, often complicated by small penecontemporaneous faults and folding (Figure no. 3g).
(e) Weak mudstone roof created by differential compaction fractures alongside or under sandstone or siltstone-filled channel (Figure nos. 3a, 3b and 3c).
(f) Efficiency of roadway drivage machinery impaired by sandstone-filled channel cutting down close above or into seam.
(g) High temperature sparking caused by machinery impacting on sandstones or siltstones.
(h) Corrosion of machinery, 'gumming-up' by softened clay-rock or flooding due to tapping water in ribbon sandstone.

(i) Increased machinery maintenance created by many of the above circumstances.

APPENDIX IV

Method for determining ribbon width from scattered borehole data.

(1) Plot, as a transparent overlay and 'by eye', a reasonable course for the ribbon, giving it a width based on experience and a length across the full prospect unless otherwise known to terminate.
(2) Plot two other ribbons with similar sinuosity and length but with widths say 33% greater and 33% smaller.
(3) Overlay one ribbon on to a base map of the same scale showing the borehole locations. Count the number of boreholes intersected by the ribbon. Repeat this in random positions say 40 times. Control the direction of the channel within reasonable limits of the direction indicated by the known borehole data but vary the position laterally. Turn the overlay over or end-for-end after each 10 trials so as to give four ribbon shapes with respect to the boreholes.
(4) Repeat this procedure for all three ribbon widths.
(5) Plot histograms of the frequency of borehole intersections for each of the three widths. The most likely true width is that which peaks at the number of borehole intersections tallying with the number of actual ribbon identifications at the one horizon in the borehole cores

For greater or lesser sinuosities, the number of 'provings' varies in direct proportion to the ribbon length, but within reasonable limits this will not significantly affect the most likely ribbon width. The aim is for the final interpretation to show a width consistent with previous experience of sinuosity and with the number of localities at which it has been located in borehole cores by chance. Modifications of this procedure suited to various circumstances are readily devised.

13

Seismic and Electromagnetic Techniques: Techniques for Premine Planning

By Joseph L. Condon
Research Supervisor
Pre-Mining Conditions Group
Denver Research Center
Bureau of Mines
U.S. Department of the Interior
Denver, Colorado, United States

ABSTRACT

PreMining Investigations are the activities necessary to acquire information for economic evaluation and engineering design of a mine, subsequent to reconnaissance exploration and prior to mining. The Bureau of Mines is doing research on methods to obtain premining information on problems such as detection of old mine workings, abandoned oil wells, channel sands, faults, and other geologic features that are undesirable from a production or safety aspect. Techniques being developed to acquire the necessary information fall into two primary categories: seismic and electromagnetic. The seismic methods include, high-resolution seismic and acoustic methods applied from the surface and inseam channel waves applied underground. The electromagnetic, or radar, methods are applied from the surface, from boreholes, and from the working face underground. All of these methods are useful under certain site-specific conditions, but none of these methods are universally applicable for all coal mining conditions.

INTRODUCTION

Few if any U.S. coal regions are uniform in geologic structure or devoid of irregularities in the coalbed or roof rock. Also, some regions have suffered more severe deformation of the original strata than other regions. The coalfields of the eastern one-third of the United States frequently have complex geologic structures. There are also some large geologic structure disruptions in the mountainous regions of the West. When mining takes place in the subsequent zones of deformation and faulting, there are usually complex problems to overcome.

One problem that plagues underground coal mining is roof stability. Many roof falls have been caused by channel sands above or in the coal bed. The hypothetical cross section in Figure 1 shows a sand lens that was deposited after the coal was formed and that may be a source of roof fall conditions. Sometimes these channel sands cut into the coal seam and completely replace the coal, resulting not only in possible roof falls but also in loss of coal. Many sand deposits are old stream channels that meandered through the coalbeds, cutting out the coal and replacing it with sand. These meandering channel sands, and to a lesser degree deltaic sands, occur quite frequently throughout coal regions and exhibit few definable patterns. Studies of known channel sands show that they range from 50 feet to several hundred feet wide and from 5 to 50 feet thick and that their location is almost unpredicable.

Another structure that is commonly found in some mining districts is the geologic fault. Faults occur in many different forms, and if they exceed about 1 meter (3 feet) in displacement, production problems will occur and the mine plan must be changed, particularly if the presence of the fault was previously unknown.

Another problem that is of interest primarily from the safety aspect is the presence of voids, or old mine workings. Many of these abandoned workings are not adequately mapped and have been filled with water or methane. Accidently minin into these old workings presents a considerable hazard and has caused numerous fatalities from inundations.

The presence of unknown oil or gas wells also presents a hazard, when continuous miners or longwall operations accidently strike the casings.

Two other types of geologic structures that occur unpredictably are various sized and shaped rolls and folds. For standard mining techniques and particularly for increasing longwall mining use, the presence of rolls and folds, like that of unknown faults and channel sands, will cause safety and economic difficulties for the industry.

Figure No. 1 is potential mining hazards.

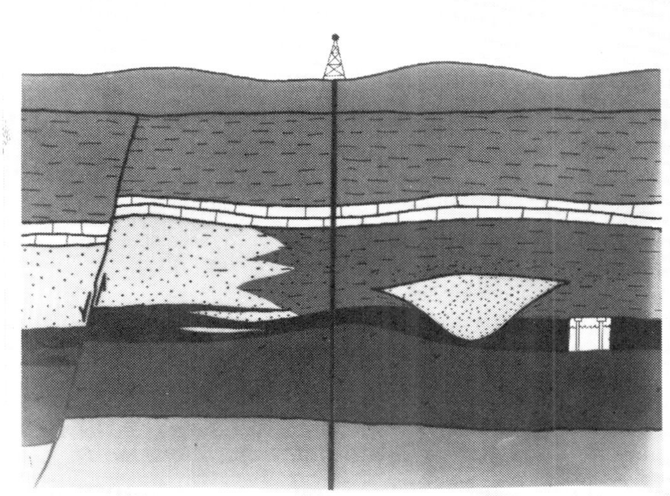

In the past, drilling has been the primary tool for pre-mining investigations. Drill core or cuttings were used by mine geologists to verify coal seam quality and extent. Drilling has typically been done on a preselected grid pattern. This method of investigation is inconclusive for detection of channel sands and voids, and sometimes for faults, because it often leaves large blocks of a deposit without definition.

It would take a drill pattern spaced possibly as close as 5-foot centers to adequately map old workings or channel sands, and such a pattern would be prohibitively expensive.

SEISMIC METHODS

The petroleum industry for years has been using reflection seismic methods for investigation of deep geologic structures to depths of 2,000 to 33,000 feet. Only recently have the U.S. Bureau of Mines and several other organizations applied this technique to shallow investigations. Although the two applications are similar in effect and in the "picture" they produce, shallow reflection methods require modified procedures. Therefore, the Bureau of Mines has initiated several research projects with objectives to minimize surface noise generated by the signal source, select optimum detector spacing, generate high-frequency source signals, and use improved data collection and data processing technology.

Conventional High-Resolution Seismic Methods

The basic technique of the shallow seismic method is shown in the schematic, Figure 2. An energy source is activated to impart a seismic signal into the earth. Seismic waves travel from the input point to the coal surface, which acts as a mirror, reflecting part of the waves back toward the surface. The geophones detect these reflected seismic waves and produce a small-voltage signal representative of the received waves. These voltage pulses are digitally processed and recorded on magnetic tape. A "quick look" analog record is made to verify that good data are being recorded and to aid in optimizing the field procedures. The analog record alone does not have the resolution required for detailed subsurface mapping, so the digital field tape is processed on a computer and a "picture" of the subsurface is reconstructed to scale. In this idealized example, the seismic signals passing through the channel sand have their waveform altered, thus revealing the presence of the sand. In other instances reflections from the sand itself may be used to determine its occurrence. In either case, knowledgeable interpretation of the data is necessary for an accurate analysis.

Many seismic energy sources have been tested. Small explosive charges are probably the most common means of creating seismic signals in excess of 50Hz at present. Weight drops, hammer blows, air guns, gas exploders, and piezoelectric stacks have been tested. Each has advantages and disadvantages. Dynoseis, originally developed by Atlantic Richfield Oil Company, is one of the seismic signal sources that can be used with this method. This system uses an exploding gas mixture inside a piston to send an impulse into the ground. It is very fast and quiet and has few adverse environmental effects so that it is very desirable for use in urban or other sensitive areas.

The reflected seismic signals are detected by electromechanical transducers called geophones that are carefully selected so that their frequency response matches the frequencies transmitted through the rock by the seismic source. Normally there are several of these geophones laid out in array and electrically connected and recorded on one channel so that noise effects are cancelled and valid signals are additive. Usually 12 to 28 channels of information are recorded in digital format on recorders that are mounted in a field van, which is used to traverse the area of interest.

Figure No. 2 is schematic of seismic techniques.

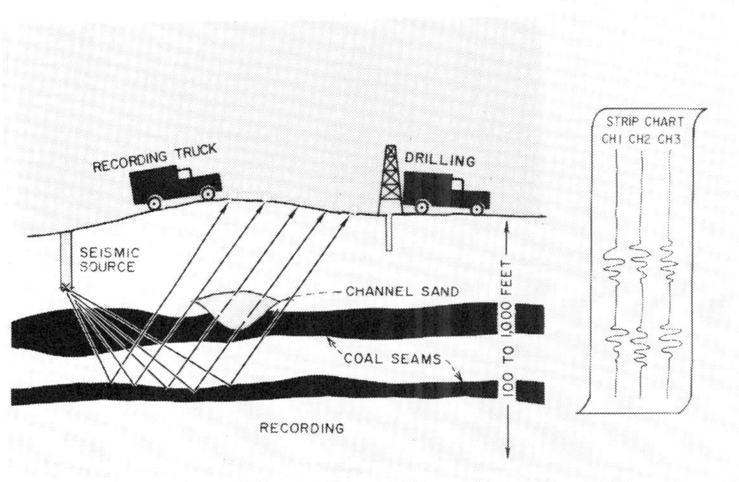

These techniques result in frequencies being recorded in the 50 to 250-Hertz range as compared to the frequencies in 5 to 50-Hertz range commonly used in the petroleum exploration business. The advantage of higher frequency sources is that you increase resolution, which means that you can detect smaller geologic features and locate them with greater accuracy. For example, at 400Hz a coalbed of about 2-foot thickness could be detected, under ideal conditions for a rock media with a 9,600-ft/s velocity. The disadvantage of a higher frequency source is that more of the energy is dissipated in transmission, thereby limiting the depth of penetration. Therefore, the conventional high-resolution seismic methods normally are useful at depths from about 100 feet down to several thousand feet.

Acoustic Methods

The Bureau of Mines has developed a very high frequency seismic signal source that has an operating range from 100 to 10,000Hz, which is in the acoustic range. This device is an electromechanical vibrator in contact with the ground that is computer controlled to vibrate a continous wave (CW) or any desired frequency, to sweep through a range of frequencies, or to generate a pulse.

The high frequencies generated result in a high rate of absorption in the ground. This is partly compensated for by repeating the signal many times at the same location and summing up the reflected signals, a process called stacking. This system is capable of stacking several hundred impulses in only a few minutes. In general, this system is useful to depths of about 1,000 feet.

Guided Seismic Waves in Coal Seams

Elastic waves launched laterally into a layered medium, as typified by stratified sedimentary geologic bedding, can generally experience guided wave propagation in those layers that have lower elastic wave velocities than their adjacent boundary layers. Guided seismic energy confinement within such a low-velocity layer is the result of total internal reflection in the layer of those seismic wavefronts that are incident upon the boundary interfaces at angles beyond the critical angle of reflection. In the case of coal layers, the elastic properties of coal are such that, in general, seismic velocities in coal seams are usually less than the velocities in typical boundary shale and sandstone rock materials. Consequently, coal seams may potentially be used advantageously as seismic waveguides for purposes of inundation hazard detection.

Guided waves in coal seams can potentially provide reflections from anomalous conditions in a seam such as vertical displacements due to faultings, seam pinch outs, channel sand cut outs, and abandoned mine openings which may represent possible water inundation hazards. Guided waves in coal seams can also be used to test the continuity of coal seams using through-transmission techniques to determine the presence of vertical displacements or other anomalies that can inhibit wave guide propagation.

Previous investigations of guided waves in coal seams have discussed and emphasized the Rayleigh-channel wave as the primary wave type of concern. Moreover, the various experimental studies and evaluations of coal seam waveguides have utilized seismic excitation sources such as explosive shots and hammer blow impulses which, because of their pressure pulse nature, tend to generate the P waves necessary to excite Rayleigh-channel waves. In contrast, while Love waves have been occasionally observed experimentally in coal seams, their merit has not been fully recognized because the source excitation methods have not optimized the generation of SH waves. To a large extent this technical focus on Rayleigh-channel waves in coal seams has been a natural result of adapting and applying conventional seismic methods and equipment to geophysical measurements related to longwall coal production in Europe. As such, the use of Rayleigh-channel waves is now emerging as a practical geophysical technique for the assessment of European longwall coal seam conditions.

Several investigators have pointed out that, on the basis of theoretical considerations, SH waves are likely to be the most efficient wave type confinable within a coal seam. Further, the present study has pointed out that shear waves are also the most reliable wave type for detecting cavities in coal seams because of the near-perfect reflections from air or water interfaces with the coal. Therefore, in the application requiring search and detection of inundation hazards from the working face in U.S. coal mines, the optimization of methods based upon guided SH waves seems justified. The Bureau of Mines has just begun an optimization effort to simplify the underground seismic equipment, speed up the underground measurement procedures, and increase the reliability of detecting the anomalous coal seam target.

ELECTROMAGNETIC TECHNIQUES

Electromagnetic methods of investigating the earth for geologic anomalies are potentially practical, provided that the probing technique and the equipment system can be optimally adapted to the earth materials and field conditions to be investigated.

The Bureau of Mines investigations utilize frequencies of about 20 megahertz to 1 gigahertz which fall into the microwave range, and these techniques are often referred to as microwave methods. These systems are referred to as radar when the electromagnetic wave is radiated by a transmitter and reflections produced by discontinuities in the medium are observed by a receiver system in the same location as the transmitter.

Penetration of electromagnetic energy into earth materials is restricted by the dissipative characteristics of the materials and by the frequency dependence of RF attenuation characteristics. The primary result of this frequency-dependent loss is the fact that the transmitted pulse waveform undergoes a progressive change with distance propagated, the effect being more pronounced for the higher frequency spectral components.

The Bureau of Mines is pursuing research in applying electromagnetic methods from three basic geometries: from the surface directed downward, from boreholes directed azimuthly, and from underground working faces in the direction of face advance.

Radar Applied from the Surface

The U.S. Bureau of Mines effort to date on radar applied from the surface has resulted in a prototype, portable field recording system. The system transmits a radio wave pulse about 25 billionths of a second in duration through an antenna laid on the surface. The frequencies in the pulse are about the same as those used in television. A second nearby antenna receives pulses reflected from mines and other geological contrasts such as faults. This received signal is recorded on digital tape through the guidance of a small programmable field computer terminal. The entire system is two-man portable and operates from batteries. Access to phone lines permits on site data enhancement with Bureau of Mines computer programs on a multitude of available large-scale computers. The data are gathered over a traverse every half-meter or closer. They are plotted, and the plot looks quite similar to a seismogram for seismic data presentation. Like seismograms, the data are often difficult to visually assess in this form and must be enhanced. The data are treated with correlation, filtering, stacking, and other approaches very similar to those evolved so successfully in the seismic industry. A schematic of the antenna layout and typical received signal is shown in Figure 3.

Although the emphasis to date has been on field instrumentation development and data enhancement techniques, some field measurements have been conducted. Well-defined reflections from a 5-foot-wide tunnel 30 feet deep in hard rock and from fault planes in the tunnel vicinity are observed in the enhanced data shown in Figure 4. These records have been verified through agreement with expected returns from the adit and with the faults detailed by three-dimensional plots of resistivity measurements. We are now in a position to apply the measurement system and the Bureau of Mines-developed computer data enhancement techniques to other radar sounding sites. In the near future work will be carried out to detect abandoned mines beneath open pit bottoms

Figure No. 3 is schematic of void detection using radar.

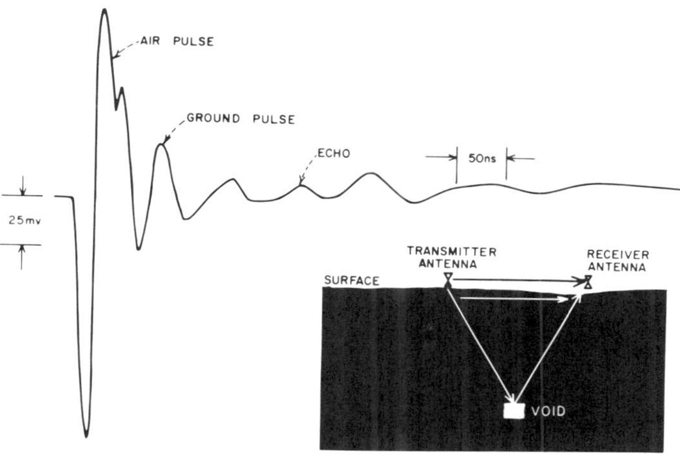

Figure No. 4 is radar reflections from an adit 30 feet deep in hard rock.

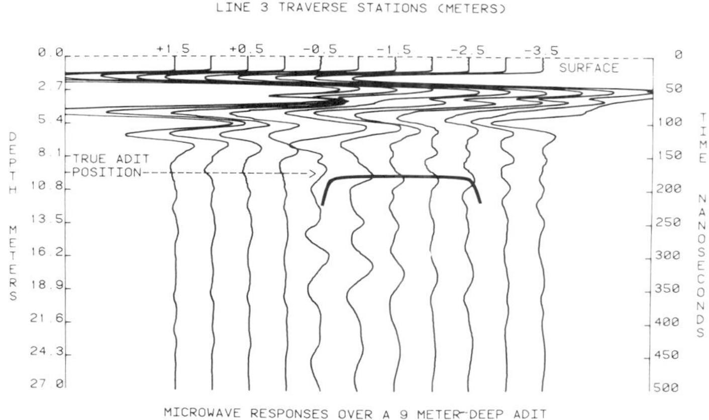

This technique will probably not be effective in areas of layered sedimentary rocks or with significant surface till, since these materials usually have dielectric properties that prohibit sufficient penetration of the electromagnetic waves.

Electromagnetic Techniques Applied from Boreholes

Since the depth of penetration of surface radar is quite limited in coal measures, a research project was undertaken to develop a radar unit that could be lowered down a borehole into the coal seam to scan the area for potential hazards in advance of mining.

An experimental pulsed electromagnetic probe for use in 6-inch vertical boreholes was designed, constructed, and tested for the Bureau of Mines by Southwest Research Institute.

The basic system was taken as a point of departure for the detailed design tasks. The three major units are: the Surface Control Unit, the Transmitter/Receiver Unit, and the Antenna Unit.

The Surface Control Unit contains the probe control circuits and also provides a termination and output connection for the radar signals and the external equipment synchronizing pulses which come up hole from the Transmitter/Receiver probe.

The Transmitter/Receiver Unit houses the bulk of the system electronics.

The Antenna Unit is attached directly to the Transmitter/Receiver module and is the radiating and receiving element of the radar.

The design of a suitable antenna was, perhaps, the most demanding task of the radar develpment. A number of configurations were considered and tested. The tests and analyses indicated that a coaxial traveling wave borehole antenna concept exhibited the necessary broad frequency response in the VHF range.

Depending upon the geologic materials involved, penetration distances up to about 100 meters can be achieved with this technique. The source waveform found to give the best overall geologic detection results, taking into account propagation losses in coal and sedimentary rock materials and reflection losses between coal and other geologic materials, was a single-cycle sinusoidal oscillation having a time duration of approximately 10 nanoseconds. In practice this pulse waveform, essentially one cycle of a 100-MHz sinewave, would be subject to bandwidth limitations imposed by the transmitter pulse-generating circuits, the transmitting and receiving antenna(s), and the receiver circuits comprising the electromagnetic system. As a realistic limit, the useful system bandwidth was not anticipated to exceed about one decade of frequency range from its low limit to its high limit. Therefore, for a frequency band having a geometric mean frequency of 100 MHz, the required antenna response would cover the range 31.6 to 316 MHz. The resulting electromagnetic pulse-echo system could be characterized as a VHF ground-penetrating radar since its operating frequency spectrum essentially covers the complete 30 to 300-MHz VHF band.

The first major set of field tests was performed at the Kemmerer Coal Co. pit mine at Kemmerer, Wyo. This is not an underground mine, and it was not anticipated that there would be the geological anomalies available for use as targets that contribute to bad roof conditions. Rather this first set of tests was specifically designed to measure the basic radar performance parameters with the downhole probe completely immersed in a homogeneous volume of coal. The 100-foot-thick No. 1 seam at Kemmerer offered this opportunity. A series of dry holes were drilled down into the coal from the exposed top of the seam. The operating radar probe was lowered into one of the holes, and a separate receiver was then placed in each of the other holes. Measurements of signal amplitude and travel time were made to characterize the electromagnetic properties of the coal. Also by making a series of measurements in a close hole with the receiver at various positions above and below the radar probe, it was possible to determine experimentally the radar antenna radiation pattern. As anticipated, the beam pattern was basically horizontal, omnidirectional, and slightly tilted downward.

One additional hole was drilled into the coal approximately 50 feet from the suspected location of an old abandoned underground mine. The location of the old mine is reasonably well indicated by at least one large collapsed area on the surface. Unfortunately, an electrosensitive strip chart recording of the radar output as a function of depth along the borehole indicates some instrumental problems with the equipment. In spite of these instrumental signals there are features in the record that can be interpreted to be geologically induced. The most notable is the return indicated on the record at approximately 35 feet below the surface and at a 50-foot range from the borehole. This is the predicted position of the abandoned underground workings, and it does provide some assurance that the radar is at least sensitive to anomalies in uniform by thick coal.

A second set of experiments were directed more specifically at underground mine situations. The York Canyon coal mine at Raton, N.M. was selected for these tests. There are both underground mining and strip mining operations at this location.

In the first experiments at this mine, vertical holes were drilled through the overburden and through coal pillars in the underground mine. The first hole was intended to pass through the 200-foot-thick pillar at approximately 50 feet from the rib of the mine entry. The second hole was intended to pass through the pillar near the center, 100 feet from either entry and 50 feet from a fault, as indicated in the mine layout.

The coal at the test location was approximately 8 feet thick. This thickness allowed the full antenna plus part of the electronics module to be completely in the coal.

The holes were probed with the radar. No clear reflections were obtained from the rib of the entry or from the fault.

The results of the experiments at the tow coal mine sites were far from conclusive and did not reveal the full potential of the concept. Uncertainties with regard to the precise location of reflecting objects and instrumental signals masking echo signals made it difficult to fully evaluate the detection capability of the radar. Hole-to-hole measurements clearly show that the transmitted energy is propagating in the coal very much as anticipated. The energy returned to the receiver is a function of the reflectivity of the illuminated objects. Additional coal mine experiments with improved elements in the radar will be required to gain a better understanding of the system response to geologic anomalies.

Future research will also concentrate on developing a focused borehole antenna to increase the range of the technique and to provide a degree of directionality for the reflected signals.

Electromagnetic Measurements from the Working Face

Hazards such as old mine workings, cased oil wells, and other undesirable geologic features may best be detected using radar measurements from the working face. The Bureau of Mines initiated a research project to develop a practical radar to penetrate at least 50 feet in advance of the working face for detection of these hazards.

Initial work was concentrated on developing focused antennas for pulsed and FM-CW radar systems. Field experiments for testing a prototype system are now underway.

CONCLUSION

In conclusion it must be stated that none of the methods discussed will be useful for all geologic and mining conditions. However, if each of these techniques is used as part of a suite of geophysical tools, a synergistic effect is observed where nore is learned about the geologic environment in advance of mining than would be expected from a simple compilation of data from the separate sources.

DISCUSSION

QUESTION: What is your current limit of detection for the thickness of the channel sands?

ANSWER: The conventional high-resolution seismic reflection technique is capable of detecting a 10 ft thick channel sand at a depth of 500 ft. At lesser depths thinner sands may be detected and conversely at greater depths only thicker sands would be detected.

QUESTION: How do you use explosives in coal holes? Isn't this against U. S. G. S. regulations?

ANSWER: The explosive shots depicted in the slide for the purpose of generating seismic energy were not fired in the coal itself, but were fired in the overburden, near the surface, a considerable distance above the coal. This is a common practice in seismic exploration and is not expressly prohibited by any government agency as long as there is compliance with other general rules for use of explosives.

QUESTION: What minimum channel-sand thicknesses are detectable by your acoustic technique?

ANSWER: The acoustic reflection technique may enable detection of channel sands as thin as 1 ft at a depth of 500 ft, but further research is needed to determine the precise capabilities of the system, under a variety of conditions.

QUESTION: What is the power required, frequency range, and driving mechanism of the variable frequency source you used?

ANSWER: The variable frequency source utilizes a 5KVA motor generator driving an electrodynamic shaker with a driving frequency synthesizer capable of a 20 Hz to 8,000 Hz range. However, the system seems to be most effective in the 1,000 Hz to 3,000 Hz range.

QUESTION: Re: Rock Springs experiment: What was depth of coal seam? geophone array? sampling method? processing techniques?

ANSWER: The Rock Springs data was from a coal seam 170 feet deep, using a Tektronix digital processing oscilloscope for sampling. Data processing was done using standard statics methods and cross-correlation with the pilot signal.

QUESTION: What frequencies are you using in your radar work?

ANSWER: The radar work was done using frequencies in the 20 megahertz to 1 gigahertz range.

QUESTION: With your array of equipment, what is your maximum depth of penetration?

ANSWER: The seismic acoustic method has a maximum depth of penetration of perhaps 500 ft and the conventional high-resolution seismic method has a depth of penetration of 5,000 ft or more. The depth of penetration is limited only by the amount of energy that can be put into the ground and by the desired resolution. The poorer the resolution that can be tolerated, the greater the depth of penetration that can be achieved.

QUESTION: Why not use acoustic energy source as a reconnaissance tool?

ANSWER: The acoustic energy source in its current stage of development is too expensive and too slow in application to be a viable reconnaissance tool. It's best used to get a very detailed, accurate look at a limited target area.

QUESTION: What is the maximum depth of investigation you hope to obtain using radar and how effective if at all do you expect it to be in coal bearing rock of a discontinuous nonuniform character?

ANSWER: The radar should be able to penetrate 50 to 100 ft in coal and perhaps several times that in hard rock. As in seismic methods the desired resolution is a major trade-off in achieving penetration. The radar should be able to detect major structures or voids even in discontinuous nonuniform coal, but probably at a cost in resolution.

QUESTION: Are there any presently practical techniques for locating abandoned underground workings from the surface?

ANSWER: It is possible in some conditions to detect abandoned underground workings using conventional surface high-resolution seismic methods. These methods are presently being used by some mining companies.

QUESTION: What do you consider is the optimum frequency for obtaining maximum resolution at depths around 2,500 ft?

ANSWER: Optimum seismic frequencies for maximum resolution at 2,500 ft depth would be approximately 500 Hz, depending upon a number of site specific parameters.

QUESTION: How adversely seriously are you affected by variations in surface topography (such as in southern West Virginia)?

ANSWER: Variations in surface topography are at best a great nuisance or in severe cases can completely prohibit surface seismic operations. Generally, the variations can be dealt with utilizing close control in surveying, recording, playback, and applying migration analysis with close manual manipulation of the data processing.

QUESTION: What are the differences in your method of storing and analyzing acoustic data when it is used in the pulsed mode versus in the swept frequency mode?

ANSWER: There are no differences since the data is cross-correlated with the pilot (input) signal.

14
Geochemical Affinities as Exploration and Classification Tools

By Alv Orheim
Geologist
Store Norske Spitzbergen Kulkompani A/S
Longyearbyen, Norway

INTRODUCTION

The coal industry is given an optimistic future by most experts. It is expected that coal must cover more of the global energy demand and also substitute oil in petrochemical industry.

This will undoubtedly influence the market prices, and the market will become more selective (1). Weight will be put on many different criteria of quality which today have little significance.

This paper deals with the chemical composition of coals, and how this may play a part in this development related to petrographic and physical characteristics.

The close relation between exploration effort and market prospects is evident, and will not be discussed.

CHEMICAL COMPOSITION OF PLANTS AND VEGETATIVE AREAS

Coal is formed under unstable basin conditions in the sense that small eustatic changes or change in groundwater level would have great influence on the deposition.

The area of peat formation was flat and just above or below sea level. A slight relative change in sea level was therefore sufficient to produce inundations or emergences (2).

The chemical conditions in the basin can be considered influenced by two components: biochemical and geochemical, and by the availability of each element in the basin. Together these variables form the biogeochemical order of the basin.

Biochemical component

It is a fair assumption that prevailing vegetation demonstrates the same chemical system as that of plants 60 or 300 million years ago. Looking at present botany can therefore give clues for evaluating the biochemical component of coal (3).

The elements can be divided into three groups: macronutrients, micronutrients and passive elements.

The ash content is normally greatest in leaves and bark, lower in the tissue of branches and least in wood structure of stems.

Macronutrients of plants are C, H, O, N, P, S, K, Ca, Mg and Fe. The first 3 represent 95-99% by weight of plant matter.

From a coal quality standpoint the following six elements call for the greatest interest (4):

Phosphorus plays a major role in plant growth. It is enriched in leaves, seeds and pollen. The plant has a mechanism for retrieving P from the leaves before they fall off.

30-40% of P in plants are in organic compounds, but only a few of these have been defined. Due to this concentration the decomposition of humus may be a decisive factor for the quantity of P in plants.

From inorganic compounds phosphorus is retrieved from phosphates of Ca, Fe and Al.

Fe and Al are insoluble at low pH, opposed to Ca phosphates, soluble under alkaline conditions.

Sulfur will normally be found as sulfates in organic amino acids. They are readily leached, but ample supplies of inorganic sulfide ions oxidize and reform amino acids with sulfate.

Potassium corresponds to phosphorus in plant distribution, i. e. leaves and spores. It is easily activated and transported by groundwater.

Calcium is concentrated in the wood structure and bark. It is fairly stable in plants. Calcium plays a major role for the total growth potential of an area, e. g. great influence on the pH value.

Magnesium will normally be found in mineral matter and enters humic acids only to a small degree. It is readily leached.

Iron represents the least important of major elements for plant growth. By some authors it is classified as micronutrient. Normally there is a surplus of iron, although arid conditions or high pH may cause iron deficiency.

Micronutrients are important elements of which the plants only need very small amounts. These include manganese, copper, boron, molybdenum, zinc and chlorine.

Those elements are normally found in humic acid or in secondary inorganic particles, often insoluble by plant acids.

Over sandy soil there might be scarcity of micronutrients due to severe leaching, while an underbed of clay will give a surplus.

Passive elements represent all elements without definite growth significance. It is believed that some of these elements may have functions, which so far have not been detected.

Some of the elements, e. g. Pb, Cd and Hg, are venomous to plants when in high concentrations. In some cases the plant rejects adsorption of these elements.

Se, Co, F and I are all elements with influence on flora. The concentration of these elements in plants are therefore of great importance.

High pH value hampers elements like B, Mg, Cu, Zn to

Element	Earth's Crust	Soil	Plant Ash	Water
Ag	0.07	1.0	1.0	0.0003
As	2	5	4	0.003
Au	0.004	0.002	0.005	0.000004
B	10	10	700	4.6
Ba	425	500	280	0.03
Be	2.8	0.8	1.2	
Cd	0.2	0.5	0.1	0.0001
Co	25	10	9	0.0005
Cr	100	200	9	0.00005
Cu	55	20	180	0.003
F	625			1.3
Fe	64,500	30,000	6,700	0.01
Ga	15	20	1	0.00009
Ge	1.5	5	5	0.00007
Hg	0.08	0.01	0.01	0.00003
I	0.5			0.06
In	0.1			
Li	20	30	2	0.003
Mg	21,000		700	1,000
Mn	1,000	850	4,800	0.002
Mo	1.7	2.5	13	0.01
Nb	20	15	0.3	
Ni	75	40	65	0.002
P	100	40	65	0.07
Pb	12.5	10	70	0.00003
Pt	0.005			
Rb	90	300	2	0.12
Re	0.0005	0.005	0.005	
Sb	0.2	0.5	1.0	0.001
Se	0.05	0.5	1.0	0.004
Sn	32	10	1.0	
Sr	375	300	30	8
Ti	5,700	5,000	2	0.001
U	2.7	1.0	0.6	0.003
V	135	100	22	0.002
W	1.5	1.0	0.5	0.00003
Zn	70	50	1,400	0.01

Table No. I. Clarke values of elements (ppm).
Compiled from (5) and (6)

enter organic compounds.

Clarke value of elements

The availability of different elements is a decisive factor for the distribution in soils or peat.

Table No. I reviews the over all clarke value in rock, water, soil and plants. As can be seen the values differ to a great extent. Some of the macronutrients are in fact only trace elements in soil or water. Abundant elements like Si and Al are of little importance to vegetation.

From Table No. II it is evident that the majority of elements are dissolved in sea water and river water with great variety. The table indicates which elements are the most interesting as salinity indicators.

Table No. II. Elements dissolved in sea water and river water (ppm). From (7).

Element	Sea water	River water
Al		0.69
B	0.01	0.00000001
Br	0.19	0.00000002
C	0.08	7.30
Ca	1.16	20.49
Cl	55.05	5.68
F	0.01	0.000001
Fe		1.44
K	1.10	2.12
Mg	3.69	3.41
Na	30.61	5.79
S	2.56	4.49
Si		5.45
Sr	0.03	0.0000005

The very large variety in distributional pattern, and the abundance of major elements, call for some distinction to be made. As in all geochemical prospecting the distribution of minor or trace elements are likely to show the most "readable" pattern.

The effort to bring order and understanding into the biogeochemistry, and make this applicable to coal exploration, should therefore be concentrated on these minor elements.

Geochemical component

The availability of a particular element in the basin is governed by several environmental factors (8). The interaction between vegetation, soil and plant nutrients is previously discussed. Below is given a summary of some other factors with effect on the chemistry of the basin.

Mobilization by weathering of soil minerals is largely a question of the stability of ions. This can be expressed by the ionic potential (Z/r) where Z is the electric charge and r is the ionic radius.

When the potential is lower than 2, ions tend to remain in solution. This is the case for Pb, Hg, Cd, Sr, Na, K.

S, P, B, Se and Ni have potential above 12. At this value oxyacid anions, i. e. sulfates and phosphates, tend to form.

Elements with intermediate ionic potential are Mg, Fe, Al, Ge, Mo, U, Ca.

The biosphere, in particular the lower organisms, is an important agency in weathering of minerals. In organic soils there is a surplus of humic acids and bacteria. Detrital minerals will therefore to a large degree be dissolved, and new compounds, chelates, will form.

Clay minerals and humus have a high adsorption capacity of ions. In clay this is a result of their sheet structure and an excess of oxygen, which gives an overall negative charge of the crystal (9).

The high adsorption capacity of humus is caused by the large and complicated carbohydroxyl molecules with several free radicals for cation bonding.

The adsorption in soil is only half of that of clay minerals, while humus has a capacity 10 times higher than clay minerals (10, 11). Ionization adsorption are increased by higher pH values.

The biochemical cycle describes the transport and concentration of elements by vegetation.

Enrichment of elements takes place in the upper half of the root system and in leaves, i. e. on soil surface (humus horizon), as the leaves wither and fall, see figure No. 1.

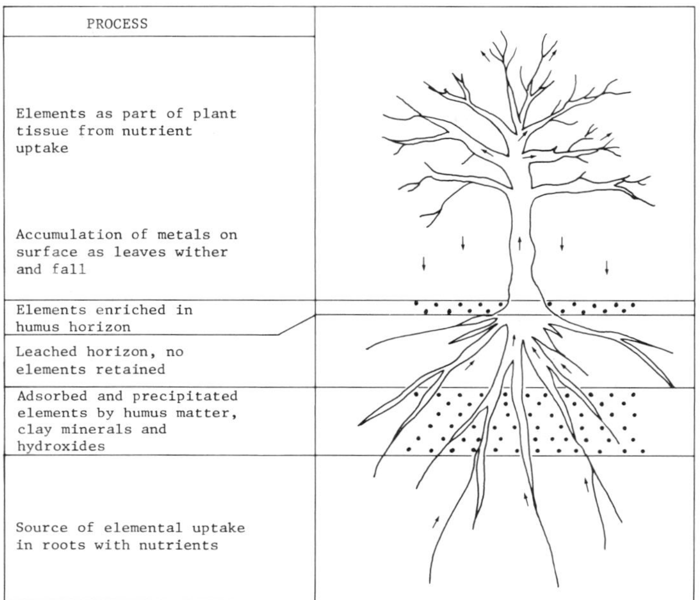

Figure No. 1. The biogeochemical cycle. After (10).

Carbonates, sulfates, phosphates and humic iron complexes are leached by water and recycled. Other compounds, insoluble hydroxides and humic complexes, are retained in the humic layers. This results in significant enrichment of As, Be, Cd, Co, Ge, Au, Pb, Mn, Ni, Sc, Ag, Sn, U, Zn.

Microorganisms represent a strong factor in mobilization and fixation of elements. It is sufficient here to mention bacteria of carbon, nitrogen, sulfur and phosphorus.

Oxidizing and reducing bacteria influence the precipitation of many bi- or multi-valence elements, e. g. ferrous/ferric iron.

Biogeochemical order.

The two components discussed, and the local availability of elements, interact to form a specific biogeochemical order of element distribution.

Normally the geochemical component will play the major role. However, in a peat forming basin the biochemical processes will have the dominating influence.

This is due to the effective humic acid and the active part microorganisms and plants play in the chemical cycles.

The organometallic compounds formed by humus tend to form colloidal structures. Each particle has an electric charge and is encircled by water molecules. The colloids have a large body compared to weight and are therefore kept "floating."

This colloidal state is unstable. Small changes in Eh and pH values disrupt the stability, and the molecules form clusters and precipitate.

When reviewing changes in trace element distribution with respect to coal exploration it is essential to have the colloidal basic chemistry of the basin in mind.

TRACE ELEMENTS IN COAL

Until recently, fairly little systematic work was done to analyze coal beyond the major elements. The reason for this is to some extent to be found in a lack of proper analytic methods.

Table No. III summarizes the results from 4 different coals and the average content of shale. It can be seen that the concentrations differ from coalfield to coalfield. This is supported by recent analytic work on US coals.

By comparing coals from Appalachian coalfields, the Illinois Basin and the western coal areas dr. Gluskoter _et al_. (13) were able to draw some generalizations for the occurrence and distribution of trace elements:

1. Elements that have relatively wide ranges in concentration include those found in coals within sulfide and sulfate minerals.
   Elements with narrow ranges or smaller standard deviation are associated with silicate minerals or the organic matter.

2. Only four elements: boron, chlorine, arsenic and selenium, show a regular enrichment in coals compared to

their clarke value. Depletion of element concentration is thus more common.

3. Bench sampling demonstrated for most elements a variable vertical distribution. However, concentrations at the top and bottom of sections are common.

4. Rock units immediately associated with the coals (roof shales, underclays and partings) show higher concentration for the majority of elements.

5. By analyzing washed samples at specific gravity fractions it was possible to establish the organic affinity of the elements.

   This can either be expressed by washability curves, histograms (see Figure No. 2) or by calculating an "organic affinity index."

   Thus the elements can be classified in four groups:
   a) organic
   b) intermediate organic
   c) intermediate inorganic and
   d) inorganic.

6. Within one basin these results can be generalized. However, correlation between larger coal areas is only partly positive.

Table No. III. Trace elements in total coal and shale. From (12). (ppm)

Element	Shale	Total coal New South Wales	Sydney Coalfield Nova Scotia	Svea seam Spitsbergen	Crown seam North Wales
Ag	0.07	0.05	-	-	0.05
As	13	-	100	-	-
B	100	60	17	145	12
Ba	580	100	35	170	63
Be	3	1	2	1	-
Co	19	4	10	2	20
Cr	90	7	5	38	31
Cu	45	15	-	3	29
Ga	19	4	-	4	8
Ge	1.6	6	5	1	-
Li	66	-	-	100	29
Mn	850	150	140	20	28
Mo	2.6	1	7	0.5	5.5
Ni	68	15	15	6	54.5
Pb	20	8	66	10	80
Sn	6	1	1	-	0.5
Sr	300	100	65	250	26
V	130	20	14	30	56
Y	26	8	-	13	-
Zn	95	100	25	-	-
Zr	160	100	-	14	23

Although these results are restricted to three coal areas, they indicate some of the conclusions that can be drawn from a rather comprehensive analytic work.

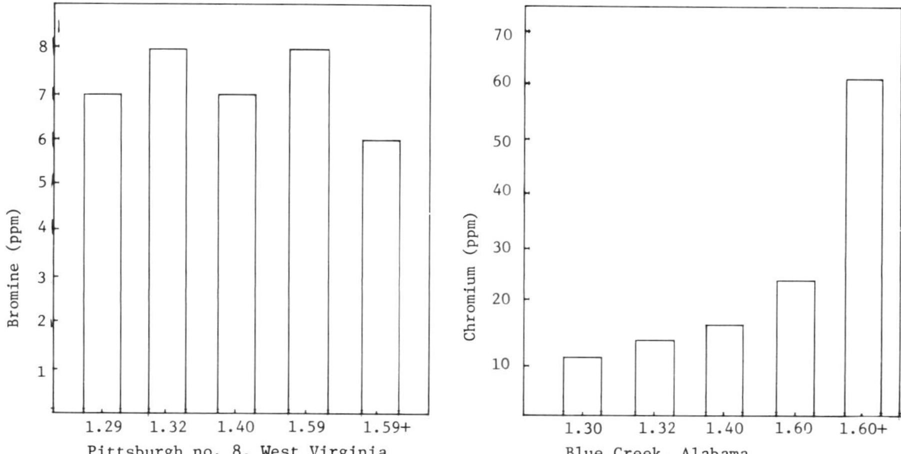

Figure No. 2 . Histograms of organic element, Br, and inorganic element, Cr. From (13).

The vertical variable distribution is by most authors taken as indicator of an epigenetic mineral deposition from solvents moving in given layers, and in particular in detrital deposits of floor or partings.

Considering the biochemical cycle as shown in Fig. 2 it is also possible to achieve preburial vertical distribution. The plants will continuously enrich the top soil of the basin. They will therefore be growing on soil with increasing abundance of certain elements.

A climate change followed by change in vegetation can disrupt this steady development, and thus bring forth variable distribution.

A distinction must be made between preburial or early diagenetic inclusions and minerals, as opposed to epigenetic reactions. Obviously only the former can be used as environmental indicators.

Likewise detrital minerals (clastic minerals deposited by surface water) can only indicate physical environmental changes.

It may be concluded that any element showing homogeneous distribution vertically is preburial or early diagenetic.

Elements showing vertical variation may be either preburial or epigenetic.

The organic affinities of Ni and Cu are low in coals rich in these elements. At low concentrations the affinities are higher.

This reflects a dual distribution pattern. The elements are primarily enriched in organometallic compounds giving a generally low value in most coal basins. Under specific conditions abundance of Ni and Cu arises in a basin, and the elements associate with inorganic compounds. The result is low organic background concentration exceeded by much higher inorganic "peaks."

Mo, Ga, Ge, As and P have a similar distribution with opposite sign, i. e. normally they are in low concentrations as inorganic matter, but show organic affinity when strongly concentrated.

The influence of local components makes it impossible to transfer results from one basin to another. A comprehensive regional analytic work must be done prior to any exploration conclusions based on trace element distribution.

In the studies done so far significant correlative factors have always been found. The apparent confusion in results from area to area should therefore not discourage a closer look at trace element distribution.

The following are suggestions to where such studies might be a useful help.

TRACE ELEMENTS IN EXPLORATION

Seam identification

Adsorption from groundwater or primary enrichment by plants are the main accumulation factors of minerals under peat formation. These factors will be relatively uniform within one peat forming cycle. The very high ion exchange capacity of humic acid, and to some extent clay minerals, favor this process.

Elements enriched thus can be boron, selenium, nickel, copper in coal, and gallium, lithium, chromium and vanadium in lattices of clay into coal.

Of these B and Li have fairly high concentrations in salt water. Influence from sea water could therefore disguise them as seam indicators.

Se is most certainly concentrated botanically, and may be the most promising indicator. However, as it is readily concentrated it is not likely to show the greatest anomalies from one peat formation to another.

Primarily a seam identification should be sought between Ni, Cu, Ga, Cr and V.

Paleosalinity indicator

The possibility of plotting changing salinity of the basin has intrigued many workers. This would of course give vital information on seawater influence and the environments of the basin.

Boron has been suggested as a good indicator and indeed in many cases it shows good correlation. Chlorine or the ratio Cl/alkali and Mg anomalies attract attention because of the burial adsorption in clay minerals, and should therefore be a possible salinity indicator.

The use of these indicators is hampered by analytic problems. How can one be sure the concentrations are due to preburial salinity changes? Discussion of these problems has so far proved that within some basins positive correlation can be established. The following two different correlations are reported:

1. Marine shales are enriched in B and Rb; Ga is abundant in fresh water deposits. Analysis of the organic fraction of the same sediments shows concentration of Ni and V in the marine shales, whereas Pb, Zn, Cu and Sn are enriched in the fresh water deposits (14).

2. B, V, Cr, Cu, Ga and Ni are significantly more abundant in marine deposits (15).

Basin configuration, paleomorphology

Certain organometallic compounds show distribution pattern depended on the distance to basin margins. This is an interesting feature when exploiting coal formations within crevasse-splay deposits.

Elements likely to precipitate as organic colloids are Pb, Zn, Cd, Hg, Cu, Ni. As previously stated this can be derived from the ionic potential of the elements. Such elements will show a laterally decreasing concentration moving away from detrital source.

It is supposed that coal at basin margin will have a more heterogeneous vertical distribution. This is attributed to the more variable conditions of weathering and erosion in the borderland (13, 16).

Sedimentary facies

Based upon the precipitation of iron bearing minerals, the coal basin can be divided into sulfide, sulfide-carbonate, carbonate and oxide facies. However, the precipitated iron minerals are particularly susceptible to diagenetic changes, and direct examination may prove impossible.

The chalcophile elements As, Co, Ni, Pb and Sb are likely to correspond to the facies. There is also reported correlation between these elements and Ge.

A correlative analysis may give values reflecting the sedimentary facies.

Rate of peat formation

This may be the most interesting aspect of trace element distribution and covers three different conditions.

1. Concentration of elements in certain plants has been known for many years. Varying vegetation reflects changing climatic conditions or other growth parameters, i. e. moisture.

   A correlation between these factors could give wide-ranging results.

2. The biogeochemical cycle results in a cumulative enrichment of As, Be, Cd, Co, Ge, Au, Pb, Mn, Ni, Sc, Ac, Sn, U, Zn in the topmost layer of the soil.

   Under peat formation subsequent plants grow on enriched

topsoil. Under uniform conditions this can have an exponential concentration effect. Abrupt changes will leave significant variations in distribution.

A particular distribution pattern or a horizon enriched on certain elements may therefore be a marker bed for simultaneous events.

3. Most coal seams are a result of both autochthonous and allochthonous processes.

In the Longyear seam of Spitsbergen, phosphorus shows a tendency to be enriched in organic colloidal suspension. Abundance of organic phosphorus at certain horizons indicates therefore precipitation of allochthonous coal within an autochthonous formation. Often this will be as channel fillings in syngenetic or early diagenetic eroded river courses. Figure No. 3.

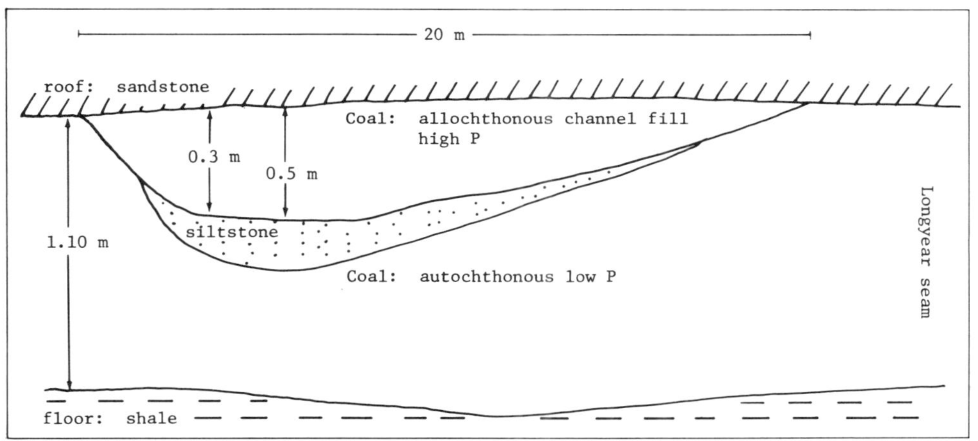

Figure No. 3. Channel fillings of phosphorus-rich coal at top of Longyear seam, Spitsbergen.

TRACE ELEMENTS IN CLASSIFICATION

A trace element may either have a definite quality in itself, e. g. pollution; or it can be associated with constituents of coal or ash with a specific quality effect.

The following are suggestions to show how this is applicable with regard to different aspect of coal technology.

Maceral distribution.

Although much work has been done to correlate elements and macerals little has evolved yet. This is, however, an interesting field as the macerals have specific technological properties, e. g. exinite may be preferred in gasification and liquefaction due to higher hydrogen content than the other macerals.

The vitrinite group may be enriched in Ga, Ca, Ge, Ba, V, Ni and Zr compared to the other maceral groups (17).

P, Cu, Mo, Ti, K are enriched in the inertinites (fusinite) P and K are also enriched in sporinite (exinite groups).

The exinite group (cutinite) may show concentration of Ag, As, Au, Be, Cd, Co, Ge, Mn, Ni, Pb, Sn, U, Zn.

Elements like Co, Fe, Se, Cd, Hg, Pb show no particular concentration in the macerals, and are likely to be found in both the vitrinite and exinite groups.

Preparation prospects

Organic compounds are considered non-extractable. Elements forming such compounds are Ge, Be, B, V, Ti and Sb. The last is of most interest as it is a chalcophile.

Abundance of Sb indicates therefore a high level of organic sulfur. Opposed to this are the inorganic chalcophiles.

Other elements reported inorganic are Zn, Cd, Mn, Mo, Fe, Ga, and Sn.

Heterogeneous distribution indicates high content of inorganic compounds and are thus in favor of a satisfactory cleaning potential.

Preburial clay, i. e. finely dispersed clay in the coal substance, contaminates the coal product. The three elements Cr, V and Li are all adsorbed in the clay lattices, but represent different time of genesis. Cr is typical postburial, while Li is typical preburial and V can be either Change in ratio between these elements may therefore be correlated to change in preparation prospects of the coal.

Trace elements as polluters

Probably the main reasons for doing trace element analysis are the environmental recommendations the coal industry will face in the future.

Besides sulfur and other major constituents the elements Hg, Cd, Be, In, U, and As are all under surveillance.

Although they represent inorganic compounds, except Be, ordinary preparation may not be able to reduce their concentrations significantly. The elements are normally concentrated at top, and sometimes bottom of the seam. One way to avoid the problems might therefore be to do selective mining and leave the upper 5 cm of the seam unmined.

In the future coal producers might be asked to produce regular run of mine analysis for "pollution elements." As they often are well below detection limits of normal analysis this could be a problem.

However, several positive correlations are suggested between trace elements and elements of higher abundance, e. g. Zn/Cd. A correlation factor from a fair number of samples may therefore solve this problem.

Several of the trace elements represent process contaminators that might have a significant market influence. Elements like Hg, Pb, Cd, As, Sb, Se, Fe and Bi are chemicals hazardous to some of the catalysts proposed in liquefaction and gasification processes.

Utilization of ash and refuses

An expansion of coal utilizaton will also involve growing deposits of ash and refuses.

To what extent this represents a problem, environmentally or technologically, is dependent on various local factors. Present use as road material, fillers, building materials, etc. will certainly be expanded.

However, both ash and refuses represent a concentration of elements by a factor of 10 to 20, and trace elements may thus be concentrated in extractable amounts. V, As, Hg, Sr, Ge, Se are reported to have concentrations worth considering (19). Although they might not represent a large positive income, they could reduce an otherwise net negative burden of how to handle deposits of ash and refuses.

Technological parameters

Several parameters regarding fusion temperatures, fouling, slagging, swelling, etc. are developed (20). They are based upon certain ratios between the major base/acid elements. How they apply from one basin to the other is debated.

Correlation factors for a major element and trace elements may indicate to what extent the parameters correspond.

This is demonstrated by the following, regarding the Svea seam of Spitsbergen.

The mineral dawsonite ($NaAl(OH)_2CO_3$) is of great abundance in the Svea seam. This mineral has distinctive features, genetically and technologically, from other sodium minerals, i. e. chlorides. However, in the commonly used "total base factor" Na both from dawsonite and chlorides will enter, and thus disrupt, the technological significance of the factor

There is a positive correlation between P, Sr, Na and dawsonite. This correlation may therefore be used to indicate when the "total base factor" is unreliable.

SUMMARY

The biochemical processes in vegetation are important factors governing the distribution of trace elements in coals. The same applies to the geochemical syngenetic and epigenetic processes.

Clay minerals and humus, in particular, have high adsorption capacity of ions. Preburial distribution of elements are therefore almost solely a result of humus activi

Only preburial or early diagenetic elements may be correlated to basin geomorphology and seam configuration. Many elements represent an analytic problem regarding distinction between preburial and postburial distribution.

Elements of high mobility are not suitable for correlative analysis as they are enriched uniformly over large areas. Significant distribution patterns are normally best establish in elements with relatively low background value.

Several positive correlations have so far been established, regarding both coal formation and coal utilization. However, the correlation is normally only valid within one coal area, as the local vegetation and availability of elements differ.

Trace element distribution and the methods described here are not opposed to other exploration techniques or classification criteria. They are only supplementary indicators.

As coal mining goes deeper and into more remote areas exploration costs will escalate. Every useful bit of information must then be extracted from boreholes and mines.

A more selective market will only tend to emphasize this.

ACKNOWLEDGEMENTS

I am indebted to my company for the opportunity to present this paper and publish some of our results. Several colleagues at Berbau-Forschung, Essen and Ruhrkohle AG Bochum, in particular Dr. M-U. Otte of the latter, Dr. Gluskoter of Illinois and Prof. Kullerud of Indiana and Prof. Laag of the Agricultural University of Norway, have supplied invaluable information and idea.

REFERENCES

1. Falcon, R. M. S., "Coal in South Africa, Part II." Minerals Sci. Engng., Vol 10, No. 1, 1978.

2. Moore, R. C., "Late Paleozoic Cyclic Sedimentation in Central United States" in Adventure in Earth History. W. H. Freeman and Company, San Francisco, 1970.

3. Goldschmidt, V. M., "The Principles of Distribution of Chemical Elements in Minerals and Rocks." Chem. Soc. Jour. Part 1, 1937.

4. Laag, J., "Soil Sciences" (Norwegian text). Agricultural University of Norway Prints, 1975.

5. Pauly, H., "Geokemi." Polyteknish Forlag, Copenhagen, 1968.

6. Berkman, D. A. and Ryall, W. R., (ed.): "Field Geologists' Manual." Austr. Inst. of Mining and Metallurgy, Victoria, 1976.

7. Op cit (4)

8. Brooks, R. R.,"Geobotany and Biogeochemistry in Mineral Exploration." Harper & Row, New York, 1972.

9. Op cit (8)

10. Hawkes, H. E. and Webb, J. S., "Geochemistry in Mineral Exploration." Harper & Row, New York, 1962.

11. Op cit (4)

12. Nicholls, G. D., "The Geochemistry of Coal-Bearing Strata" in "Coal and Coal-Bearing Strata," Oliver & Boyd, Ltd., London, 1968.

13. Gluskoter, H. J. et al, "Trace Elements in Coal: Occurrence and Distribution." Illinois State Geo. Survey, Circular 499, 1977.

14. Degens, E. T., Williams, E. G., Keith, M. L., "Environmental Studies of Carboniferous Sediments. Part I." Bull - Am. Assn. Petrol. Geol. 41, 1957.

15. Potter, P. E., Shimp, N. F., Witters, J., "Trace Elements in Marine and Freshwater Argillaceous Sediments." Geochim. et Cosmochim. Acta 27, 1963.

16. Zubic, P. et al, "Distribution of Minor Elements in Coal of the Eastern Interior Region." U. S. Geol. Surv. Bull. 1117-B, 1964.

17. Otte, M. U., "Spurelemente in einigen deutschen Steinkohlen." Chemie der Erde 16, 1953.

18. "Stach's Textbook of Coal Petrology, 2nd Edition." Gebruder Borntraeger, Berlin-Stuttgart, 1975.

19. Kullerud, G., Purdue University, Indiana, personal comm.

20. Schmidt, R. A., "Consumer Coal Criteria as a Guide to Exploration" in "Coal Exploration Symposium Proceedings," London 1976, Miller Freeman Publ., 1976.

Australia: Methods and Results

15. Hunter Valley Coal: A Successful Australian Exploration Project

by George E. Edwards and Brian W. Vitnell

16. Efficient Australian Practice in Exploration and Evaluation of Coal Prospects

by David Svenson

15

Hunter Valley Coal: A Successful Australian Exploration Project

By George E. Edwards, Manager Marketing (Technical),
and Brian W. Vitnell, Chief Geologist,
Coal and Allied Industries Limited
Sydney, New South Wales, Australia

SUMMARY

This paper outlines the exploration and testing of an area of some 900 million tonnes of measured plus indicated coal reserves near Warkworth in the Hunter Valley of New South Wales, Australia, where coal production will commence in mid-1979.

More than 350 slim cored holes (51 millimetre (mm) diameter) have been drilled in the exploration programme, together with almost 500 rotary open-holes (114mm diameter) to delineate oxidation by analysis of chip samples and 23 large diameter (203mm) cored holes for bulk sample testing. A 2 metre x 1 metre shaft was sunk for further testing and to produce bulk samples for assessment by potential consumers.

Comprehensive chemical, physical and float/sink analyses of coal bore cores, large and small, have been undertaken throughout the programme and all data obtained has been encoded for computer playback of plans, sections and selected tabular and mining information.

Geophysical surface exploration, comprising ground magnetometer and resistivity surveys, has also been undertaken to provide greater detail of faults and igneous intrusion in the initial operations area at Hunter Valley No. 1 Mine, to the north of the Hunter River. Operations will also take place at Hunter Valley No.2 Mine, south of the Hunter River, in the future.

The thirteen seams occurring in the two areas contain high volatile coals of both coking and steaming potential, and marketing related studies in both areas of utilisation have been undertaken on the various samples produced during the exploration and testing programmes.

INTRODUCTION

The fastest developing coal area in Australia is centred around Warkworth, some 100 kilometres west of Newcastle in the Central Hunter Valley of New South Wales, Australia (Fig. 1).

Twenty years ago the nearest working coalfield of the Central Hunter Valley was located some 15 kilometres to the north at Ravensworth-Liddell and the sole knowledge of the vast potential of the Warkworth coalfield came from isolated reports by landowners of thick coal seams occasionally encountered in water bores.

Following some spasmodic private company exploration drilling in the 1960's the Geological Survey of the N.S.W. Department of Mines commenced sinking "scout" boreholes throughout the area on a 4 kilometre spacing, controlled to a degree by surface mapping, in the early 1970's.

Coal & Allied Industries Limited, a major N.S.W. coal producer was allocated a 34.6 square kilometre Authorisation to explore for coal over part of the Warkworth coalfield in early 1974. Operations commenced in May 1974 and will be completed in March 1979. The area finally explored was 47.6 square kilometre after having been extended northward twice.

At a late stage of the exploration the Government of the State of New South Wales reviewed the right of the Company to mine the resource proved up before agreeing to issue a Mining Lease. The Government decided on the allocation of two separate leases to be known as Hunter Valley No.1 Mine and Hunter Valley No.2 Mine, situated north and south of the Hunter River, which flows through the exploration

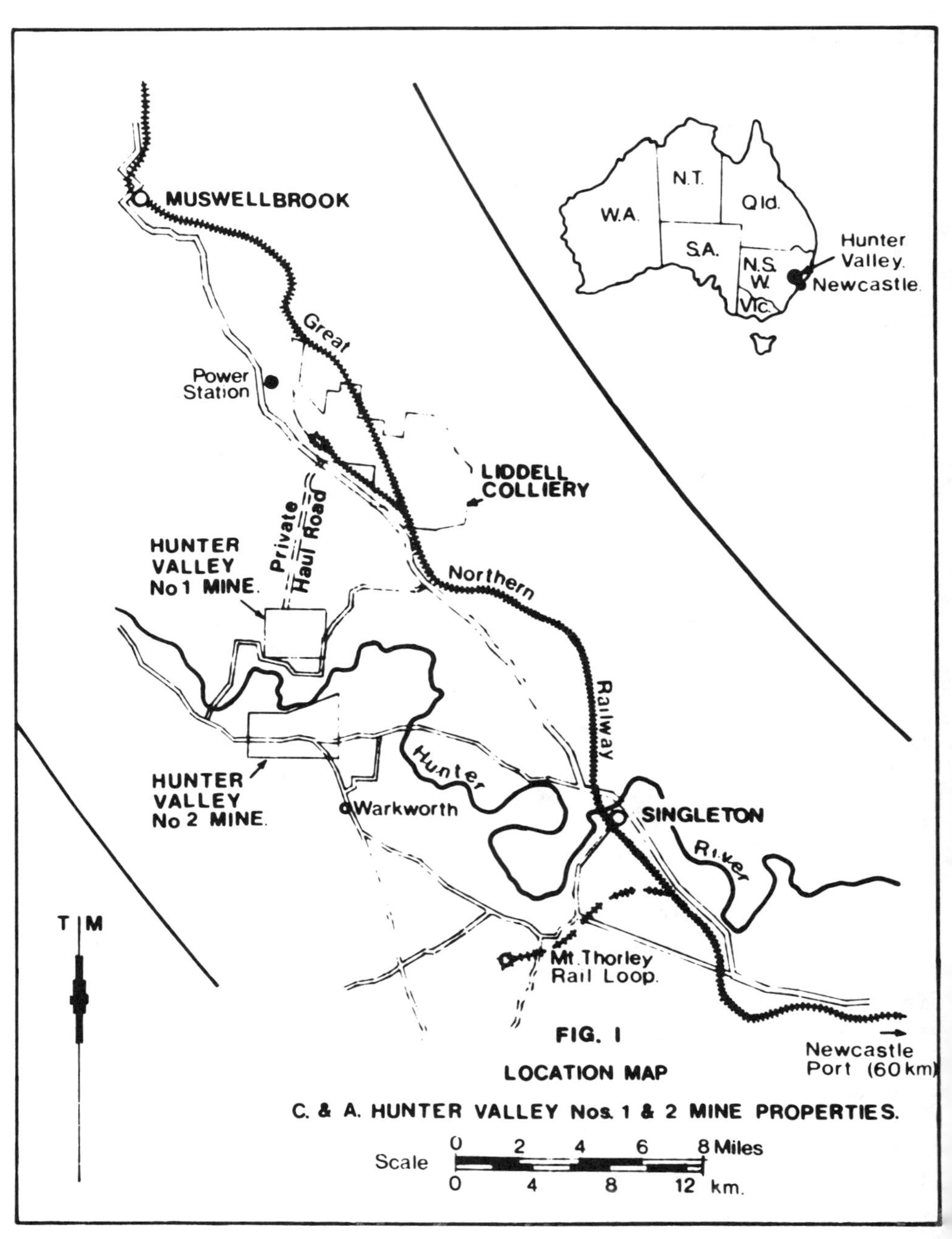

FIG. I

LOCATION MAP

C. & A. HUNTER VALLEY Nos. 1 & 2 MINE PROPERTIES.

Authorisations. The former will contain 56 million tonnes of unoxidised coal in three seams, the latter 350 million tonnes in thirteen seams, with working thicknesses ranging from 1 to 7 metres. Total reserves proven had totalled some 900 million tonnes. It is planned to work all coal in the two mine areas by open-cut methods. The seams in both areas will yield high volatile coking and steaming quality coals.

Coal production will commence in mid-1979 at the Hunter Valley No.1 Mine and later at the Hunter Valley No.2 Mine. Production will begin as a truck and shovel operation using scrapers and dozers, with the use of draglines in the No.2 Mine where the seams are deeper. Production will commence at the rate of 1.5-2.0 million tonnes per annum (m.t.p.a.) and increase to about 3.5 m.t.p.a. by 1981. Further increases to 7 m.t.p.a., and possibly 10 m.t.p.a., will be achieved as markets develop in the 1980's.

EXPLORATION STRATEGY & METHODS

The suggested "Code for Calculating and Reporting Coal Reserves" issued by the Standing Committee of Coalfield Geology of N.S.W. (1), which closely follows the earlier U.S. Bureau of Mines standard, by inference, determines the spacing of drillholes in a new coalfield where other data is not available.

With modifications to allow for the adoption of the Systeme International (modified metric) in Australia in 1970, these are:-

Inferred or Assumed Reserves: 4 kilometre (2 mile) grids
Indicated Reserves: 2 kilometre (1 mile) grids
Measured Reserves: 1 kilometre (½ mile) grids

using fully cored (51mm) drillholes.

A number of Government and private enterprise groups have been using this concept in coalfield exploration in Australia for some years, and it is not considered necessary to modify it where exploitation of the reserves so proved is to be carried out by sub-surface (underground) mining. Additional drilling, or geophysics, may be undertaken in areas of special interest so proved, but the equipment in use and the patterns of mining development themselves become exploration tools.

However, with the present necessity to prove up new areas in which the geology and initial drilling (plotted at plan scale 1:20 000) indicates large reserves of single or multi-seam deposits where open-cut mining may be possible, many practicing coal exploration geologists consider closer spacing of drillholes necessary.

The suggested code currently does not cover this situation. However, in 1956 the Joint Coal Board (2) closely defined the drilling grids necessary to prove up open-cut coal where overburden type and limits, and high coal extraction quantities, are critical to maximum exploitation and cost benefit.

This standard, with modifications, has been used in the Coal & Allied Warkworth Authorisations. Because of the closer tolerances and selectivity suggested this Standard takes into account a certain "recovery" factor, and gives closer information on departures from "average" conditions, thus allowing a better planning exercise for a mining method which recovers large tonnages from quite small areas. This standard commences where the existing suggested code ceases, i.e. at 1 kilometre (1000m) grids, from which previously defined "Measured reserves" revert to "Inferred Open-Cut reserves". Further drilling, also fully cored, then reduces data spacing by an approximate factor of three to 300 metre grids for computation of "Indicated Open-Cut reserves". Assessment of these results allows the drawing of preliminary open-cut plans at scale 1:10 000 on which seam outcrops and limits of full seam working thickness (i.e. non-oxidised) coal are delineated (by intersection of seam base and topographic contours). When the resource, which by now is considered an economically viable open-cut reserve, is to be developed, closer drilling grids are considered necessary.

A further reduction by a factor of three reduces data points by drilling to 100 metre grids specifically for the delineation of any partly or fully oxidised section of the seam. Depending on the methods used, and the sampling interval, "subcrops", "limits of oxidisation" or "limits of saleable coal" can be delineated to varying orders of accuracy.

At Warkworth the exploration method adopted was rotary open hole drilling (of hole diameter 114mm) using air blast circulation, and "chip" sampling at one metre intervals. Prior to this work being carried out, lithotype (brightness) logging of the coal seams in cored holes had allowed a detailed evaluation of seam morphology. On this basis, relatively inexpensive crucible swelling number tests were carried out on the samples collected from the open holes. Successful refinement of the method, based on detailed seam knowledge, allowed some selectivity by omitting stone bands and duller (low swelling) sub-sections from the samples collected. Empirical interpretation of the swelling numbers obtained yielded quite definitive information on oxidation, having regard to the sampling interval and sub-section contamination.

In general, this final stage, leading to the computation of "Measured Open-Cut reserves" of coal, could be restricted to that area of the resource where initial operations are planned, then relying on "in-pit" quality control for follow up work. However, because of constraints imposed on the company as to the size of the resource to be allocated in the area of initial operations, it was necessary to exclude the major proportion of oxidised coal and a large number of open holes were drilled and tested in the manner described. Although the tests were indicative of oxidised coking coal, the method can safely be used to delineate oxidised steaming coal because intrinsic specific energy values fall more slowly. At zero crucible swelling index the specific energy value is then more significant, provided always that the coal structure has not been completely destroyed.

In certain sub-areas where larger coal samples were required for small scale washability testing, taking account of sizing and a range of float/sink densities, larger cores (203.2mm) have been taken. Almost always, on a cost basis and for correlation with even smaller scale float/sink tests undertaken on the 51mm cores, they have been drilled alongside a suitable slim cored hole.

All drilling undertaken was by contract and performed by an organisation with a long record of successful coal exploration performance in the N.S.W. coalfields, particularly those of the Hunter Valley.

Diesel powered hydraulic feed diamond coring rigs were used to drill all holes down to 300 metre grids. Coring equipment used was NM/LC which is a derivative of the international NM size gear. Hole O.D. is the same as NM at 76mm diameter but core size at 51mm, slightly larger than NM, is considered to give better recovery in bright, friable coal seams.

Open hole drilling at 100 metre grids in the Hunter Valley No.1 area was carried out by diesel powered Fox Mobile rigs with integral mounted air compressor. Tungsten set blade bits were almost always sufficient in the soft formations drilled; rarely was it necessary to switch to tricone bits in hard sandstone formation.

Altogether a total of over 350 slim cored bores (51mm diameter) have been sunk in the whole exploration area, together with almost 500 rotary open-holes (114mm diameter) in the Hunter Valley No.1 Mine area. Also 23 large diameter cored bores (203mm diameter) have been sunk at various locations in the exploration area to date.

GEOLOGY

General

The geology of the Singleton-Muswellbrook District is described in detail by Britten (3) with special reference to the area concerned on pp.195-203. Exploration within the Company's Authorisation was confined to the sequence of seams of the Jerry's Plains Sub-Group of the Wittingham Coal Measures, which form the lower of two coal measure groups of the Singleton Coal Measures (Super Group) of Upper Permian Age (Fig.2).

The property is traversed in its northern section by the axis of the Muswellbrook Anticline, which structure controls the dips of all seams encountered. With the exception of a small area to the extreme north, where there is a south-easterly dip incorporating a shallow syncline with low dips, the complete sequence dips south-westerly, grades ranging from 1 in 15 in the south-west to 1 in 30 and less in the north-east.

Igneous intrusion is confined mainly to sparse doleritic dykes, which are believed to be of Tertiary age, and there is a small sill stratigraphically located around the top splits of the Blakefield Seam to the west centre of the property. Faulting has been encountered in drilling and has been interpreted as slight to moderate normal and reverse with displacements generally less than 20 metres.

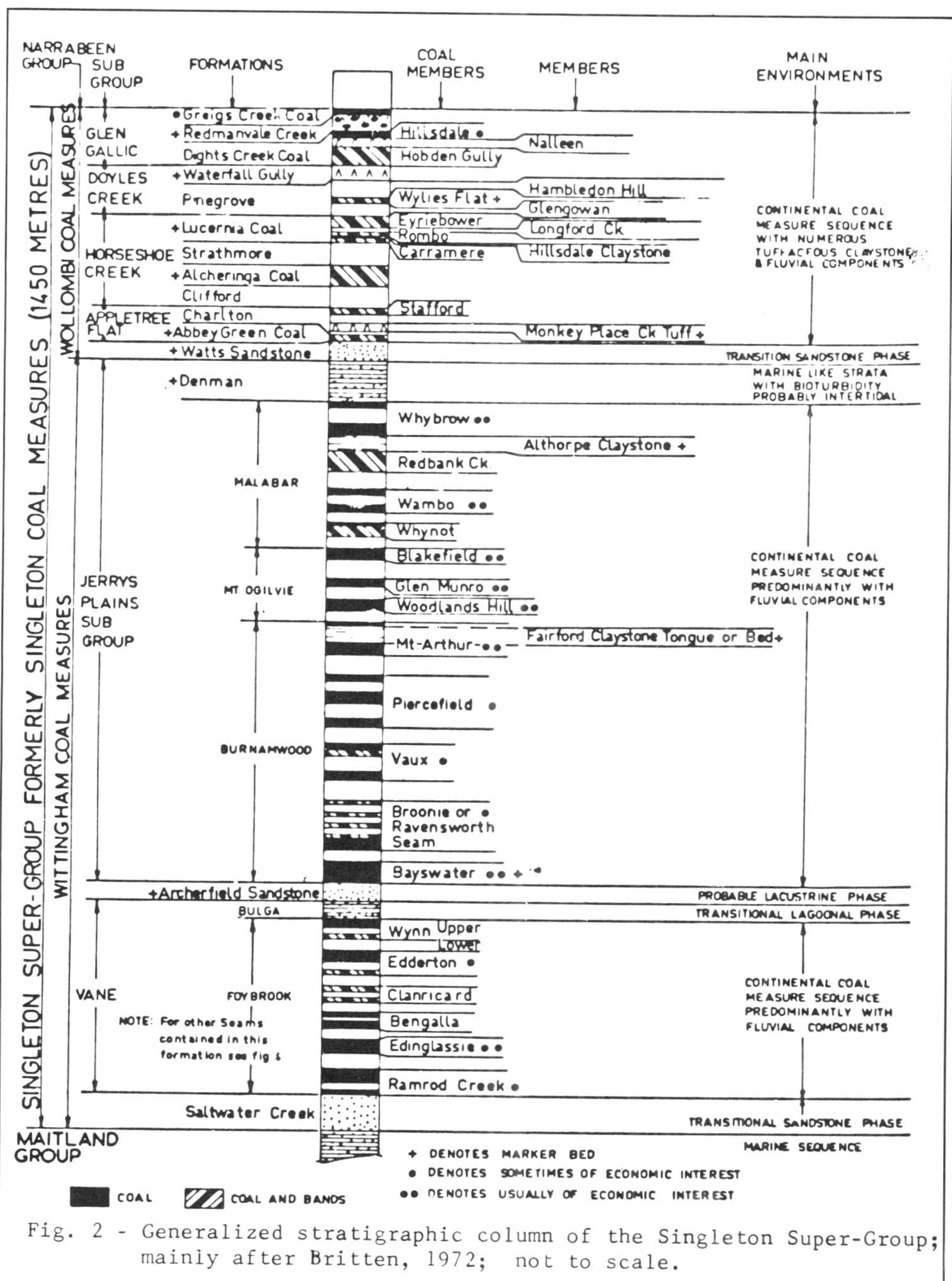

Fig. 2 - Generalized stratigraphic column of the Singleton Super-Group; mainly after Britten, 1972; not to scale.

Coal Seams

The first cored bore was drilled in the south-west corner of the property commencing at the top of the Denman Formation and penetrating all seams of the Jerry's Plains Sub-Group. The stratigraphic interval approximates 400 metres from the top of the Whybrow Seam to the base of the Bayswater Seam. The plan of seam outcrops in relation to the two mines to be developed is shown in Fig. 3; only those seams above the Bayswater Seam will be available for exploitation. A cross-section of the measures from north-east to south-west is shown in Fig. 4.

The coal seams to be worked in Hunter Valley No.1 Mine to the north of the Hunter River occur towards the base of the sub-group and comprise the Mt Arthur, Piercefield and Vaux Seams. They occur in a stratigraphic interval of 17 metres in which the intervals between them total 4 metres only. The deepest bore in the centre of this mine area was 48 metres and contained 13 metres of coal. The centre part of the area is a low plateau with minor stream dissections about its flanks which progressively remove the overlying coals by dissection of the slopes producing the general semi-circular outcrop pattern shown.

The Hunter River (a major Australian east coast stream) and its flood plain traverse the area between the Hunter Valley No.1 Mine and Hunter Valley No.2 Mine areas. At shallow depth beneath this river the Mt Arthur, Piercefield and Vaux Seams dip gently to the south. A total of more than one hundred million tonnes of coal occur in this area, the maximum extraction of which could be obtained by straightening the river from the loop shown in Fig. 3. The N.S.W. Government is investigating the feasibility of mining coal generally in this area and the environmental implications of the various feasible mining strategies.

Hunter Valley No.2 Mine area contains thirteen seams ranging in thickness from 1 to 7 metres. The seam outcrops, as will be observed in Fig. 3, reflect gradually increasing south-westerly dips intersecting a gently undulating land surface.

In the north-eastern corner of the No.2 Mine area the Mt Arthur Seam outcrops followed by the Warkworth, Woodlands Hill, Glen Munro, Blakefield, Unnamed Seams A, B, C, D (probably splits of the Blakefield) and the Whynot and Wambo Seams. Conglomerate wedges thicken up the section

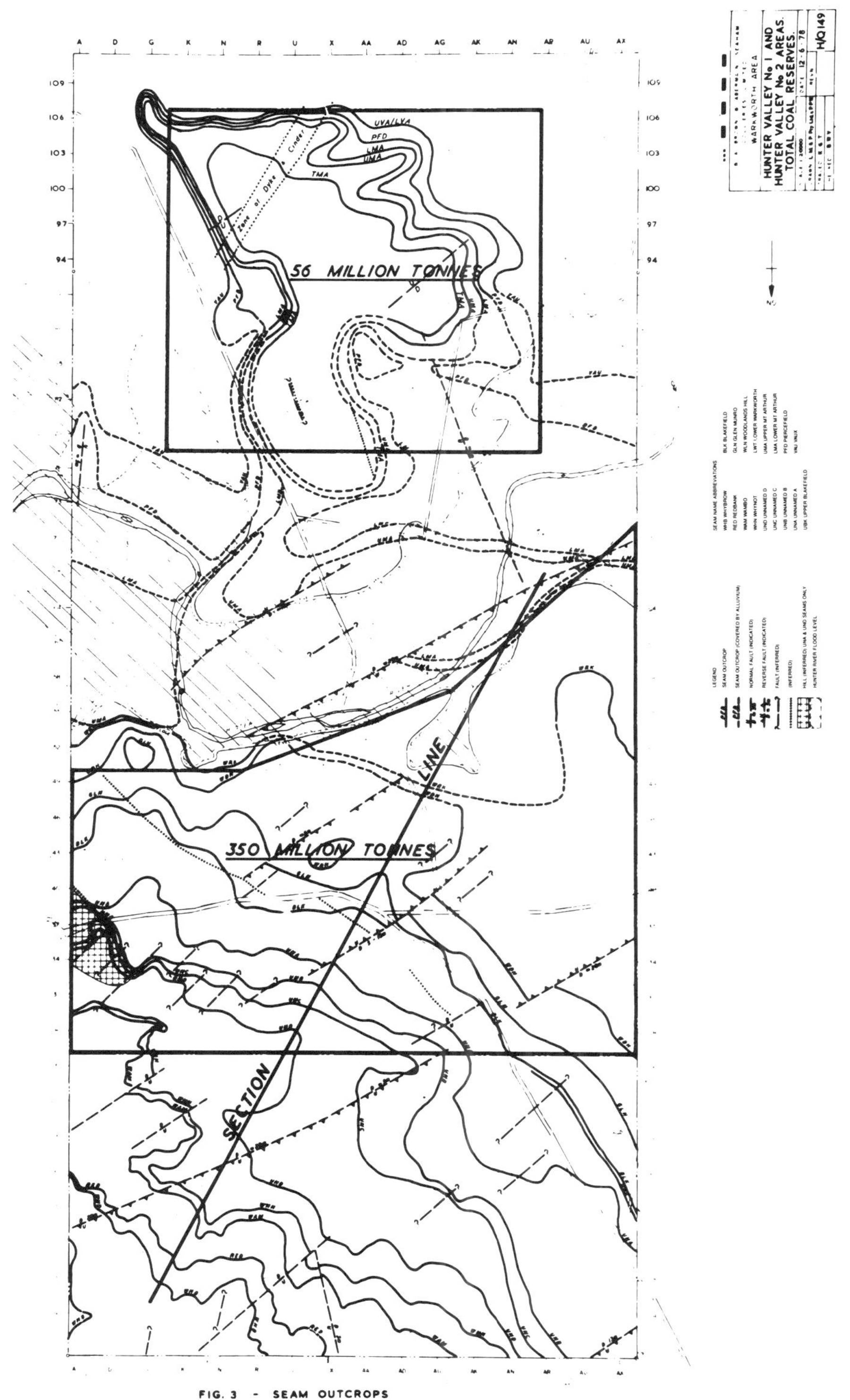

FIG. 3 - SEAM OUTCROPS

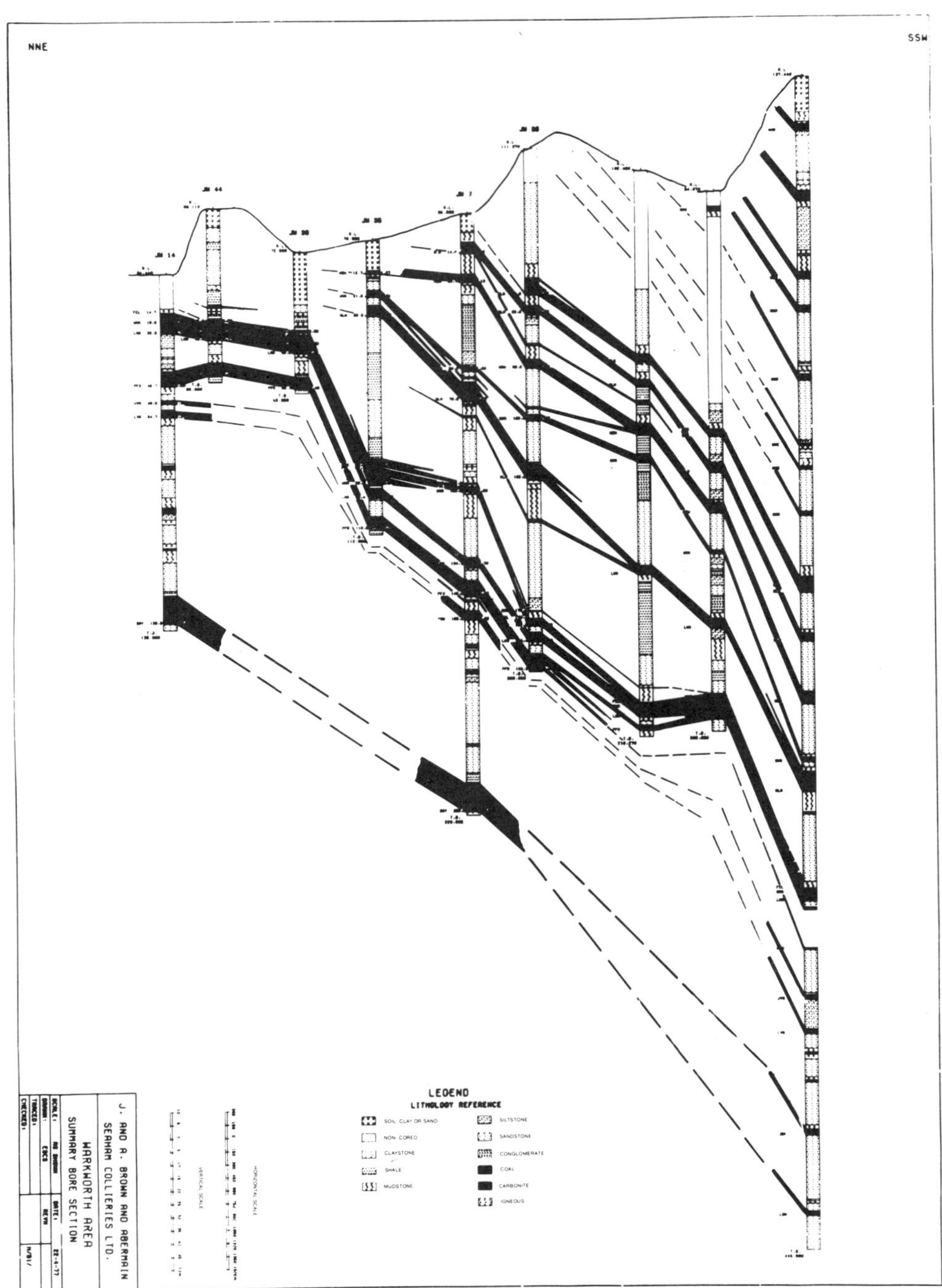

Fig. 4 - Seam Section

to the east between the Mt Arthur, Warkworth and Woodlands Hill Seams.

Near the south-west corner of Hunter Valley No.2 Mine a bore which encountered the Wambo Seam at depth 26 metres penetrated the base of the Lower Vaux Seam at 292 metres. The workable coal content of this bore was 30 metres.

ANALYSIS AND TESTING METHODS

Slim Bores

Coal seams in excess of 1 metre thick from cored drill-holes were subjected to extensive analysis and testing according to the relevant I.S.O., Australian, British or A.S. T.M. Standard, by an independent laboratory. Accepted procedures (international or domestic) were used for analysis and testing not covered by published Standards.

The coal cores were transported to the laboratory after having been logged for brightness content in the field. As work progressed beyond the early holes drilled, the field geologists became more adept at uniformly logging sub-sections or plies within the one coal seam from bore to bore.

This very important attribute has allowed a close correlation of coal seams, or sub-sections of them, throughout the entire area drilled, even to the extent of correlating the morphology of some coal plies which split away from the "main" seam and transgressing the section between, join with other seams above or below. It has also allowed uniformity of analysis within each seam.

Coal cores from the 800 metre grid drillholes have been subjected to the most intensive analysis detail.

Combining, where necessary, thinner plies or sub-sections into groups, the initial ply analysis undertaken was float/sink separation of the 25.4mm by 3.18mm and the 3.18mm by zero size fractions at nominated densities of 1.40 and 1.60. These separate samples, which allow various working section recombinations of coal, middlings and reject products, together with crucible swelling numbers determined thereon, were used as a first guide to the assessment of marketable coal potential. Because of the small top size of sample (25.4mm), the results are not used as a basis for coal preparation plant design.

Following part recombination, the coal plies within each seam were analysed for Moisture, Ash, Volatile Matter and Fixed Carbon (Proximate Analysis), together with apparent Relative Density, Specific Energy (Calorific Value), Crucible Swelling Number and Total Sulphur. Stone bands within the coal seam were analysed for apparent Relative Density and Ash only.

It was on this basis that variations of possible working sections in each coal intersection were considered. Predetermining working sections within a coal seam for analysis, although cost saving, often leads to a short-fall of information when detailed mine planning follows.

Those seam plies which were considered as the most likely working section were then, in combination, analysed chemically as above on an "ex-bands" and a "Floats 1.60 R.D." basis. The selection of a nominated density as a basis for this and further analysis was made with a general understanding of the coals being tested. There is a restricted middlings fraction in most of the coals of the area and although some minor upgrading of caking and steaming characteristics was obtained at "Floats 1.40 R.D." for the same sizing, it did not compensate for the reduction in yield. In addition to the analysis detailed above, the following analyses were also carried out on the combined "Floats 1.60 R.D." samples of the possible working section:-

Ultimate Analysis
Analysis of Ash Constituents
Ash Fusion Temperatures
Hardgrove Grindability Index
Chlorine
Phosphorus
Maceral Analysis
Vitrinite Reflectance
Gray-King Coke Type
Giesler Plastometer
Audibert-Arnu Dilatometer
Roga Caking Power

This comprehensive assessment of the coals of the Warkworth Area has given a very clear indication of their marketing potential.

The 300 metre grid cored drilling was undertaken on groups of seams to upgrade the data for reserves and quality. These groups were:-

(a) Vaux, Piercefield, Mt Arthur and Warkworth Seams,
(b) Woodlands Hill, Glen Munro, Blakefield and Unnamed A, B, C, D Seams,
(c) Whynot, Wambo, Redbank Creek and Whybrow Seams.

As a function of economics and to enable integration with subsequent pit quality control, analysis and testing from seams in these drillholes was restricted to ply-by-ply moisture, ash, relative density and crucible swelling number for coal and relative density and ash for stone bands. Composite analysis of the potential working sections was restricted to Moisture, Ash, Volatile Matter and Fixed Carbon (Proximate Analysis) together with Crucible Swelling Number, Specific Energy and Total Sulphur.

Large Diameter Bores

At ten locations a total of 23 large diameter cored bores (203mm diameter) were sunk alongside slim bore sites to give larger samples of all the seams occurring in the exploration area. These sites were chosen to intersect all seams and their splits at points where they were typical of an area, to give a complete coverage of the exploration area. Depending on the amount of coal at each location and on the purpose for the bore hole, between one and five bores were sunk at any one location.

In each case the core was taken, after logging, to the laboratory for testing and analysis. The cores for any one seam, or seam split, at any one location were bulked, subdivided into two bulk samples and the bulk samples broken by dropping from a height of 3.05 metres until the top size of one sample was 101.6mm and the other 31.8mm, to give a natural breakage distribution. These top sizes were chosen to allow assessment of liberation effects on yield and quality.

After screening the coal was tumbled in water with steel cubes in accordance with Australian Standard AS1661-1977 (4) and then screened on a 0.5mm wedgewire screen, the minus 0.5 mm material being tested using a continuous laboratory froth flotation rig and the plus 0.5mm being float/sink tested at relative densities from 1.30 to 2.00 in size fractions.

Float/sink results were used to prepare yield/ash curves after applying Ecart Probable Partition Factors to allow for the imperfections of commercial separation in the washing equipment to be used in the coal preparation plant. A composite simulated washed product was then produced using this data and results of the froth flotation tests.

The simulated washed product was subjected to detailed analysis for all of the tests listed above for the "Float 1.60 R.D." slim bore samples plus the following analyses:-

Forms of sulphur
Fluorine
Arsenic

In addition, several tests were separately done on the coals of coking and steaming potential.

Coals of coking potential were coked in various blends, some containing other coking coals produced and exported by the Company. Most of this work was done in a 7 kilogram (kg) pilot coke oven, with some confirmatory work in a 230 kg pilot coke oven (5). The cokes were tested to yield Japanese Drum, Micum, Irsid and ASTM Tumbler indices. Also, reflectance histogram (or vitrinite reflectogram) was determined to allow calculation of indices such as Composition Balance Index and Strength Index (6) used in blending assessments by some coking coal consumers.

Coals of steaming potential were subjected to a wider range of additional analyses. The Yancey, Gear and Price test was done to assist in assessments of hardness for crushing and grinding. Also a number of fly ash electrostatic precipitation tests (7) were carried out in the laboratory to allow comparisons to be made with other Company coals used in known power station situations. Also trace element studies (8) were carried out using both atomic emission spectrographic and neutron activation techniques to assess the levels of some 55 trace elements.

Samples of simulated washed products were also sent to potential consumers of both coking and steaming coals around the world for their assessment and, wherever possible, this data has been obtained to correlate with data generated in Australia and to expand knowledge of the coal's characteristics. These samples have range from a few kilograms to almost a tonne.

Shaft Samples

A 2 metre x 1 metre shaft has been sunk, using contractors, in the Hunter Valley No.1 Mine area to provide about 35 tonnes of coal for several purposes. The coal at this location was considered typical of the "average" coal quality to be supplied from 1979 onwards.

Primarily the shaft was sunk to provide bulk samples for testing around the world by potential consumers and/or their consultants; these samples range in size from a few kilograms to several tonnes. Again, results of these tests have been obtained whenever possible to improve Company knowledge of the coals. Larger samples may be required in some cases as a means of final assessment and these will be provided from the initial mining operations, as they may range from 100 tonnes to as much as 50,000 tonnes.

These bulk samples were prepared by washing the coal in a pilot coal preparation plant (9) through commercial equipment and therefore represent a truly simulated "commercial" washed product. Previous samples had all been prepared using float/sink in organic liquids, as is widely practised.

The second reason for sinking the shaft was to determine the optimum top size to crush the Mount Arthur seam prior to washing for liberation. Bore core testing had shown a reasonable proportion of middlings material to be present in the coal and that it was better to separate the coal at 31.8mm top size than at 101.6mm top size. A representative portion of Mt Arthur seam was subdivided into four representative samples and reduced to each of 101.6mm, 76.2mm, 50.8mm and 31.8mm top size before float/sink analysis and production of yield/ash curves, with allowance for commercial washing imperfections. Assessment of this data allows a top size to be selected which takes advantage of liberation and which maintains the largest size to reduce total moisture of the product and minimises the percentage of fine coal which feeds to the coal preparation plant.

COMPUTER ASSESSMENTS

When the Stage I (800 metre grid) exploration was about half complete and Stage II (300 metre grid) exploration had commenced on seams in the south-west of the property, it was realised that so much information would be obtained that computerisation of base data for retrieval and plan drawing was desirable.

Accordingly, after initial assessment was made of the Company's own computer facilities and they were found not to be entirely suitable for the purpose, a contracting organisation, who had expertise in the subject was engaged. Bore log and analytical data was initially encoded from field logs and laboratory reports, but as work progressed field logging was written direct onto encoding sheets made up for the purpose, taking into account the Company's needs.

The data from the bore log and encoding sheets was processed using the COALBOR system developed by the contractor.

After checking for coding validation, initial output from the system consisted of -

a) narrative bore log print outs suitable for report presentations;
b) a variety of geological presentations of the drilling data in the form of graphic logs of bore lithology and seam brightness, plotted automatically on a high resolution Calcomp plotter. These sections were produced directly from the data base.

Modifications to seam definition and correlation were performed directly to the data base and, when desired, updated logs and graphics were produced.

When the data acquisition phase was complete, data was extracted from the data base on a seam by seam basis for the automatic production of contour plans and their associated grids. The contour plans included a variety of digitised information including seam crops and cadastral information as well as designated bore names and values.

The grids from topography, structure, isopach, ash, density and other relevant analytical data are stored to form a integrated digital model of the deposit.

These grids were then manipulated for the generation of overburden maps, cumulative overburden/coal ratios etc., as well as to form the basis for reserve and overburden calculations.

Reserves and volumes were calculated from grids using modified areas nominated by digitising outlines of them (e.g. by land owner, crop line or mining area). Reserves outputs were produced as tables classified by overburden and coal thickness with ash and other constraints as appropriate. The tonnages were also presented as block maps showing the features of the coal and overburden on a block by block basis over the deposit.

These grids were also used as the model for strip simulation and production scheduling for the deposit.

Complete suites of sections and iso-plans by seams for Hunter Valley No.1 and No.2 mines include the following:-

- Surface Contours & Cadastral Boundaries
- Seam Outcrops, Faults, Dykes and Split Lines (exceeding 0.6 metre)
- Bore Locations & Key to Section Lines
- Graphic Borelog Sections
- Individual Seam Sections
- Limit of Saleable Coal
- Depth to Roof
- Cumulative Coal Thickness to Floor
- Net Cumulative Overburden/Coal Ratio
- Thickness of Seam
- Thickness of Seam Split
- Structure Contours on Floor
- % Thickness Stone Bands
- Ash % (db) - Ex Bands
- Volatile Matter % (daf) - Ex Bands
- Crucible Swelling Number - Ex Bands
- Gray-King Coke Type - Ex Bands
- Total Sulphur % - Ex Bands
- Specific Energy (MJ/kg) (daf) - Ex Bands
- Interval to Next Seam

COMMERCIAL QUALITY

Computer drawn iso-quality maps from the bore core testing, data generated from the large diameter bore cores and the shaft samples were taken into account, in conjunction with the mining plans, in deciding on commercial specifications. As Coal and Allied has been a coal producer since the last century and operations have traditionally been based on long term contracts it was important that the specifications could be supplied over a long term, e.g. over the life of a power station.

Specifications were reached with confidence, as the whole of the area to be mined had been systematically explored and tested and the data generated had been fully recorded and processed. These specifications are listed in Tables I and II for coking and steaming coals respectively. The specifications are conservative and will accommodate variations in the proportions of the thirteen seams occurring in the exploration area in the commercial products.

The coking coal specification represents a high volatile coal of low ash, sulphur and phosphorus which would be a valuable component in a blend for the production of metallurgical coke by conventional means or using briquetting techniques. This coal is particularly low in ash compared with other Australian high volatile coking coals (10).

The steaming coal specifications represents a high quality, low ash, low sulphur, high specific energy (calorific value) coal suitable for electricity generation, cement manufacture, formed coke and a range of general industrial applications.

CONCLUSIONS

An area of some 900 million tonnes of measured and indicated coal reserves has been systematically explored and tested near Warkworth, N.S.W., Australia.

Mining operations will commence in this area in 1979 in the Hunter Valley No.1 Mine and later in the Hunter Valley No.2 Mine. These mine areas contain 406 million tonnes of the original exploration areas.

Coals produced will be high volatile coking and steaming coals of high quality from the thirteen seams occurring in the area.

REFERENCES

1. STANDING COMMITTEE ON COALFIELD GEOLOGY OF N.S.W., Code for Calculating & Reporting Coal Reserves. Records, Geol. Surv. of N.S.W., Vol. 16 Pt. 1, March 1975.

2. JOINT COAL BOARD OF N.S.W., Estimation of Reserves - Open Cuts. Geological Standard 521 (unpublished), September 1956.

3. BRITTEN, R.A.,
Singleton - Muswellbrook District.
in Economic Geology of Australia and Papua New Guinea.
2. Coal Monograph Series No.6.
The Australasian Institute of Mining & Metallurgy 1975.

4. STANDARDS ASSOCIATION OF AUSTRALIA,
Float and Sink Testing of Hard Coal AS1661-1977

5. BURKE, L.O.,
Assessing Coals and Blends for Coke Making using
Small Scale Coke Ovens and a Modified Micum Drum Test.
ACIRL Report P.R. 75-1, April 1975.

6. JOINT COAL BOARD AND AUSTRALIAN COAL INDUSTRY
RESEARCH LABORATORIES LTD.,
Blending Australian Coals with Japanese Indigenous
Coals to meet the Metallurgical Coke requirements of
Japanese Steel Mills 1978.

7. BAKER, J.W. and SULLIVAN, K.M.,
A Strategy for Assessing the Requirements for Trapping
Power Station Fly Ash.
CSIRO Conference on Electrostatic Precipitation,
August 1978.

8. PORRIT, R.E. and SWAINE, D.J.
Mercury and Selenium in Some Australian Coals and Fly
Ash,
The Institute of Fuel (Australian Membership) 1976
Conference, Paper 18.

9. POLLARD, F.,
Developments in Coal Preparation Research,
Mine and Quarry Mechanisation 1969.

10. EDWARDS, G.E.,
Marketable Resources of Australian Coal.
Australian Black Coal Symposium, Aus.I.M.M., 1975.

COKING COAL

This coal represents a high quality, low ash, high volatile coking coal suitable as a blend component for the production of metallurgical coke by conventional means or using briquetting techniques.

Specification: Size 50 mm x 0

Total Moisture (%)	8.0
Air Dried Analysis (%):	
Moisture	3.0
Ash	7.0
Volatile Matter	35.0
Fixed Carbon	55.0
Sulphur	0.5
Crucible (Free) Swelling Index	5

Indicative Qualities:

Gray-King Coke Type	G
Roga Index	65
Audibert-Arnu Dilatometer:	
Initial Softening Temp. (°C)	365
Max. Contraction Temp. (°C)	430
Max. Dilatation Temp. (°C)	460
% Max. Contraction	30
% Max. Dilatation	−10
Gieseler Plastometer:	
Initial Softening Temp. (°C)	395
Max. Fluidity Temp. (°C)	425
Resolidification Temp. (°C)	450
Max. Fluidity (dial div/min)	100
Petrographic Analyses:	
(a) Maceral Analysis (% by volume)	
Vitrinite	70
Exinite (i) Resinite	1
(ii) Sporinite	1
(iii) Cutinite	trace
Micrinite	4
Semifusinite	22
Fusinite	trace
Minerals	
(i) Clay	2
(ii) Carbonate	trace
(iii) Pyrite	trace
(iv) Quartz	trace
(b) %Mean Max. Reflectance of All Vitrinite (Measurements at 546 nm light wavelength in oil of 1.518 refractive index)	0.71

% Phosphorus (ad)	0.01
Hardgrove Grindability Index	58
Ash Fusion Temp. (°C)	1600
Gross Calorific Value (kcal/kg)	7,400

Analysis of Ash Constituents (%):	
SiO_2	67.0
Al_2O_3	24.0
Fe_2O_3	3.0
TiO_2	2.5
CaO	0.9
MgO	0.5
Na_2O	0.3
K_2O	1.0
P_2O_5	0.5
Mn_3O_4	0.1
SO_3	0.2
SiO_2: Al_2O_3 ratio	2.79

Ultimate Analysis (% d.a.f.):	
Carbon	84.0
Hydrogen	5.3
Nitrogen	1.7
Oxygen (by difference)	8.5
Sulphur	0.5

Table I. Hunter Valley Coking Coal Quality

STEAMING COAL

This coal represents a high quality, low ash, low sulphur, high calorific value steaming coal suitable for electricity generation, cement manufacture, formed coke and a range of general industrial applications.

Specification: Size 50 mm x 0

Total Moisture (%)	8.0
Air Dried Analysis (%):	
Moisture	3.5
Ash	11.2
Volatile Matter	34.0
Fixed Carbon	51.3
Sulphur	0.6
Gross Calorific Value (a.d.)	7,075 kcal/kg 12,737 Btu/lb

Indicative Qualities:

% Phosphorus (ad)	0.02
% Chlorine (ad)	0.05
Crucible (Free) Swelling Index	2-3
Hardgrove Grindability Index	55
Yancey, Gear and Price Index	68
Abrasive Index (mg/kg)	17
Forms of Sulphur (% ad):	
Pyritic	0.10
Organic	0.45
Sulphate	0.05
Ash Fusion Temperatures (°C):	
(a) Oxidising Atmosphere	
Deformation	1550
Spherical	1580
Hemisphere	1600
Flow	+1600
(b) Reducing Atmosphere	
Deformation	1500
Spherical	1540
Hemisphere	1570
Flow	+1600
Petrographic Analyses:	
(a) Maceral Analysis (% by volume)	
Vitrinite + Exinite	70
Inerts	30
(b) % Mean Maximum of All Vitrinite	0.70

Analysis of Ash Constituents (%):	
SiO_2	68.6
Al_2O_3	24.0
Fe_2O_3	2.7
TiO_2	1.1
CaO	0.7
MgO	0.6
Na_2O	0.5
K_2O	1.1
P_2O_5	0.3
Mn_3O_4	0.1
SO_3	0.3

Ultimate Analysis (% d.a.f.):	
Carbon	82.2
Hydrogen	5.1
Nitrogen	1.8
Oxygen (by difference)	10.5
Sulphur	0.4

Trace Elements:		
Mercury	<0.05	ppm
Arsenic	1.0	ppm
Cadmium	0.05	ppm
Vanadium	15	ppm
Lead	10	ppm

Table II. Hunter Valley Steaming Coal Quality

DISCUSSION

QUESTION: Were electric logs made and if so how do they match your logging system?

ANSWER: No, the were not made. As a matter of fact, every hole, apart from some initial scout hole was cored as soon as possible after collaring. There is too much coal.

QUESTION: How many degrees of "brightness" of coal were logged in the field, and what other features were frequently found to have correlation value?

ANSWER: I think I might have answered this in the paper: 10 degrees. Other things which were of assistance were the drill core radiographs. In other words, the coal seam is X-rayed before our log is taken--is transfered from the core barrel to aluminum split cylinders, taken directly to the laboratory and a direct X-ray film is made of the core. This often aids in proving if subdivision or analytical subdivision is correct. It does also provide information about possible washability or breaking characteristics to the coal, and the distribution of ash within the coal, i. e. whether it's discrete or finely divided in the coal. Other things that have been used are analytical data. At times the swelling number; in Australia, I think it's used a lot. Generally, you get a straight line relationship between brightness and swelling number, When it doesn't, we think someone has made a mistake in logging the coal. Sulfur, at times, was used. We have a range from 0.3 to 0.8. But within a seam, the range is less, plus or minus about 0.15. And so, we could use this as a correlation tool. Analysis of the ash constituents is sometimes used. This may have indications of what was going on at the time of deposition. But these additional features were very secondary when compared to the field data of the coal bore log.

QUESTION: Were petrographic determinations utilized in your study of in-seam characteristics? That is, were "brights determined by visual examinations or by maceral examination?

ANSWER: No. Maceral determinations were only done on selected samples of Floats 1.60 R.D.

QUESTION: Could you comment on the expected product quality of Hunter Valley coal?

ANSWER: Well, besides the normal slim coal drilling, something like twenty or thirty 200 millimeter holes were drilled in the area. The mass of the sample used was considered to be consistent with the recommendations of the Standards Association of Australia. On this basis the marketing figures as presented in the paper were determined. We feel quite confident that these are right. In addition we have just completed, a 1 by 2 meter shaft through the middle of Hunter Valley No. 1, which will produce I think a five-ton sample for additional work.

QUESTION: Do you incorporate drill hole washability data from the various seams into reserve calculations and annual production plans?

ANSWER: We are, as a matter of fact, in the process. When I left, the planning engineer for Hunter Valley No. 1 was just starting on this exercise. But he got a very strongly worded warning that those washability figures from slim cores are very, very doubtful, but it will give him a figure of reserves that is better than nothing. As soon as there is any development in the mine, there will be further tests to revise these figures.

QUESTION: What is considered to be oxidized coal and how is it controlled and projected?

ANSWER: We've taken a very "off-the-cuff" determination of the limits of oxidized coal. If we get a 2-unit drop-off in our swelling number in our chip sample (which in some cases might just be due to contamination) we decide that this change is due to oxidized coal, i. e. this is only on the coking basis. From coal analysis people we have spoken to, the steaming characteristics or the Btu value of the coal are not affected until well after the coking properties deteriorate. Generally the control and projected position of the Lox line, of the seam is based upon 100 meter center drill holes, which are drawn up in section lines both north-south and east-west, and the weathering profile is produced by plotting the depth of the top of unweathered coal in each hole and projections are drawn between bore holes.

16

Efficient Australian Practice in Exploration and Evaluation of Coal Prospects

By David Svenson
Projects Manager
R. W. Miller and Company Pty. Limited
Brisbane, Queensland, Australia

ABSTRACT

Over the past 10 years, the Australian Coal Industry has enjoyed an unprecedented growth resulting from the remarkable advance of one of Japan's major industrial bases - the Steel Industry. The consequent exploration effort of our coal industry as a whole, has been sustained with a high standard of professionalism in field, laboratory and the presentation of the co-ordinated results to the world markets for both coking and fuel coal. The industry's exploration effort has been solidly technically supported by both statutory authorities and the industry's own research unit (ACIRL). Advances in field exploration and laboratory evaluation techniques and equipment have been and will continue to be essential in sustaining the effort in the future. In the next decade steam coal will become increasingly important in terms of exploration and evaluation, but as surface mine reserves of quality coking coal become fully developing, attention will again be directed to reducing the costs of exploring the underground potential.

INTRODUCTION

This paper hopefully provides a concise review of standard Australian *black* coal exploration and prospect evaluation practice and the various factors influencing or affecting that practice. The author unashamedly admits to borrowing extensively from previously published papers including those of some notable Australian coal *explorationists* and *evaluators*. Here at the beginning of the paper, the author emphasizes a preference for such terms. All executive members of any group involved in coal exploration should have a sound knowledge of and familiarity with coal science, mine development planning criteria and mining economics, if they are to be effective members of an interdisciplinary evaluation team.

The paper briefly points to, and discusses the sample pretreatment procedures developed in Australia to enable closer simulation of drill core laboratory test data with results achieved in full scale coal production.

THE EASTERN AUSTRALIAN COALFIELDS

The distribution of the major coal bearing sedimentary basins and coal mining areas in Queensland, New South Wales (NSW) and Victoria is shown in Figure 1.

The history of exploration and development of the coal resources in these States and fairly detailed descriptions of the regional geological settings and of the individual coalfields are given in Volume 2 *(Coal)* of the *Economic Geology of Australia and Papua New Guinea* (1, 2). Broad reviews of these coalfields and type/rank variations in Australian Coals are also given in *Australian Black Coal* (3, 4, 5, 6). Thus the geology of these coalfields is not discussed in detail herein.

The oldest and most significant economic coal measures in Australia occur in the Permian Bowen (2, 7) and Sydney Basins (4, 8). These basins contain practically all of the coking coal resources in Australia. Extensive reserves of low rank high volatile coal have been discovered in the Permian Galilee Basin (9), a western extension of the Bowen Basin.

Mesozoic coal bearing sedimentary basins provide

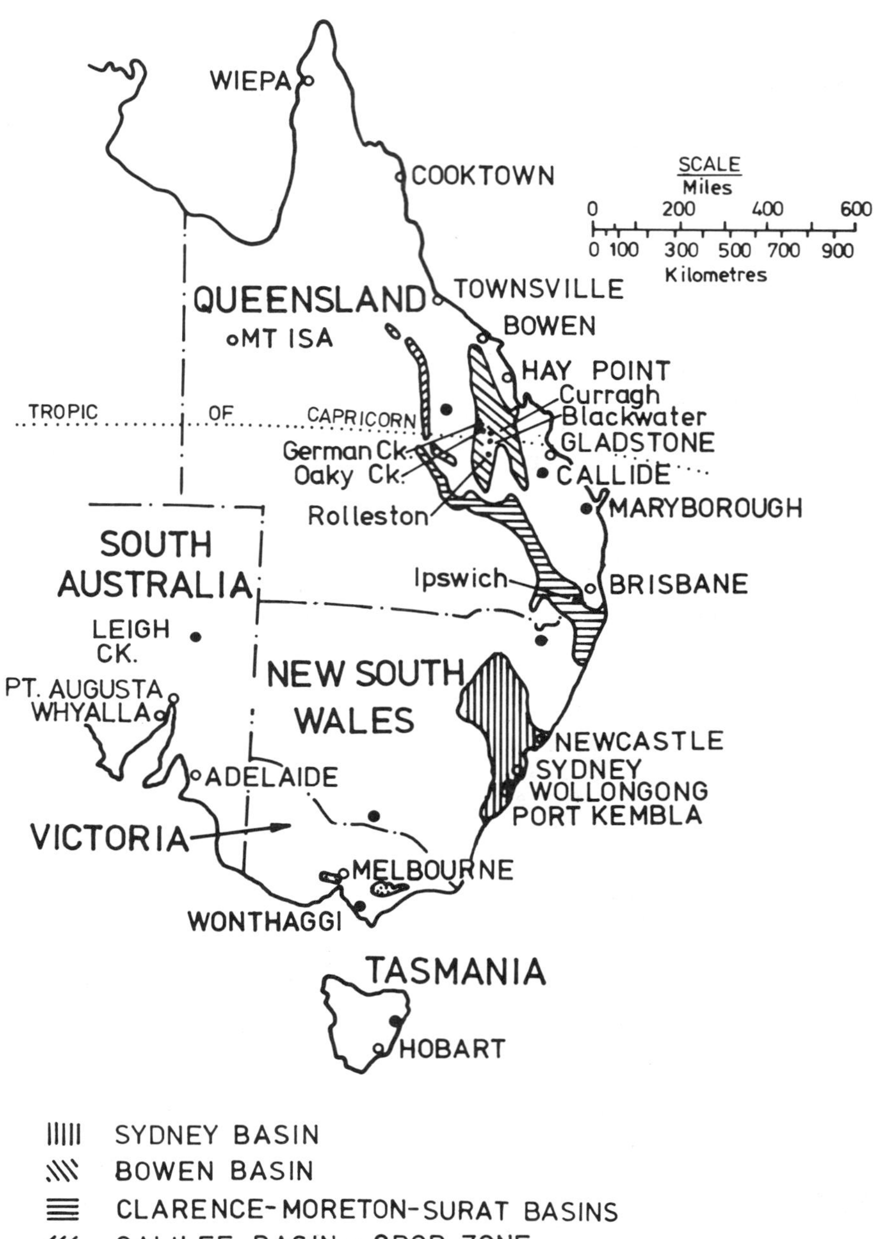

Figure 1 - Eastern Australian Coal Deposits

significant reserves of low rank coal peripheral to their margins. These basins include the Surat (10), Callide (11) and Maryborough (12) Basins in Queensland and the Clarence-Moreton Basin (13, 14) which extends through the NSW-Queensland border. Minor underground reserves of Mesozoic bituminous coal occur in Victoria (15). These, once mined to provide railway steam coal, are at present not economic.

Tertiary coals of current economic significance are confined to the Gippsland Basin and minor extensions of the Gippsland Basin in southern and south-eastern Victoria (16, 17, 18, 19). These coals are extremely low rank brown coals or lignites; however, the very large reserves and shallow disposition in thick seams make these resources a major source of power generation and alumina smelter fuel with a current total annual production exceeding 24 000 000 tonnes. The author is not familiar with coal exploration in Victoria and Victorian practice is not discussed herein.

GEOLOGIC-GEOGRAPHIC INTER-RELATIONS

The effects of topography on both coal mining development and exploration is well illustrated in the central and southern parts of the Sydney Basin where the Permian Coal Measures are overlain by a thick sequence of Triassic strata, mainly massive sandstones and open cut development is thus precluded. Most of the mining development has thus occurred from drift entries along the steeply dissected valley walls. In these conditions, exploration is inhibited by the need to penetrate relatively hard sandstones to depths exceeding 300 metres (1 000 feet) and by the relative inaccessibility of the ground for drill rigs or seismic crews.

In the Hunter Valley, the topography is mature and gentle, formed on the coal measures which have been stripped of their harsh cover of Triassic sandstone. Here several coal bearing sequences provide opportunities for both surface and underground mine development, with exploration tailored to suit the needs of each type of mining. For many years, surface mine development was discouraged in the Hunter Valley, as a matter of Government policy. The prime purposes of such policy were to protect the economic viability of existing underground mines which could not have withstood competition from low cost open cut development, and to maintain a high rate of employment in the coalfields. Thus when open cut pits scarred the countryside in the late nineteen forties to overcome fuel coal shortages, they were promptly closed when the emergency ended and not for reasons of environmental concern! Government policy has since abruptly changed, to meet the changed circumstances brought about by rising underground mining costs and the need to ensure that coal reserves were not wastefully *sterilized* by inappropriate mining methods. Since a revised policy statement (20) was issued in 1975, surface mining is actively encouraged (subject to constraints imposed by current political concern for the environment), to ensure that coal mining recovery is maximal.

In the larger sedimentary basins of Queensland, the topography along the basin margins, in the crop zones of the coal measures, tends to encourage rather than inhibit exploration for surface mine coal reserves although other factors provide impediments to drilling. In the Bowen Basin these include:

. extensive areas of *black soil* which become impassable for vehicle movement in the summer wet season,

. areas of Tertiary gravel sands either unconsolidated or silicified which make drilling difficult and expensive,

. areas of surface basalt cover which increase overburden depths beyond economic limits,

. areas where coal measures are extensively intruded and coal seams irregularly destroyed, making the need for close drilling imperative, while increasing drilling time and cost.

The Triassic, Callide Basin in Central Queensland exhibits relatively rugged topography also due to erosion resistant sandstone members and although the thick coal seams tend to counter the adverse effects of such conditions on surface mining potential, drilling conditions are relatively difficult. In the Triassic Coal Measures of the Ipswich Coalfield near Brisbane, residential and industrial development has tended to prohibit surface mine development although surface mine potential is limited. Exploration likewise tends to be constrained, but in most Australian coalfields, such constraints to exploration are rare.

MINING LEGISLATION & STATUTORY AUTHORITIES IN NSW & QUEENSLAND

In Queensland all of the coal resources are the property of the *Crown*. Coal mining and exploration is controlled by the State Government through the Department of (the Minister for) Mines under the *Mining Act of 1976-78 and Regulations* and the *Mining Royalties Act of 1974*.

Currently mining royalty is payable at rates of:

. 4 per cent of the value of coal mined for export by underground methods,

. 5 per cent of the value of coal surface mined for export, and

. 5 cents per tonne for coal mined for consumption within Queensland.

The State Government also owns and controls the railways and sets freight rates for rail haulage of all coal. The mining company has to provide all of the capital required for new or upgraded rail lines and all rolling stock and this is repayable at an annual 8 per cent interest rate.

Coal mining tenements include Authorities to Prospect and Coal Mining Leases or Special Coal Mining Leases. The Authority to Prospect for Coal is subject to granting by the Minister who has wide powers under the Mining Act, for determining period of tenure, area and expenditure requirements. However, current policy generally limits the area to 300 nautical square mile *Sub Blocks* (1 Sub Block is bounded by 1 minute of latitude and longitude), to be held for 3 years with or without relinquishment (usually a minimum of one-third the total area held per year) and exploration expenditure arbitrarily fixed generally according to the area's economic potential. Coal mining leases can only be issued, over ground already held as an Authority to Prospect for Coal, to holders or assignees of an Authority to Prospect. Coal mining leases exceeding 300 hectares (1 square mile) are issued as special coal mining leases. The tenure of a coal mining lease is usually for 21 years but is renewable.

The growth of the Japanese steel industry, its search

for relatively cheap sources of coking coal, and the success of Utah Development Company in outlining very large reserves of high quality coking coal, led to an unprecedented *rash* of applications for exploration rights by many companies (mostly genuine explorers!) in Queensland. To rationalize coal exploration and development and to provide a sound inventory of the State's coal resources, the Queensland Government, in the early 1970s, reserved the most prospective coal areas and intensified its own program of exploratory drilling. Selected areas have been released from reserve and awarded to companies, on the basis of detailed tenders. The granting of an area depends on the quality of the tenderer's proposed exploration and development program and the amount of expenditure proposed for exploration and development, and not on a competitive cash bid basis. The Queensland Government actively encourages overseas investors, as a matter of policy, to foster economic development, diversification and decentralization. However, to maintain local equity as high as possible, the Australian (Federal) Government may use its export control powers to limit overseas equity in resource development.

To ensure compliance with the terms and conditions of Authorities to Prospect for Coal, a substantial cash interest bearing deposit is usually required to be submitted with yearly rentals payable in advance to the Department of Mines. Rentals on new Authorities are currently $15/Sub Block/year. Very substantial security deposits are usually required from the holders of a coal mining lease to ensure compliance with lease conditions, especially the requirements regarding reclamation and environmental protection.

The Geological Survey of Queensland (GSQ) has a relatively large and active coal geology group and has provided an excellent service to both the coal industry and the electric power generation authorities which are State owned public utilities throughout Australia (2). This *Coal Section* of GSQ was formed in 1950, following the recommendation of Powell Duffryn Technical Services in a monumental report to the Queensland Government in 1949.

A continuous assessment of the State's coal resources is being made by GSQ through its monitoring of private company exploration and the conducting of its own reconnaissance exploration, aided by the Drilling Branch of the Department of Mines.

The scope of Departmental coal exploration drilling in Queensland is indicated by the fact that between 1950 and 1976, some 3 689 holes were drilled, for a total meterage of 603 044, most of which was core drilling.

In conjunction with the Australian Government Bureau of Mineral Resources, Geology and Geophysics (BMR), the GSQ and the NSW Geological Survey have conducted regional mapping programs for the production of 1:250 000 map sheets with explanatory notes and these have provided the regional geological framework for the planning of detailed coal exploration programs by both the State organization and private enterprise.

Although not directly concerned with coal exploration, BMR geologists have provided useful data in the form of major reports and monographs on regional geological aspects and a BMR published Map of the Bowen Basin which although now out of date, continues to be widely used.

In NSW, while most of the in situ coal is owned by the Crown, significant areas of coal are privately owned and mining companies have to pay the royalty (currently $1/tonne) to the private owner. Exploration rights in the form of Licences, prospecting or permit areas may be granted by the Minister for Mines to an established coal mining company to replace a mined out, relinquished or depleting area, or tenders may be called over areas released following a Government exploration program. While the successful tenderer must propose a comprehensive and rational exploration and development program, in NSW success is evidently gained on the basis of the estimated maximum cash return to the Government in the form of cash and super royalty payments.

Although not as large as the GSQ coal section, the NSW Geological Survey's coal group is responsible for definition of the regional coal geology of the State and for providing the necessary background data for determining NSW Government policy in relation to exploration and development of coal resources. The NSW Geological Survey have mounted exploratory drilling programs throughout the State in the continuing assessment and evaluation of NSW coal resources. One such drilling program has recently confirmed significant surface mine reserves at shallow depth near Gunnedah in northern NSW - the most northerly extent of Permian Coal Measures in NSW.

Another feature of the NSW coal mining environment is the presence of the *Joint Coal Board* (JCB), a joint Federal-State Government body established under Coal Industry Acts 1946 of the Australian and NSW Parliaments, responsible jointly through the appropriate Ministers to each Government. The functions and powers of the JCB include:

. to ensure coal production to meet the requirements of all local markets and to encourage export,

. to ensure optimum conservation and development of NSW coal resources to the best advantage in the public interest,

. to control production quantities, coal quality and sales prices in the public interest, and

. to promote the welfare of the coal industry work force.

In exercising its functions the JCB has the power to veto or impose special conditions in the granting of coal mining leases, which are usually more difficult to finalize in NSW than is the case in Queensland. Substantial security deposits are also usually required in NSW to ensure control of lease conditions and especially compliance with environmental standards.

While the Queensland Government adopts a pragmatic approach to environmental problems, in NSW both local government and the environmental protection authority can impose difficult conditions, and in both States the Australian Government can impose its will in this area, if the development involves export and it so wishes, through its overiding export control.

The JCB,to fully realize all of its functions in the NSW Coal Industry, established coal geology and fuel technology arms shortly after its creation in the late 1940s. Under the guidance of its first Chief Geologist, K.G. Mossher, and his successor, J.B. Robinson, the JCB has provided a service to the coal industry by making available at cost, the services of its highly trained geologists, particularly for planning and executing exploration programs including the supervision of drilling and the standardized logging of cores, designation and correlation of seams on a regional basis. The JCB geological arm has also proved invaluable in providing (philosophically!) an excellent training ground for coal geologists recruited into the private sector of the Australian Coal Industry.

The Queensland Coal Board (QCB) is constituted under the Coal Industry (Control) Acts of 1948 to 1965. It was designed principally to ensure that the conservation and development of the State's coal resources was to the best advantage in the public interest, to promote the welfare of the work force and to encourage maximum productivity through co-operation between management and labour.

The QCB is financed mainly by levy of the coal owners. Its principal activities involve maintenance of comprehensive production and safety records and statistics, advancing loans for capital equipment purchases and welfare projects, sponsoring research projects and, of major importance, the fixing of coal prices for coal produced for consumption within Queensland.

COAL FOR ELECTRIC POWER GENERATION

In NSW and Queensland, all power is generated by statutory public utilities - The State Electricity Commission of NSW (Elcom) and, in Queensland, regional electricity generating boards, with planning of power developments by The State Electricity Commission of Queensland (SECQ).

Elcom through their subsidiary, Elcom Collieries Pty. Ltd., operate seven underground mines and two open cuts are mined by contract for Elcom.

The NSW Government have reserved relatively large areas of economic coal potential for future steam coal production, and active exploration programs are pursued to ensure a continuing supply of coal for future power generation demand. The exploration programs are usually planned in consultation with other Government departments and conducted with the assistance of JCB geological staff.

In Queensland, all coal for power generation is mined privately, under supply contract arrangements between the various generating boards and individual mining companies. The SECQ, responsible for overall power generation planning, has used the services of the GSQ to evaluate the steam coal resources of the State and to assist in evaluating tenders submitted by private mining companies for supply of coal to new or existing power stations. Since 1976 SECQ has funded more detailed exploration by GSQ in a Departmental Reserve Area at Curragh, north of Blackwater (Figure 1), to evaluate available surface mine reserves for use in the State's major power plant at Gladstone. However, this work is being done preparatory to the production of both coking and steam coal at Curragh by a private mining group, under an agreement between the private mining group and the statutory authorities.

Except in the Ipswich Coalfield, Queensland, practically all black steam coal in Australia is supplied as raw coal, with relatively small proportions of middlings steam coal produced as a by-product of coking coal production. Ash content generally ranges from about 10 to 22 per cent at total moisture contents of 5 to 15 per cent and specific energy in the range 20 to 28 megaJoules/kilogram (8 700 to 12 000 British thermal units/pound). Very few problems are recorded with respect to deleterious properties of Australian black steam coals. Most have low sulphur, chlorine and nitrogen contents. However, fly ash precipitation is generally relatively difficult with NSW coals and the Permian Bowen Basin coals have very variable and irregular ash fusion properties, commonly with reducing atmosphere flow temperatures as low as 1 200°C or less.

In exploring and evaluating steam coals, particular attention is paid to the pattern of ash fusion temperature distribution and to the prediction of ash precipitation properties, which for meaningful results requires a relatively large bulk sample, costly to obtain by core drilling.

SERVICE ORGANIZATIONS AIDING THE AUSTRALIAN COAL INDUSTRY

Commonwealth Scientific and Industrial Research Organization (CSIRO)

The CSIRO, an Australian Federal Government body, has provided practical research assistance in exploration, development and utilization of coal resources, through its Coal Research Division and Geomechanics Sections. It conducts active research into the physics and chemistry of coals and aspects of coal utilization, maintains an excellent standard-setting coal petrological laboratory and ensures that the industry is well informed through its publication of the periodical *Coal Research in Australia* and its regularly distributed updated listing of relevant publications.

Australian Coal Industries Research Laboratories Ltd. (ACIRL)

This organization was founded originally as the Australian Coal Association Research Ltd. in 1956 by the Australian Coal Association (Coal Mine Owners) to provide the industry with a scientific and specialist technical service. In 1965 it was reconstituted in its present form with the additional support of the Federal and State Governments and it can now claim the distinction of being among the world's foremost coal testing authorities. It conducts vigorous research activities especially in the fields of coal liquefaction and applied rock mechanics for overcoming difficult mining conditions, and it provides comprehensive laboratory testing facilities for the industry at North Ryde and Maitland in NSW, at Ipswich and Rockhampton in Queensland.

JAPANESE INFLUENCE IN COAL EXPLORATION AND EVALUATION

The meticulous professionalism of the Japanese Steel Industry deserves mention herein as an influence for maintaining high standards of exploration, testing and evaluation on the part of the Australian Coal Industry, to ensure that markets can be negotiated with mutual respect, on the basis of adequate factual data, objectively presented.

BASES FOR PLANNING EXPLORATION PROGRAMS (24)

All major coal bearing sedimentary basins in Australia have now been outlined and private enterprise companies have confined their attention to these major basins in selecting areas for exploration.

Such areas have a pre-existing data base of variable quantity and quality. This includes published topographic and cadastral maps, geological maps, and geological reports, aerial photography, data gathered from literature research and local knowledge including reported coal occurrences, reports of previous exploration activities, water bore records etc.

Obviously a geological basis for selection of any coal prospect is needed, but unless the area selected has been previously explored, the existing data base will need to be expanded before fully effective exploratory drilling programs can be mounted. This further collection of data will involve geological reconnaissance mapping, using both photogeological studies and ground surveys, to enable drilling targets to be selected on the basis of geological criteria.

The initial stage of the exploration program - the *planning* or *research* stage involves only the collation and evaluation of the data base, and this work may be completed before the prospective area is acquired.

Assuming that the evaluation provides sufficient incentive, a second or *reconnaissance* stage is undertaken, whereby:

(i) A topographic/geological base map is prepared of the area from all available information including aerial photogeological studies.

(ii) Geological mapping is conducted to improve the quality of the base map and ensure that all significant data is recorded.

(iii) An interpretation of the geological structure is made, as thoroughly as possible, from the surface data and any additional subsurface information, and information extrapolated from beyond the area boundaries.

In this stage, stratigraphic drilling programs have been conducted, particularly in areas of inferior outcrop due to deep or extreme weathering, or extensive superficial cover. Other exploratory techniques also have been used, including geophysical methods - seismic refraction or ground electrical resistivity surveys, particularly for gauging depth of weathering or superficial cover. If substantial deep underground reserves of good quality coals can be confidently predicted, seismic reflection surveys may be warranted.

During the reconnaissance stage, sufficient drilling is conducted to broadly outline the extent of potentially economic coal deposits, indicate the number, lateral continuity, and geological structure of the coal seams, and lithotypes and properties of the coal. Initially, this drilling is non-core, unless or until core drilling is justified when coal of economic significance has already been revealed and sufficient reserves have been outlined in structural conditions suitable for economic mining operations. However, for reconnaissance drilling in areas of potential underground development, core drilling is often the most appropriate, particularly where coal measure strata are little known and definitive data is required, or the strata are relatively hard and drilling costs for non-core drilling are relatively high. The accuracy of logged formation depths and thicknesses reduces with increasing depth of non-core drilling, due to the time-lag for return of drill cuttings. The record of a non-cored hole can be further complicated if undetected caving of the hole occurs during drilling. Thus, unless wire-line, geophysical logs are made, the hole record may be unreliable.

The reconnaissance stage will be completed when the potentially economic coal deposits have been located and approximately outlined, and the data base becomes sufficient for either a decision to terminate drilling, or plan the third stage of exploration in the area(s) of economic potential.

The third stage of exploration - the *prospecting* or *preliminary mine planning* stage is planned on the updated data base with the objective of confident, conservative assessment of coal reserves, coal quality, mining conditions, and (for potential surface mines) overburden quantities for preliminary mine and production planning, costing studies, and market surveys. In this stage, the density of the drilling pattern is generally dictated both by geological structure, including the uniformity of seam thickness distribution, and by the lateral variation in coal quality in the area. For areas of underground mine potential, holes are spaced at centres not more than 2 kilometres apart, with all seams cored for measurement of reserves to at least *indicated* status. However, a more closely spaced drilling pattern is used locally when needed for adequate definition of geological structure, or to obtain sufficient data for definition of overall coal quality and lateral variation in coal type and properties.

For areas of surface mine potential, the pattern and spacing of the drilling in this third stage is controlled by several factors, including complexity of structure, lateral variation in coal type and properties of individual seams, and, locally, topography. Drill spacing is such as to enable calculation of coal reserves within the definition of reserves to *measured* status. The drill spacing is such as to enable confident interpretation of the updip limit of economically minable coal seams, define the depth of weathering with sufficient accuracy for confident preliminary mine design and production planning, and to enable a preliminary specification of the overall quality of the proposed coal production. The coring of coal seams is usually spaced from 200 to 400 metres to ensure adequate statistical evaluation of coal quality. Depending on the complexity of structure and the uniformity of seam thickness, non-cored holes for definition of structure are spaced down to 50 metres apart. Recently wire-line logs are being more frequently run to ensure reliability of seam depths and thickness.

In this third stage, the drilling is usually conducted on cross strike traverse lines. Drilling costs are controlled by coring only the coal seams either by drilling to the roof of the coal seam at a depth previously established by non-core drilling (target coring), or by the *touch-coring* of the coal seam. The latter involves running the core barrel as soon as cuttings from a readily identifiable marker horizon in the roof appear at the surface, or in the absence of a marker, cuttings from the top of coal seam. Obviously, this technique is only suitable when drilling at relatively shallow depth using high velocity air circulation, or when there is a recognizable marker formation, as it has often involved unrepresentative sampling through loss of the top section of the coal seam. Some fully cored holes are now usually drilled, to ensure that the engineering properties of the overburden formations can be objectively assessed. These are located to provide cores fully representative of the various strata, in both weathered and unweathered states. In this stage, particularly in coking coal exploration, large diameter cores are drilled to enable thorough evaluation of coal quality through testing of bulk samples in the proposed production size range.

In the fourth, *detailed mine planning* stage, the drilling program is based on the preliminary mine plan developed during the previous stage.

For the development of an underground mine area, this stage may extend for almost all of the mine life, with drilling conducted in advance of the development of each mine district. However the initial program includes core drilling to assist in the selection of shaft and mine entry sites, and the scope and intensity of the drilling program depends upon the geological structure and variations in coal quality as indicated from the existing data base.

In the final stage of exploratory drilling of a surface mine area, the program is best based on the preliminary mine plan, to check the validity of the assumptions made in the plan, and to define the open cut low wall limits as accurately as possible. The latter task will involve careful planning of drilling to determine the *soot* and *oxidation* lines or updip limits of coal of satisfactory quality, and to define the low wall location. This involves *chip* (non-core) drilling in traverses across the strike of the originally interpreted oxidation line with two holes per traverse, one on each side of the line. Spacing between traverses will depend upon the regularity of the oxidation line, and a third hole may be required for accurate definition. Figure 2 shows an oxidation line traverse section for determination of the limit of oxidation for a coking coal, using both visual signs of weathering and swell No.s determined in the field. The *soot* line is the line updip of which the coal is reduced to soot and of no economic use.

GEOLOGICAL LOGGING AND SAMPLING

Required Information

The more experienced Australian coal explorers realize that detailed geological logging of all exploratory drill holes is essential in all stages of drilling, if optimum economy is to be achieved, and most organizations now engaged in coal exploration have standard procedures for

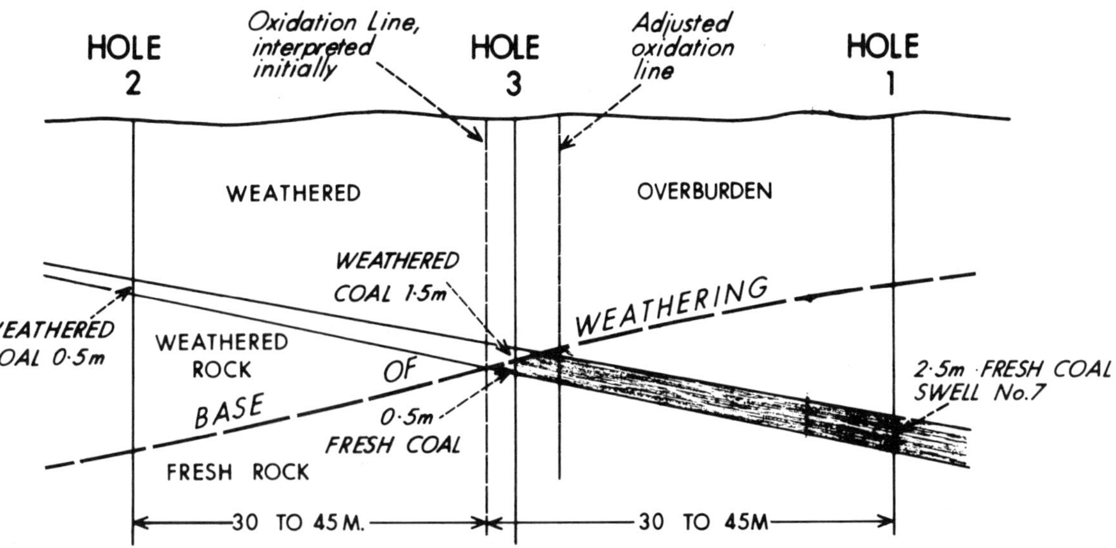

Figure 2 - Oxidation Line Drill Traverse Section (24)

recording drilling data and geological logging and sampling of cuttings and cores.

The important information recorded includes:

1. <u>Drilling Data</u>

   . hole No.,
   . date of drilling,
   . hole location co-ordinates and elevation,
   . depths of various types of drilling (i.e. blade bits, hammer and/or core drilling),
   . drilling penetration rates and bit usage,
   . type of core barrel and core bit,
   . type of circulation fluid and depths and degree of loss of circulation during drilling,
   . hole, bit and core diameters,
   . details of hole casing and stand pipe,
   . core recovery and reasons for core losses.

2. <u>Formation Rock Types and Properties</u>

   . colour,
   . hardness and tenacity,
   . grain size and grain sorting,
   . bedding and other sedimentary structures,
   . nature and attitude of joints, cleavage planes, sheared zones,
   . degree of weathering,
   . degree of consolidation, lithifaction and nature of grain cementation,
   . conditions of core (stick lengths and nature of breaks).

3. <u>Coal Seam Lithotypes and Structures</u>

   . coal lithotype vertical distribution using standard terminology,
   . apparent truncation or repetition of seam,
   . the nature and attitude of structures including bedding or banding, cleats,
   . the mode of occurrence of secondary minerals, particularly pyrite.

4. <u>Geohydrological Data</u>

   . ground water levels and water inflows,
   . reasons for fluctuation in ground water levels, losses of circulation or water inflows,
   . water quality, if large water inflows are apparent.

Although most of the above geological data can only be obtained from continuous cores, careful examination of drill cuttings enables a reasonably reliable assessment of depositional environments using sedimentological principles (25), and engineering properties. For the latter, accurate records of drilling penetration rates and bit usage provide useful information for mine planning engineers in assessment of overburden drilling and blasting operations and costs. Careful description of the physical properties of the strata in standard terminology (26) understood by both the explorer's geologists and engineers will provide a useful framework for engineering geological assessment and geomechanics studies which are now being frequently used for assessment of surface mine slope stability or roof and floor conditions which largely control productivity in underground mine operations.

Sampling Techniques

The prerequisite for accurate sampling of drill cores is disturbance-free retrieval from the core barrel, and secure storage and packaging of cores and samples in the intervals between sampling and testing. Thus, core should be drilled using a split inner tube barrel to deliver core in *in ground* condition, and for the most efficient logging and sampling, ideally the core should be logged in the split tube and sampled immediately. Where this is not practicable, the core can be removed after initial logging, by:

- placing a PVC (or aluminium) half tube over the core exposed in the bottom split of the barrel inner tube,
- carefully inverting the PVC tube and replacing the barrel tube with another PVC half tube,
- carefully binding the PVC tubing and plugging both ends,
- sealing the tube by encasing the bound tube in clear polyethylene tubing, before inserting in the core box to achieve a close fit.

As coal cores are generally lost in testing, a photographic record of the core is desirable, to further aid the mine planning process. Fortunately core photography is becoming more frequent and is virtually a standard procedure in the core libraries of the State Geological Surveys.

Selection of coal seam core sample intervals depends upon several factors, including:

- the scope of the required testing (in terms of minimum sample mass meeting universal standards for particular tests),
- the lithotype profile and variation in ash content between seam plies or sub-sections,
- the proposed utilisation of the coal and the desired production quality specifications for various uses of the coal.

As vertical variation in important coal properties within a coal seam may not be precisely gauged from visual examination, particularly in the early stages of exploration, the coal seams are usually sampled and analysed in plies or sections. The initial sample intervals should be as small as is consistent with the standard minimum test sample mass. Where the lithotype profile is particularly heterogeneous, with interbanded coal and stone, the sampling intervals are selected to enable the determination of an economic mining section, based on the relationship of the properties of contiguous sampled sections, in various combinations, with required coal production quality specifications. In this regard, it is essential to sample all of the coal seam, including stone and dirt bands. There are numerous past instances where intra-seam non-coal bands have been discarded or not subsequently tested and this has resulted in uncertain assessment of run-of-mine coal quality. Unless the intra-seam non-coal banding can be readily selectively mined because of its thickness, hardness and lateral continuity, or it can be left in the floor, its sampling is necessary. While it is necessary to sectionally sample all coal seam intersections of heterogeneous profile, seams with homogenous or characteristic profiles persistent over large areas should not require intensive sub-sectional sampling once the lithotype profile/quality relations have been established, as such sampling leads to unnecessary expense in terms of both testing and result processing costs. However selection of sampling intervals is usually no longer left to unqualified or inexperienced personnel.

Where large diameter cores are taken to provide bulk samples for comprehensive coal quality evaluation, the sites for the cored holes are carefully selected, on the basis of the results of previous slim core drilling, to ensure that the samples will represent the range in proposed production coal quality. A slim core pilot hole should be drilled at each bulk sample location to ensure that the large cores will yield the required sample size and quality. Detailed sub-sectional sampling of the pilot 50mm core should avoid the need for sub-sectional testing of the required large diameter core intervals.

The same principles apply in the logging of the large cores as in the 50mm core logging. Although some explorers neglect to log large cores, such logging is considered essential if results are to be fully assessed and correlated with the data from the 50mm core drilling program. Moreover presentation of detailed core logs together with the results of the bulk sample testing should engender confidence on the part of the prospective market.

SUPERVISION OF EXPLORATION

The people responsible for supervision of the exploratory drilling programs are now mainly experienced senior professionals, well qualified in all phases of coal mining geology and thoroughly familiar with drilling equipment and techniques, and appreciative of the engineering significance of geological criteria. They are responsible for drilling contract supervision and co-ordination of the work of assistant geologists, surveyors and geophysical contractors. Each drill rig should be systematically supervised and the logging and sampling is now being conducted more frequently by graduate geologists. The use of unqualified personnel has been satisfactory if they are well trained and possess a sense of responsibility. However, at the present time, there is no shortage of graduates who can be rapidly trained to use their knowledge and initiative for recording and interpretation of the important data. There is ample evidence that discontinuous drilling supervision and/or a lack of adequate recording procedures has resulted in the loss of or unreliability of data.

PRESENTATION OF RESULTS

The simplest and most effective method of presentation of a core log is in the graphical, histogram form, whereby

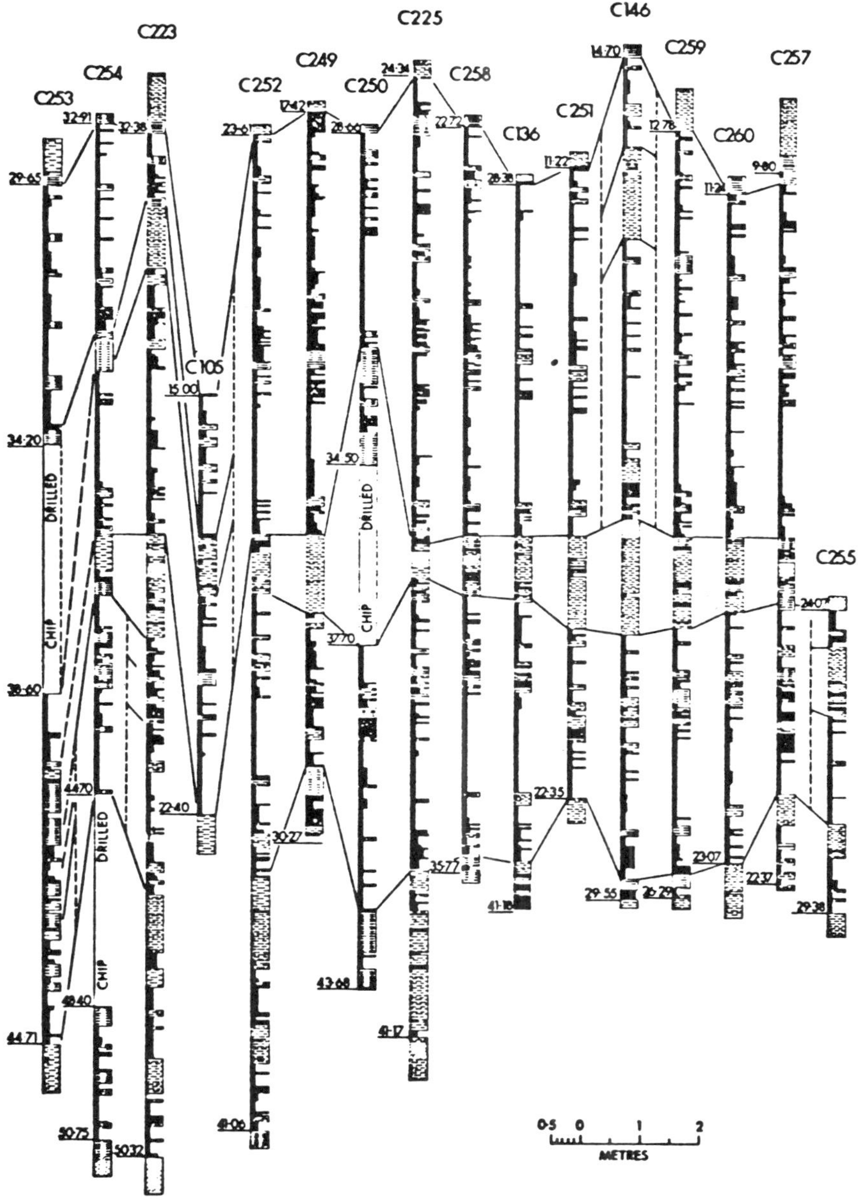

Figure 3 - Coal Seam Lithotype and Correlation Diagram (24)

the coal lithotype block is drawn (it can be computer plotted) with the lithotype profile proportional to the percentage of bright coal (visible vitrain). Presentation of these profiles in lateral succession enables correlation of coal seams on the basis of distinctive profiles as shown in Figure 3. Where geophysical logs have been run in both cored and non-cored holes the geophysical log plots drawn alongside the graphic lithotype logs enable confident extension of the correlation through non-cored seam intersections.

QUEENSLAND EXAMPLES

Surat Basin

Before 1973, the primary geological data base in the Surat Basin was limited by the lack of outcrop over much of the region and incentive for coal exploration was provided only by high volatile coal occurrences in water bores and in some long abandoned prospective coal mines. No systematic stratigraphic interpretation had been made and scout drilling was largely a hit-and-miss affair. However, a well planned program of stratigraphic drilling by the Queensland Department of Mines (27, 28) enabled the regional stratigraphy to be outlined and the more prospective horizons for coal deposition to be indicated or inferred. Brigalow Mines Pty. Ltd. by reconnaissance drilling, located the basinal distribution of the coal bearing horizons. By applying sedimentological principles to the study of previously located coal deposits in the Basin, a model of coal deposition was constructed to establish environmental characteristics. Further economically significant coal deposits were located using drilling programs planned on this basis.

Central Queensland

In the area shown in Figure 4, exploration first commenced in 1969, encouraged by the occurrence of thick coal intersections in water bores and shallow stratigraphic holes drilled by the BMR in Upper Permian Blackwater Group strata. The drilling of four scout holes confirmed the presence of high volatile low ash coal at shallow depth.

Regional stratigraphy and geological structure had been outlined by the GSQ and BMR in a joint regional mapping

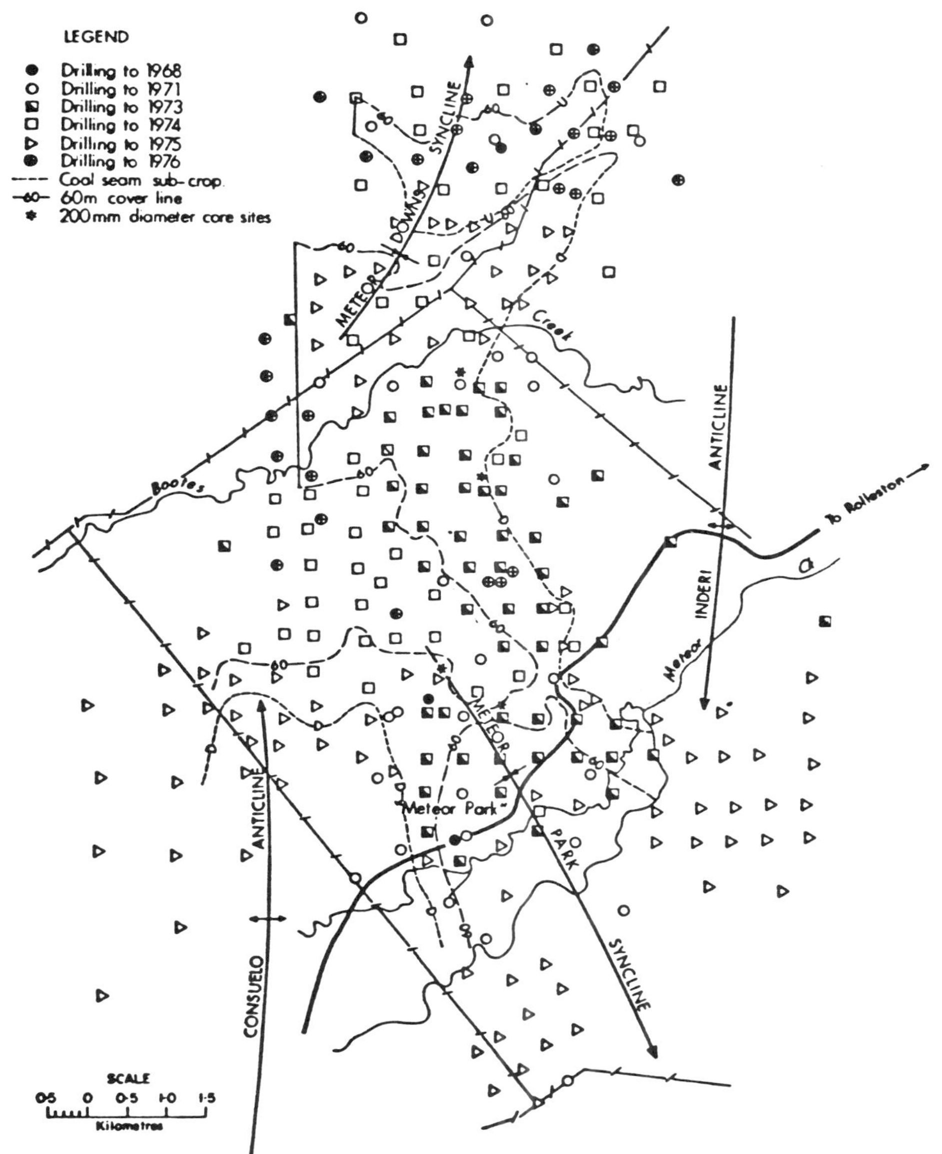

Figure 4 - Plan of Staged Exploration Showing Coal Seam Structure (24)

program, but much of the area was covered by Tertiary basalt flows which obscurred the Permian structural framework. However, the area had been explored, in the search for petroleum, by extensive seismic reflection surveys. Structure contours on the Aldebaran Sandstone, the shallowest and most persistent marker horizon in the area, interpreted from the seismic reflection data, were projected upwards to indicate the possible structural form of the Blackwater Group. The map obtained, although lacking accuracy in detail, outlined three areas of potential coal deposits at shallow depth. One of these areas, the Meteor Park Prospect, has been subject to staged exploratory drilling with the third or preliminary mine planning stage now completed. The layout of successive drilling programs is shown in Figure 4 together with the geological structural elements. The drilling has been supplemented by geophysical logging and ground electrical resistivity surveys. Here, drilling has been conducted with *Mayhew 1000* truck mounted rigs using high volume auxiliary air compressors. Most of the drilling has involved non-core drilling of the superficial cover and overlying coal measures with selective coring of coal seams using target depths obtained in pilot holes, or by touch-coring, a method necessitated, in the earlier stages of drilling, by the extensive occurrence of hard olivine basalts and superficial bouldery gravel deposits which impose additional expense in pilot hole drilling. With expansion of the data base through more reliable structural interpretation, coal seams can be fully cored by conservative calculation of the required depth for commencement of coring. The touch-core technique has led to samples being not fully representative of the seams. However comparison of analysis data between fully cored and touch-cored samples shows no significant difference in indicated coal quality parameters in this case. Moreover the reduced seam thickness tends to reduce calculated *in situ* coal quantities and increase overburden ratio - an added safety factor, in this regard.

TOPOGRAPHIC, CADASTRAL MAPPING, SURVEY CONTROL

In the NSW and Queensland coalfields, cadastral maps showing real property boundaries, prepared by the State *Lands Departments* are generally adequate for exploration purposes. However, until recently, statutory topographic maps were sometimes of indifferent quality except in the relatively small areas in or adjacent to highly urbanised areas in NSW. With co-operation between the Federal *Division of National Mapping* and the State Lands and Survey Departments, this situation is rapidly being remedied with an increasing cover of 1:100 000 sheets with 10 metre contour intervals and mapping to scales down to 1:25 000.

In the more recent times, the value of accurate mapping has become increasingly appreciated and once the ultimate economic potential of a prospect is indicated, it is now usual for explorers to produce the required topographic maps in modern aerial photo stereo plotters operated by specialist companies.

Ground control is usually difficult in the less accessible parts of Central Queensland where topography tends to inhibit triangulation methods. However, lack of triangulation has been overcome by modern electronic telemetric survey techniques.

At the present time the survey standards of the Departments of Mines in NSW and Queensland differ in terms of established map grids and statutory mine plan scale requirements, but the SI (metric) system is now well established. The lack of care in accurate location of drill sites in some prospects in the past on the part of both private companies and the Government authorities has led to a considerable mass of useful geological and coal quality data being inapplicable, and drill sites are now usually surveyed as soon as the sites are pegged or drilled.

DRILLING EQUIPMENT AND TECHNIQUES (29)

Most exploratory drilling in the NSW coalfields has been conducted using conventional diamond drill rigs of modern design, usually truck or trailer mounted. Conventional diamond drill rigs have also been used extensively by the Queensland Department of Mines for coal exploration. However almost all of the exploratory drilling of surface mine coal prospects throughout Queensland has been conducted using rotary rigs typified by the *Mayhew 1000* truck mounted unit. The rotary rig has proved capable of both rapid non-core drilling and core drilling with high core recoveries when modern core barrels have been used by experienced drillers.

For deep or continuous core drilling, wire-line equipment is now in general use. Both wire-line and conventional coring is effected using stationary split inner tube barrels, for minimum core disturbance, maximum core recovery, ease and accuracy of logging and sub-sectional sampling. Diamond core bits are generally required in NSW, with water circulation. In Queensland, the generally softer coal measures strata can be successfully cored using tungsten carbide bits and air circulation is the preferred method where possible.

Most of the non-core drilling is accomplished using blade bits, with down hole hammers used for penetrating hard formations, being preferred, wherever possible, to the relatively slower and more costly roller *tri-cone* bit.

The exploratory drilling arm of the Australian mining industry is becoming increasingly competitive, requiring modern equipment, techniques and skilled, dedicated drillers to meet the clients' demands for quantity and quality of performance. The success of drilling programs is largely dependent on co-operation and good relations between client and drilling contractor, and a comprehensive specification for exploratory drilling contracts is considered essential by this author to ensure this. Most of the major companies in Australia now insist on drilling being conducted accordingly.

GEOPHYSICS

A comprehensive review of geophysics and its increasing role in Australian coal exploration is beyond the scope of this paper but mention should be made of:

- the increasing acceptance of, due to success achieved in both simple and multi-mode wire-line logging including density, high resolution density, gamma ray, neutron, sonic, and resistivity logging for interpretation of stratigraphy and structure and very accurate positioning of coal seams in rapid non-core drilling programs,
- wide use of ground resistivity methods in conjunction with drilling for structural interpretation, particularly displacement of coal

seams and the definition of displacement direction by faulting,

recent apparently successful application of seismic reflection and refraction methods as developed in the United Kingdom for structural interpretation at depths previously considered to be too shallow for the application of the reflection method.

COAL QUALITY EVALUATION

In the advanced stage of exploration, testing may be conducted on samples obtained from exploratory shafts or trial adits to supplement data from core testing. Sometimes a trial open cut has been used to obtain trial shipment samples of 10 000 tonnes for evaluation by the overseas users.

The growth of coal exploration in Eastern Australia has led to a rapid expansion of ACIRL laboratory facilities but the testing loads have been such as to encourage the establishment of properly equipped laboratories by private companies other than BHP. This company, the sole Australian steel producer, has long maintained up-to-date research and comprehensive coal and steel laboratory testing facilities at Shortland, near Newcastle, as well as at the steel plants at Wollongong - Port Kembla and Whyalla in South Australia. Independent testing authorities responsible for monitoring coal shipment quality have expanded their facilities to cater for a demand from explorers for rapid testing of coal cores and several of the larger coal mining companies operate their own laboratories for both production quality control and exploration quality evaluation. The latest of these, recently opened by R.W. Miller at Carrington, near Newcastle NSW, is capable of performing most of the standard and special tests required for evaluation of coking and steam coals and coal liquefaction potential. A National Association of Testing Authorities (NATA) registers laboratories meeting the required standards for consistent, reliable performance.

For pilot scale preparation plant process testing of bulk samples, ACIRL maintain facilities at Rockhampton in Queensland and Maitland in the Hunter Valley, NSW.

Pretreatment of Core Samples

In recent years ACIRL has successfully endeavoured to obtain effective and reliable test results from coal cores that reflect, as closely as practicable, the physical and chemical properties of the salable coal products, particularly the coking coals. Test results from bore cores prepared in the traditional manner doubtfully reflect with sufficient accuracy the proposed product quality and yield, mainly because of the difference in the particle size consist between the crushed core sample and run-of-mine or the clean coal products. *Fumble* factors have thus been applied to such results, to ensure that projected yields and specified ash contents are conservatively predicted and this has led to some uncertainty on the part of both the producer and the coal buyer.

A.J. Le Page and B.W. Proudfoot of ACIRL (30, 31) have developed a method of pretreatment of bore cores to ensure that the particle size distribution conforms as nearly as possible in a bore core sample to that obtained in a sample of the same coal treated through a commercial preparation plant. The method was developed through continuing research using a range of different coals and is considered to be particularly important for the relatively high rank, high vitrinite content coals of the Bowen Basin which have Hardgrove grindability indices of 90 or more, tending to produce excess fines - around 30 per cent passing the 30 mesh (500 micron) screen.

The theory and experimental work which led to development of the pretreatment method has been discussed by the researchers and is beyond the scope of this paper. However, the essentials of, and basis for the procedure are outlined below.

It is known that Australian coals as mined and commercially processed generally conform to the Bennett (32) equation for particle size distribution of crushed coal on a Rosin Rammler graph, but crushed core samples generally do not. A method was devised whereby drill core of a coal of known or predicted Rosin Rammler size distribution, was crushed to a nominated top particle size and the particle size distribution of the core sample initially adjusted by dry tumbling to conform to the predicted ROM size distribution. Thence, to simulate wet processing, it is wet tumbled to ensure breakdown of weak clay particles as would occur in commercial processing.

The apparatus consists of a metal drum of 200 litre capacity. The coal sample is rotated in the drum at 20 revolutions/minute for a period directly related to the Hardgrove grindability index (31). The maximum size of sample is 50 kilograms. Steel cubes of 50 millimetres side are loaded in the drum to simulate impact breakdown of the coal, the number of cubes ranging from eighteen for a 50 kilogram sample to two for the minimum size sample of 6 kilograms. For pretreating slim core samples (diameter 50 millimetres), a seam sample thickness of at least 2.4 metres is required. The dimensions of the apparatus are given in Standards Association of Australia AS 1661-1977.

For most new projects run-of-mine size distribution is not known, but it is reasonably predictable from a relationship between the Rosin Rammler distribution constant n and the Hardgrove grindability index (33). Moreover testing has shown that for some Australian Permian Coals there are good correlation coefficients for n and Hardgrove grindability index (HGI), n and $\bar{R}_o$ max. vitrinite, and HGI and $\bar{R}_o$ max. vitrinite:

$$\text{HGI} = 35.5(n) - 1.54$$
correlation = 0.91, samples = 21

$$\text{HGI} = 73.66(\bar{R}_o \text{ max.}) - 2.26$$
correlation = 0.98, samples = 21

$$\bar{R}_o \text{ max.} = 0.39 - 1.38 \ln(n)$$
correlation = 0.88, samples = 21

The method is by no means accepted by all Australian explorers but it has proved effective in the Bowen Basin where coals of moderate rank and high Hardgrove grindability index tend to produce an excess of passing 500 micron fines. The effect on the particle size distribution of a 150 millimetre coal core is shown in Figure 5.

Froth Flotation Testing

Laboratory froth flotation testing of the ultrafines (passing 500 micron) has gained more widespread acceptance since Le Page (34) developed a laboratory process for continuous steady state conditions in a laboratory flotation cell and showed that the results of

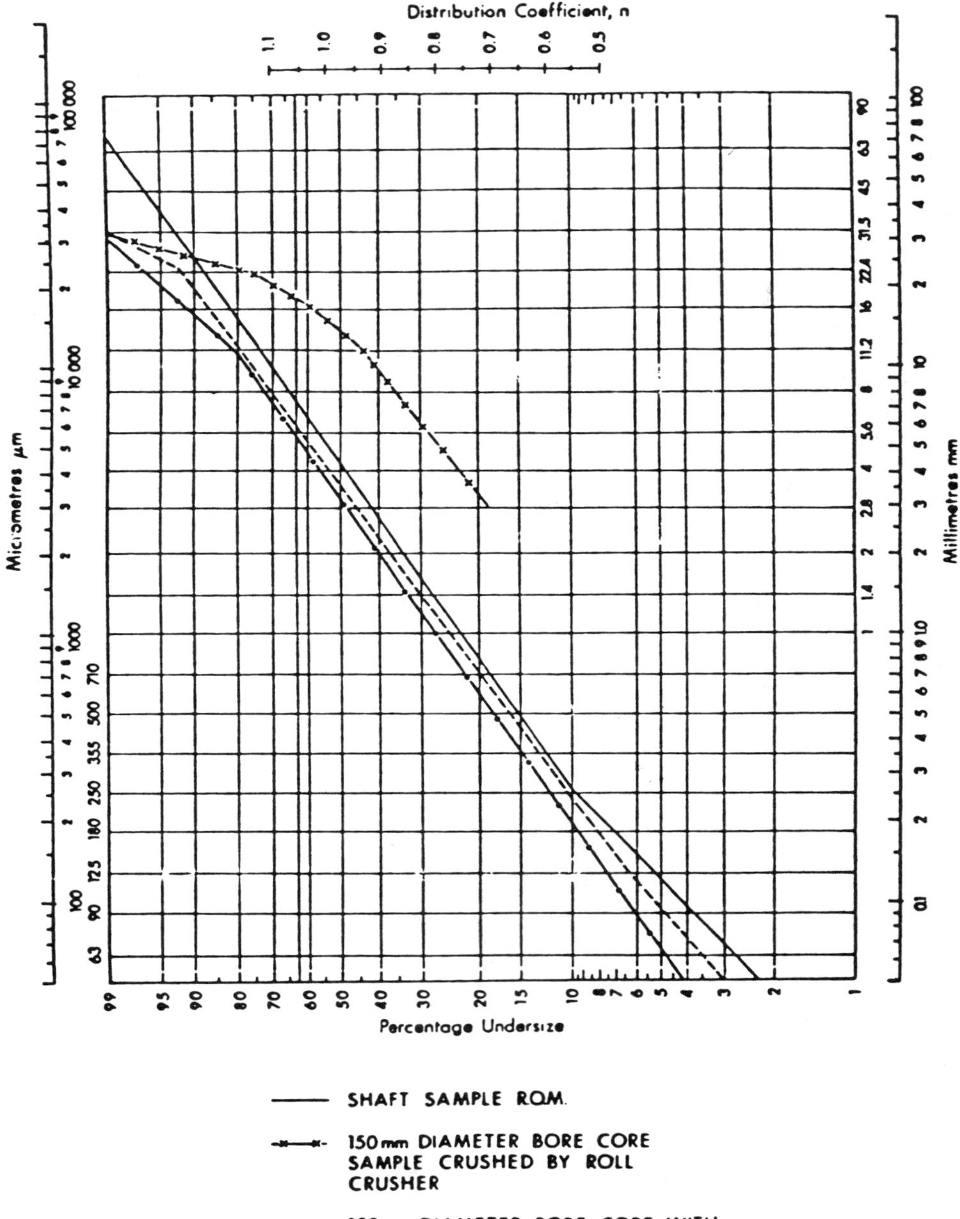

Figure 5 - Rosin Rammler Curves Showing Pretreatment Effects on 150mm Core in Relation to ROM and Salable Coal Gradings (31)

testing under these conditions conformed more closely with actual pilot plant froth yield and ash content than the batch flotation test result.

Coking Coal Evaluation

Following pretreatment, core samples including nominal 50 millimetre diameter (NQ - NM), 60 - 63 millimetre diameter HQ and large cores (100 to 200 millimetre diameter) are normally float/sink tested. The slim cores are usually tested at three to four separation densities after crushing to passing ½ inch mesh (12 - 13 millimetres) in the size fraction ½ inch x 0 or ½ inch x 30 mesh (500 micron) with froth flotation cell (continuous) testing of the ultrafines. Key coking quality parameters will then be determined for an apparent optimum separation density selected on the basis of yield, ash and coking quality as reflected by swell or plastic properties, rank and reactive maceral content. Alternatively, a *practicable* ash content is selected and yields of clean coal for that ash content are recorded and plotted. In these ways coal quality is related to mine production scheduling and a viable production quality specification determined statistically with some confidence.

Large diameter core testing is used to run full scale washability tests and produce washability curves over a range of separation densities from 1.30 to 2.00 and for several size fractions with froth flotation testing of the ultrafines.

Research by several evaluators has shown that variations in coking quality with size fraction may be important, particularly in predominantly dull inert rich coal with only thin bands of vitrinite distributed throughout the seam. The latter tends to report in the finer size fractions as exemplified at Curragh where in one seam, a high quality coking coal can be produced in the passing 10 millimetre fraction by separation at a relatively low density, whereas the coarse fraction shows no coking potential. When tested in the full size range, the seam shows little if any desirable coking properties.

Since the parameters of Gieseler maximum fluidity, rank in terms of $\bar{R}_o$ max. vitrinite and reactive maceral content have become important in terms of comparative price comparisons of Australian coking coals on the Japanese market. These parameters tend to be more frequently measured as a matter of routine to encourage favourable market reaction to the coal offered for sale. 7 kilogram pilot coke oven tests are also frequently conducted and as these tests do not always reflect maximum possible coke strength, 230 kilogram coke oven tests are relatively frequently conducted on selected combined large core samples representative of overall production quality to re-inforce the quality evaluation based purely on coal properties.

Australian Coking Coal, Comparative Cash Value

A renewed vigor in exploration of better quality coking coal has been spurred by the Queensland Government's granting of Authorities to Prospect in areas highly prospective for high quality coking coal at German and Oaky Creeks. This has been in spite of a marked worldwide down turn in the demand for coking coal. The competition for the available markets has led to comparisons between current Australian coking coal prices with prospective prices based on the price quality relationship developed in Japan and known in Australia as the *NKK Formulae.*

Prices calculated using these formulae are given in Table I, using two USA Reference coals and Itman (USA) with South Blackwater coal. The prices calculated as shown below tend to refute R.J. Bennett's conclusion (35) that: *It is doubly impossible to measure the cash value of coking properties in any systematic way.* The calculations also show that the quality of the recently explored inland coking coal resources is favourable, relative to currently exported Queensland coals.

The formulae used in the price calculations are:

$$BP\ (US\$) = C_1(\bar{R}_o max)(1-I) + C_2(I) + C_3(\log MF)$$

in which: BP = Base Price

C_1, C_3 = \$ value constants calculated from sales price of two reference coals by solving simultaneous equations of the above formula

C_2 = Current sales price of Hongay Coal (anthracite) (\$30)

I = Inert maceral content, volume % including inertinite + 2/3 semi-inertinite

logMF = Log Gieseler maximum fluidity, dial divisions/minute

and MP = Merit (Sales) Price

MP (US$) = (BP + 2(a) + 10(s) + 0.3(vm))tm

in which: a = difference* in ash % between R & X

s = difference* in total sulphur between R & X

vm = difference* in volatile matter between R & X

tm = % adjustment for difference in total moisture content between R & X

$$\frac{(100-TMX)}{(100-TMR)}$$

TM = total moisture

R = reference coal

X = coal for price determination

* negative (-) when R value is lower

To solve the above equations using:

Itman Sale Price FOBT @ 1.11.77 = US$68.42, and
Pittston Sale Price FOBT @ 1.11.77 = US$63.65

US$63.65 = (BP + 2(0) + 10(0.12) - 0.3(9.5)1.018
BP = 63.65 - 1.22 + 2.90/1.018
= 64.03
Itman BP = US$68.42 = C_1(1.52)(0.87)+30(0.13)+C_3(1.903)
Pittston BP = US$64.03 = C_1(1.10)(0.888)+30(0.112)+C_3(3.602)
68.42 = C_1(1.3224) + 3.66 + C_3(1.903)
64.03 = C_1(0.9768) + 3.36 + C_3(3.602), and
C_1 = 40.56, C_3 = 5.84 (C_2 = $30)

For example, the calculation for Curragh coal is:

BP = 40.56(1.31)(1-0.30) + 30(0.30) + 5.84(2.0)
= $57.86

MP = (57.86-2(0.5) + 10(0.23) - 0.3(2.1))0.978
= $57.24

Table No. I

Comparison of Coal Quality Specifications, Coking Quality Parameters, Calculated Prices (*NKK* Formulae) Australian and U.S.A. Reference Coals

COAL	ITMAN USA	PITTSTON USA	GREGORY	NORWICH PARK	GERMAN CREEK	HAIL CREEK	SOUTH BLACKWATER	MOURA	BULLI SEAM BURRAGORANG NSW	OAKY ** CREEK	CURRAGH
Contract & ● FOBT Price	68.42#	63.65#	52.00ϕ	-	-	-	49.50#	49.44#	49.32#	-	-
Calculated * Sales Price	-	-	49.28[1] 52.06[2]	54.22[1] 53.50[2]	64.24[1] 65.81[2]	45.67[1] 45.63[2]	49.07[1] 50.67[2]	50.46[1] 53.03[2]	47.40[1] 49.71[2]	56.97[1] 59.61[2]	57.24[1] 58.08[2]
$\bar{R}_o$ Max.	1.52	1.10	1.0	1.6	1.40	1.32	1.0	1.12	0.98	1.25	1.26
Inerts * Vol. %	13.0	11.2	23.0	29.0	15.0	44.0	30.0	33.4	32.0	15.0	30.0
Max. Fluidity dd/min	1.903 (log)	3.603 (log)	2500	10.7	1000	6.5	200	3.05 (log)	620	5000	100
Ash %	6.5	6.5	8.5	10.3	8.5	9.5	7.3	8.6	10.0	8.5	7.0
Total Sulphur	0.80	0.68	0.80	0.65	0.80	0.40	0.50	0.45	0.45	0.85	0.57
Volatile Matter	19.5	29.0	31.5	17.0	21.5	20.0	28.5	30.4	28.0	26.0	21.6
Total Moisture	7.7	6.0	10.0	10.0	10.0	10.0	10.0	9.5	7.0	10.0	10.0

ϕ Proposed as at 13.5.1977
● Prices in US$ per <u>tonne</u>
** Assuming production from both surface & underground mines

* Using NKK formulae for base and merit prices
1. Reference Coals Itman and Pittston
2. Reference Coals Itman and South Blackwater (Prices as at January 1978)
\# Prices as at 1.11.1977

COAL RESERVES ASSESSMENT

Standards for calculation of coal reserves and definitions of categories of reserves *measured, indicated* and *inferred* have been set by the statutory authorities and the Australasian Institute of Mining and Metallurgy, based on those specified by the United States Geological Survey.

Experience has shown that such categories and definitions should be considered with objective caution, particularly in areas of underground reserves and/or where irregular igneous intrusions occur. A bank of experience is developing for Central Queensland surface mine reserve estimates and, for normal conditions, not more than 5 per cent pit losses need to be accounted in converting in situ to surface run-of-mine reserves. Dilution factors due to contamination by floor material or slumped overburden are usually conservatively estimated at 1 per cent. Although slumps are common and wet season conditions may cause floor contamination, the overall dilution factors are very low. In spite of the praiseworthy advances in preparation plant yield prediction, the calculated yield factors are usually reduced by about 5 per cent, if not by the mining engineer estimator, by the potential and cautious Japanese buyer! A further conservative factor is usually applicable in that coal quantities are calculated on an *air dried* basis, mainly because the actual *bed moisture* of the *in situ* coal is difficult to assess. Product total moisture estimates are based on experience of production of similar coals. In Bowen Basin production, total moisture is about 10 per cent (about 2 per cent higher than the initial estimates of the 1960s). Most Australians would note the relative ease of computing quantities in the metric system and commend it to our American colleagues!

ELECTRONIC DATA PROCESSING

This will undoubtedly become universal for routine storage, retrieval, statistical processing and plotting of geological and coal quality data and calculations of quality related quantities. Several systems have been offered for data storage and retrieval and simple sectional plotting and a considerable effort is current in developing suitable comprehensive systems for use in exploration and mine development. However, as yet, it must be said that the large computer is not yet fully effective in Australian coal exploration or indeed the coal industry generally. Nevertheless its time rapidly approaches!

ACKNOWLEDGEMENTS

The author wishes to thank R.W. Miller & Company Pty. Limited for permission to publish this paper and his many friends in the Australian Coal Industry for their help and forbearance over the past 10 years.

REFERENCES

1. Traves, D.M. & King, D., Editors, Economic Geology of Australia and Papua New Guinea, Volume 2, Coal Monograph Series No. 6, *Aus.I.M.M.*, Melbourne 1975.

2. Branagan, D.F., *Distribution and Geological Setting of Coal Measures in Australia and Papua New Guinea* in Economic Geology of Australia and Papua New Guinea, Volume 2, Coal Monograph Series No. 6, *Aus.I.M.M.*, Melbourne 1975.

3. Milligan, E.N., *The Geology of the Bowen & Galilee Basin Coalfields* in Australian Black Coal, *Aus.I.M.M., Illawara Br.*, Wollongong 1975.

4. Blayden, I.D., *Geology of Australian Coalfields - Some Other Basins* in Australian Black Coal, *Aus.I.M.M., Illawara Br.*, Wollongong 1975.

5. Robinson, J.B. & Shiels, O.J., *The Permian Coal Deposits of New South Wales* in Australian Black Coal, *Aus. I.M.M., Illawara Br.*, Wollongong 1975.

6. Cook, A.C., *The Spatial and Temporal Variation of the Type and Rank of Australian Coals* in Australian Black Coal, *Aus.I.M.M., Illawara Br.*, Wollongong 1975.

7. Edwards, G.E., *Marketable Resources of Australian Coal* in Australian Black Coal, *Aus.I.M.M., Illawara Br.*, Wollongong 1975.

8. Hawthorne, W.L., *Regional Coal Geology of the Bowen Basin* in Economic Geology of Australia and Papua New Guinea, Volume 2, Coal Monograph Series No. 6, *Aus.I.M.M.*, Melbourne 1975.

9. Stuntz, J., *Regional Coal Geology of the Main Coal Province of New South Wales* in Economic Geology of Australia and Papua New Guinea, Volume 2, Coal Monograph Series No. 6, *Aus.I.M.M.*, Melbourne 1975.

10. Carr, A.F., *Galilee Basin, Q.* in Economic Geology of Australia and Papua New Guinea, Volume 2, Coal Monograph Series No. 6, *Aus.I.M.M.*, Melbourne 1975.

11. Swarbrick, C.F.J., *Surat Basin, Q.* in Economic Geology of Australia and Papua New Guinea, Volume 2, Coal Monograph Series No. 6, *Aus.I.M.M.*, Melbourne 1975.

12. Svenson, D. & Hayes, S.W., *Callide Coal Measures, Q.* in Economic Geology of Australia and Papua New Guinea, Volume 2, Coal Monograph Series No. 6, *Aus.I.M.M.*, Melbourne 1975.

13. Koppe, W.H., *Maryborough Basin, Q.* in Economic Geology of Australia and Papua New Guinea, Volume 2, Coal Monograph Series No. 6, *Aus.I.M.M.*, Melbourne 1975.

14. Mengel, D.C., *Ipswich Coalfield, Q.* in Economic Geology of Australia and Papua New Guinea, Volume 2, Coal Monograph Series No. 6, *Aus.I.M.M.*, Melbourne 1975.

15. Cranfield, L.C., McElroy, C.T. & Swarbrick, C.F., *Clarence Moreton Basin, NSW and Q.* in Economic Geology of Australia and Papua New Guinea, Volume 2, Coal Monograph Series No. 6, *Aus.I.M.M.*, Melbourne 1975.

16. Knight, J.L., *Wonthaggi and Other Cretaceous Black Coalfields, Victoria* in Economic Geology of Australia and Papua New Guinea, Volume 2, Coal Monograph Series No. 6, *Aus.I.M.M.*, Melbourne 1975.

17. Gloe, C.S., *Latrobe Valley Coalfields, Vic.* in Economic Geology of Australia and Papua New Guinea, Volume 2, Coal Monograph Series No. 6, *Aus.I.M.M.*, Melbourne 1975.

18. Knight, J.L., *Gelliondale Coalfield, Vic.* in Economic Geology of Australia and Papua New Guinea, Volume 2, Coal Monograph Series No. 6, *Aus.I.M.M.*, Melbourne 1975.

19. Knight, J.L., *Altora-Bacchus Mara, Lah Lah, Wensleydale, Binwerrin and Deans Marsh Districts, Victoria* in Economic Geology of Australia and Papua New Guinea, Volume 2, Coal Monograph Series No. 6, *Aus.I.M.M.*, Melbourne 1975.

20. George, A.M., *Anglesea Coalfield, Victoria* in Economic Geology of Australia and Papua New Guinea, Volume 2, Coal Monograph Series No. 6, *Aus.I.M.M.*, Melbourne 1975.

21. Hartnell, B.W., Smith M.J. & Wybrow, K.G., *Annual Report 1974-1975* Joint Coal Board, Sydney 1975.

22. Mengel, D.C., *The Role of the Geological Survey of Queensland in Exploratory Coal Drilling* Symposium on Coal Borehole Evaluation, *Aus.I.M.M., Southern Queensland Branch*, Brisbane 1977.

23. Cook, G.W., Preuss, O. & Thorpe, G.T., *26th Annual Report* The Queensland Coal Board, Brisbane 30 June 1977.

24. Leblang, G.M. & Svenson, D., *Planning of Exploratory Drilling Programs, Logging and Sampling of Exploratory Drill Holes For Coal*, Symposium on Coal Borehole Evaluation, *Aus.I.M.M., Southern Queensland Branch*, Brisbane October-November 1977.

25. Conybeare, C.B.E. & Crook, K.A.W., *Manual of Sedimentary Structures*, Bull. 102, *Bureau of Min. Res., Geol, Geophys.*, Canberra 1968.

26. McMahon, B.K., Douglas, D.J. & Burgess, P.J., *Engineering Classification of Sedimentary Rocks in the Sydney Basin*, Aust. Geomech. J. 65(1)5L, 1975.

27. Swarbrick, C.F., *Stratigraphy and Economic Potential of the Imjena Creek Group in the Surat Basin*, Geol. Surv. Qld. Rep. No. 79, Brisbane 1973.

28. Power, P.E. & Devine, S.G., *Surat Basin, Australia - Subsurface Stratigraphy, History and Petroleum*, Bull. A.A. P.6, Vol. 54 No. 12, 1970.

29. Price, A.G. & Svenson, D., *Drilling Equipment and Techniques in Australian Coal Exploration*, Symposium on Coal Borehole Evaluation, *Aus.I.M.M., Southern Queensland Branch*, Brisbane October-November 1977.

30. Le Page, A.J. & Proudfoot, B.W., *Use of Laboratory Data for Preparation Plant Prediction and Assessment*, ACIRL Report PR74-3, Ipswich Queensland 1974.

31. Proudfoot, B.W., *The Treatment of Bore Cores to Provide Reliable Data For Coal Preparation Plant Design*, Symposium on Coal Borehole Evaluation, *Aus.I.M.M., Southern Queensland Branch*, Brisbane October-November 1977.

32. Bennett, J.G., *Broken Coal*, J. Inst. Fuel 10, 1936.

33. Callcott, T.G., *Size Distribution and Single Particle Breakage*, BHP Tech. Bull. 10, 2, 26, 1966.

34. Le Page, A.J., *Laboratory Froth Flotation of Coal on a Continuous Basis*, ACIRL Report PR75-4, 1975.

35. Bennett, R.J., *The Commercial Evaluation of Coal Properties*, Australian Black Coal Symposium, *Aus.I.M.M., Illawarra Branch*, Wollongong 1975.

Western Canadian coal exploration is difficult because of the extensive folding and faulting of the coal beds, an example of which is shown above. To completely outline and delimit a coal bed is both costly and difficult.

Exploration in Western Canada

17. Problems and Solutions for Rocky Mountain Coal Exploration in Canada

by Eric W. Beresford

18. Exploring the Metallurgical Coal Deposits of the Canadian Rocky Mountains

by I. P. Dyson

17

Problems and Solutions for Rocky Mountain Coal Exploration in Canada

By Eric W. Beresford
Manager, Coal Division
Union Oil Company of Canada Limited
Calgary, Alberta, Canada

INTRODUCTION

The coal bearing measures of the Rocky Mountain region of Western Canada cover an area approximately 750 miles long by 30 miles wide and have been the scene of many exploration programs since the re-surge of metallurgical coal demand from the mid 1960's, to the present time. However only a small proportion, approximately 25 percent of this area has been explored in any great detail leaving the rest relatively unexplored.

Some 130 companies presently hold coal leases in the mountain and foothill regions of British Columbia and Alberta with lease blocks varying in size from 50 square miles to 270 square miles. (See Figure I - Western Canada major lease blocks).

These leases are at varying stages of coal exploration from basic field mapping through to completely proven properties by extensive drilling, sampling, and coal testing.

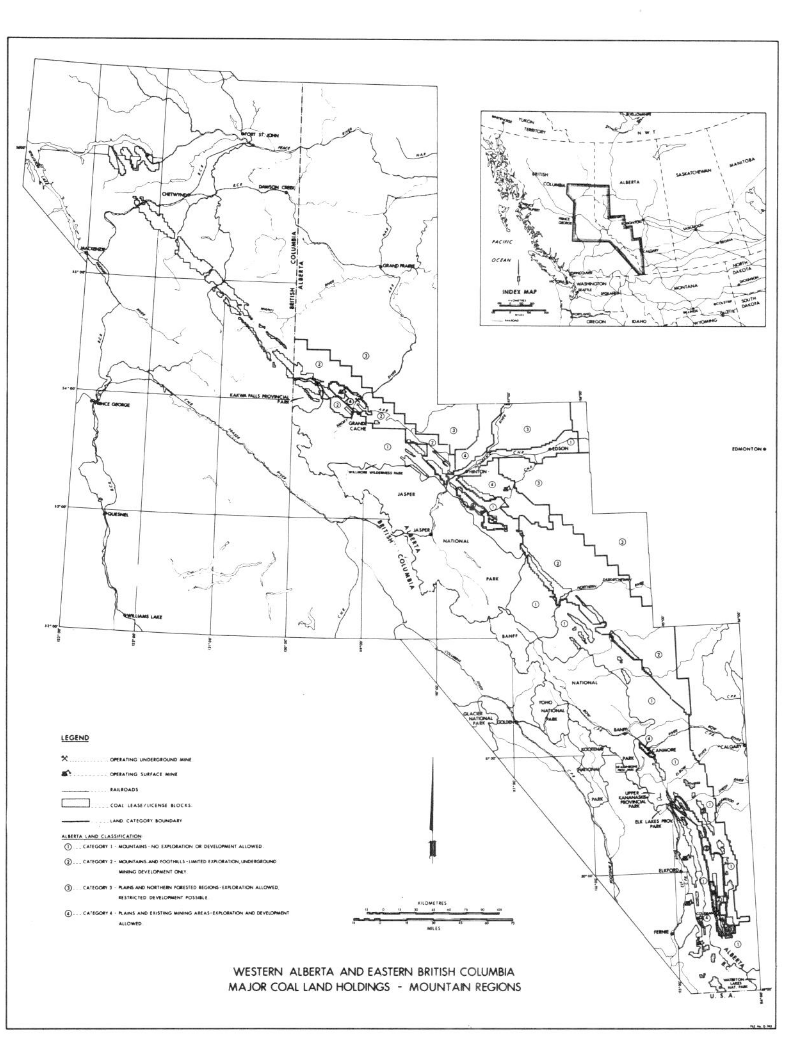
LEGEND
OPERATING UNDERGROUND MINE
OPERATING SURFACE MINE
RAILROADS
COAL LEASE/LICENSE BLOCKS
LAND CATEGORY BOUNDARY
ALBERTA LAND CLASSIFICATION
① ... CATEGORY 1 - MOUNTAINS - NO EXPLORATION OR DEVELOPMENT ALLOWED.
② ... CATEGORY 2 - MOUNTAINS AND FOOTHILLS - LIMITED EXPLORATION, UNDERGROUND MINING DEVELOPMENT ONLY.
③ ... CATEGORY 3 - PLAINS AND NORTHERN FORESTED REGIONS - EXPLORATION ALLOWED, RESTRICTED DEVELOPMENT POSSIBLE.
④ ... CATEGORY 4 - PLAINS AND EXISTING MINING AREAS - EXPLORATION AND DEVELOPMENT ALLOWED.
KILOMETRES
MILES
INDEX MAP
PACIFIC OCEAN
BRITISH COLUMBIA
ALBERTA
SASKATCHEWAN
MANITOBA
WASHINGTON
OREGON
IDAHO
MONTANA
NORTH DAKOTA
SOUTH DAKOTA
WYOMING
FORT ST. JOHN
DAWSON CREEK
CHETWYND
GRAND PRAIRIE
KAKWA FALLS PROVINCIAL PARK
GRANDE CACHE
WILLMORE WILDERNESS PARK
PRINCE GEORGE
QUESNEL
WILLIAMS LAKE
JASPER
JASPER NATIONAL PARK
EDSON
EDMONTON
BANFF NATIONAL PARK
YOHO NATIONAL PARK
GLACIER NATIONAL PARK
KOOTENAY NATIONAL PARK
CANMORE
CALGARY
UPPER KANANASKIS PROVINCIAL PARK
ELK LAKES PROV. PARK
ELKFORD
FERNIE
WATERTON LAKES NAT. PARK
U. S. A.
WESTERN ALBERTA AND EASTERN BRITISH COLUMBIA
MAJOR COAL LAND HOLDINGS - MOUNTAIN REGIONS

FIGURE NO. 1

In the mountain region there are only 7 operating companies producing 13 million tons of clean coal per year with 85 percent by surface mining methods and the rest by underground methods.

Before these mines came into production some 4 to 5 years of intensive exploration work had been carried out on the lease areas, leading through Feasibility Studies and more recently Environmental Impact Studies.

Upwards of $10.0 million dollars of high risk capital money may have to be spent on a property before the production decision is made and my paper today will identify some of the problems facing companies in the mountain region of Western Canada.

The geology of the region will be described in other papers but it is sufficient for me to say that generally the coal seams are intensely deformed by severe folding and faulting with varying thickness and seam quality, often over a very localized area, caused by tectonic movement during the mountain building process. (See Figure 2).

This makes the task of fully defining a coal deposit an extremely difficult and costly operation if no major surprises are to be encountered when actual mining has commenced.

The object of the majority of these exploration projects to date has been to find a deposit of low to med-volatile, high rank metallurgical grade coal hopefully within reasonable range of a railway line and a town.

PROBLEMS

The problems of coal exploration in the Rockies can be divided into the following basic groups:

1) Environmental restrictions.

2) Governmental restrictions.

3) Physical restrictions.

4) Coal quality determinations.

5) Reclamation of exploration activities.

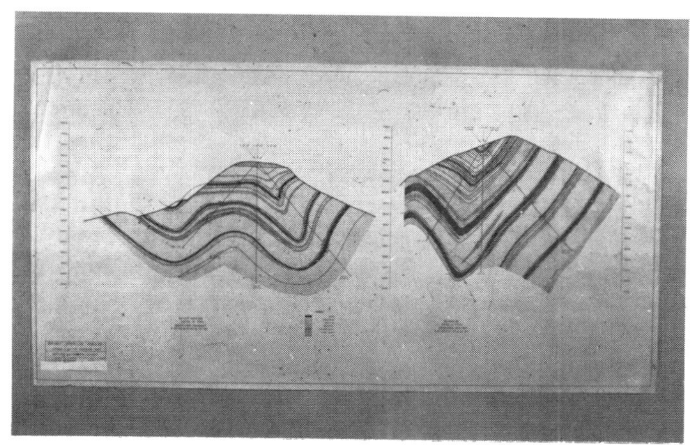

FIGURE NO. 2 - TYPICAL FOLDING IN MOUNTAIN STRUCTURE.

FIGURE NO. 3 - EXPLORATION ROAD DISTURBANCE.

FIGURE NO. 4 - EXPLORATION ROAD AND DRILL SITE.

1) Environmental Restrictions

Coal exploration, needless to say, suffers from the many environmental constraints imposed by governmental zoning of areas designated environmentally sensitive.

Most of the problems and concern experienced are damage to the fragile mountain alpine regions, siltation of water courses from roads and adits, and the opening up of previously pristine areas to the general public through the use of new exploration roads. (See Figure 3).

To minimize the amount of land disturbance and thus reduce the cost of reclamation a strong emphasis has been placed upon devising exploration techniques which are related to Environmental protection.

Since the implementation of an exploration program does not guarantee mine development a strong emphasis should be placed on environmental protection during the early exploration phase of a project.

Perhaps the most significant results have been gained by the use of hand trenching, as opposed to bulldozer trenching, to obtain detailed data from surface exposures of coal seams. In most exploration programs use of surface seam data is tied into data points from drill cores and in multi-seam strata numerous seam exposures are necessary to classify seam reserves into a proven category.

This intensive exploration of surface outcrops, when conducted by bulldozers, disturbs large areas not only in the trenching operation but also to provide access roads. (See Figure 4 and 5).

Hand trenching where field crew support is provided by helicopter dramatically reduces the amount of surface disturbance but is limited when depth of surface material overlying the seam is over 6 feet.

The cost of hand trenching is very low when compared to bulldozer or backhoe trenching especially when the cost of reclamation is taken into account.

The use of helicopters to move and support drills is also used in areas of high sensitivity and when long access roads are required in steep terrain and drill holes at say 2 to 3 mile intervals.

In the early stages of exploration of a lease area this method could be useful until a target area had been defined for sufficient coal reserves to justify construction of a new road system. With helicopter assisted programs the drill rig is sectionalized into about 7 unit loads and the men and materials airlifted to the rig each shift change from a base camp. The logging unit can also be transportable by helicopter to each drill site. (See Figure 6).

On a recent program involving helicopter assisted drilling on 8 drill sites spaced from 1 to 3 miles apart the additional cost amounted to $7.95 per foot of drilling. If a conventional access road had been built amounting to 12.5 miles and subsequent reclamation costs added, the cost would have worked out to $14.40 per foot of drilling.

Naturally there are some problems with using helicopters such as their vulnerability to high winds, visibility problems, steet terrain and sudden changes of weather patterns when flying in mountainous country.

It should not be the intention to use helicopter programs to replace conventional exploration techniques in all cases but rather to minimize the initial impact of exploration disturbance on a relatively virgin area.

2) <u>Governmental Restrictions</u>

In Western Canada coal measures are found both sides of the Continental Divide and both British Columbia and Alberta have different philosophies, legislation, and guidelines for coal exploration.

The Province of British Columbia recently opened up new areas in the mountain regions for coal licencing but has since frozen all claims until an assessment can be made. Designated Parks and Wilderness areas are naturally restricted but otherwise no zoning restrictions are imposed and all licence areas carry equal sensitivity.

Present legislation requires applications for a "Notice of Work" and a "Programme for Protection and Reclamation" to be made to the Chief Inspector of Mines under the Coal Mines Regulation Act prior to commencement of work.

The B.C. Government issued "Guidelines for Coal and Mineral Exploration" in 1977 and they form a comprehensive

FIGURE NO. 5 - COAL OUTCROP EXCAVATED BY BULLDOZER.

FIGURE NO. 6 - HELICOPTER MOVING PART OF DRILL RIG.

FIGURE NO. 7 - AUGUST SNOWSTORM AT DRILL SITE.

practical guide for carrying out exploration activities in that province.

In Alberta, following the Provincial Coal Policy announced in June 1976, the province has been divided into land classification categories which eliminate certain areas altogether for coal exploration activity and outline other areas for future developments.

These categories are summarized as follows:

Category 1 - Mountains and other areas of high environmental sensitivity:- no exploration or development allowed.

Category 2 - Mountains and Foothills - moderate environmental sensitivity - few infrastructure facilities:- limited exploration allowed but development restricted to underground mining methods.

Category 3 - Plains and Northern Forested regions in which land use conflicts must be resolved - mainly agricultural related:- exploration allowed and restricted development possible.

Category 4 - Mainly Plains and settled regions:- exploration and development allowed subject to normal procedures.

The net result of this is a virtual moratorium on exploration and development on new metallurgical coal properties, although some activity does continue, with existing operations in the mountains receiving expansion approvals.

There is no likely solution to these restrictions until changed circumstances in resource management policies make coal development essential to the welfare of the Province.

3) Physical Restrictions

These can be defined as terrain and climate with each one complimenting the other to make exploration difficult if not impossible in severe conditions.

The sphere of activity usually varies in elevation between 4,500 feet to 8,000 feet, often in relatively undisturbed mountain topography. Slopes are usually heavily forested up to the 6,000 foot level and require pre-logging prior to road construction through the area. Wildlife populations of elk, goats and sheep inhabit southwest facing grassy slopes which are important for winter forage and as sensitive areas, these should be avoided whenever possible.

Heavy snowfall, sub-zero temperatures and high winds make winter exploration extremely expensive and difficult and as such limits the "normal" exploration season to the 5 months between June and the end of October. Detailed surface geological mapping so necessary for interpretation of structure patterns and seam configuration is often curtailed by snowstorms in this period. (See Figure No. 7).

The job of physically keeping roads open in snow conditions, hauling water to drill rigs, freezing problems, and generally fighting the elements cause most companies to curtail field exploration activities in the mountains during the winter and spring "break-up" time.

However, I have worked on various exploration programs throughout the winter, usually to prove coal reserves required for development during the coming year, and that becomes exploration at any price.

To overcome freezing problems on diamond drill rigs, the rig is completely winterized having a propane heated drill shack combined with the mud-tanks and water storage tank. Water is hauled by a tracked Nodwell type vehicle usually from a river source and requires heating in transit. Usually one or two bulldozers of the D.6 or D.7 size equipped with ice lugs on the tracks are required for road work and snow ploughing.

4) Coal Quality Determinations

One of the most serious problems facing property evaluation is one of inadequate information concerning coal quality.

All too often as a result of poor coal core recoveries and a casual approach to adit and bulk sampling procedures. Coal Preparation Engineers have been mislead, resulting in expensive modification to Coal Preparation Plants when in production.

Often some of the problem rests in not being able to obtain a representative sample of the coal which will be mined and the fact that the geological seam boundary often bears little resemblance to the mining or recoverable coal zone when in production.

Extreme care should be taken therefore in testing all parts of the seam including allowance for seam dilution to arrive at an estimated raw coal feed to the plant.

All too often core recovery is poor and the old problem, when comparing the mechanical logs to the actual core recovered, is one of "what have we lost, coal or shale?". The variety of conditions encountered in rocky mountain coal exploration have made it very difficult to establish a set of drilling procedures that will consistently maintain both good core recoveries and satisfactory production.

Examples of typical core recoveries of 42 percent and 93.6 percent are shown in Figures 9 and 10 plotted from the geologist's measurements against the Sidewall Density log.

This gives an indication of where the core is lost from, in an effort to assess accurate quality determination of the seam.

Core recovery is dependent to a great deal on the condition of the coal seam at the point where it is intersected by the bore hole, with faulting, varying degrees of hardness through the seam, and the presence of water are some of the problems to overcome.

Coal in a semi or saturated state is difficult to retain in the inner tube, however, improvements to the wireline drill by way of hydraulically powered drill units, recorders for rotational torque, bit weights, fluid flow and pressure gauges allows the drill operator to rely on instrumentation more than the old "listen and feel" methods.

HQ wireline drilling tools recovering a 2½ inch core with the shorter 5 feet length core barrel has been found to be one of the most consistent methods for better core recovery, although PQ coring (3½ inch) has recently shown great success.

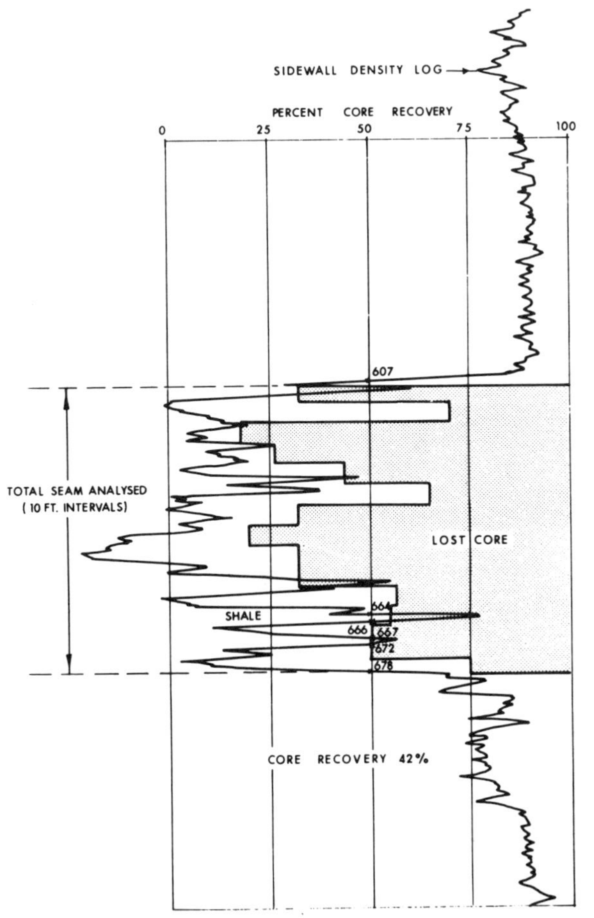

FIGURE 9
CORE RECOVERY RELATED TO SIDEWALL DENSITY LOG

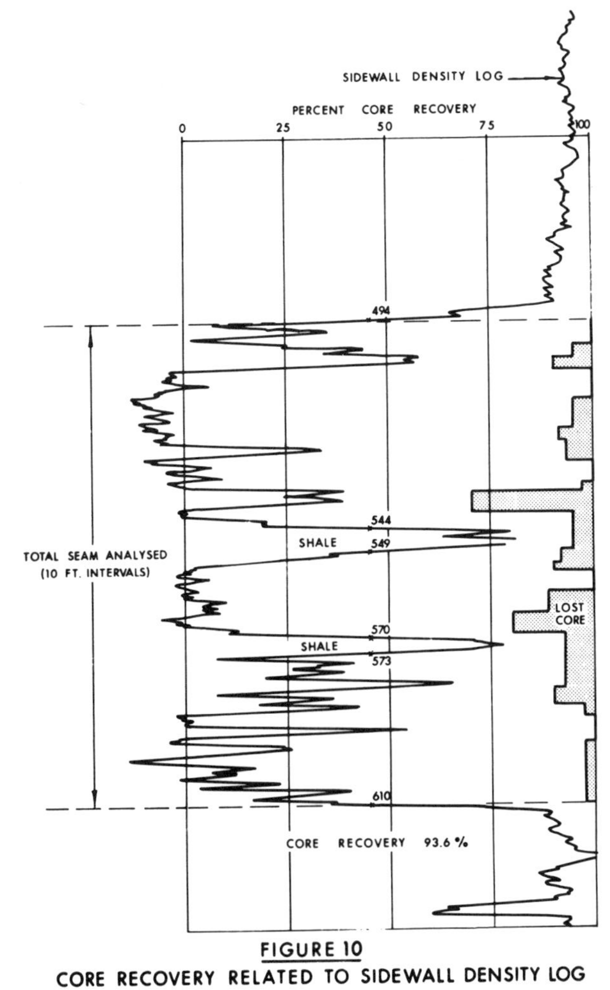

FIGURE 10
CORE RECOVERY RELATED TO SIDEWALL DENSITY LOG

The use of triple or split inner tube equipment has not improved core recovery but has proven to be a significant aid in minimizing loss in the handling of core.

Various coal drilling bonus schemes have been tried in an effort to motivate drillers to better core recoveries at the expense of penetration rate and to some extent have proved successful.

Additional payments of $10.00 per foot recovered in excess of 90 percent core recovery is now common practice on some programs.

Drivage of a 6 foot by 6 foot adit into the seam is often the only way to obtain a bulk sample for washability tests and also check the oxidisation rate against depth of cover. (See Figure No. 11).

Drivage costs are in the order of $85.00 per foot and this does not include reclamation costs or transportation of surplus coal material away from the adit site.

One of the chief problems in adit siting is to find a reasonable graded slope, away from a watercourse and where depth of cover is reached fairly quickly.

5) Reclamation of Exploration Activities

Until about 1974 the general approach to coal exploration in the mountains was by extensive access road construction followed by seam tracing of outcrop by bulldozer, trenching, test pit and adit sampling in conjunction with skid mounted diamond drill rigs requiring large drill sites.

Partly due to increased legislation affecting coal exploration and a general public awareness to environmental disturbance in mountain regions, reclamation planning and techniques have been developed to suit almost all situations.

The use of a large backhoe or Gradall machine working in conjunction with a D.6 type bulldozer can effectively reclaim exploration roads on hillside slopes up to 35 degrees (See Figures 12 and 13). The material cast over the side during road construction can be brought back into the road surface and reseeded to prevent further erosion and consequent degredation of water quality.

FIGURE NO. 11 - ADIT ENTRANCE IN COAL SEAM.

FIGURE NO. 12 - GRADALL MACHINE AND BULLDOZER RECLAIMING ROAD.

FIGURE NO. 13 - ROAD RECLAMATION IN PROGRESS.

These reclaimed access roads make excellent game trails after a few years when exploration has moved through an area. (See Figure 14).

The cost of reclaiming exploration roads as shown including seeding and fertilizing is in the order of $4,900 per mile, and the following example is given of a typical reclamation cost breakdown for a 15 foot wide exploration road.

COST BREAKDOWN

1) Gradall machine (Backhoe) with 80 foot long reach.
Production rate = 105 feet completed per hour.
Rental rate with operator = $55.00 per hour.

$$\frac{5.7 \times 5280 \times \$55}{105} = \qquad \$\ 15,765$$

2) D.6 Crawler Tractor.
Production rate including cross-ditching - 220 feet per hour.
Rental rate with operator = $32.50 per hour.

$$\frac{5.7 \times 5280 \times \$32.50}{220} = \qquad \$\ 4,446$$

3) Seeding - (mixture of 5 grass types).
Application rate at 100 pounds per acre.
34.6 acres x $0.90 per pound = $ 3,114

4) Fertilizer
Application rate - 200 pounds per acre.
34.6 acres x $0.10 per pound = $ 692

Labour

5) Seeders' wages.
2 labourers at $7.18 per hour + 37 percent fringe benefits.
Spreading rate of 3.1 hours per acre.
34.6 acres x 3.1 hours = $ 2,102

6) Supervision - transportation to site $ 2,000

Total Cost $ 28,119

Reclamation cost per mile $ 4,933

FIGURE NO. 14a Restored road cut

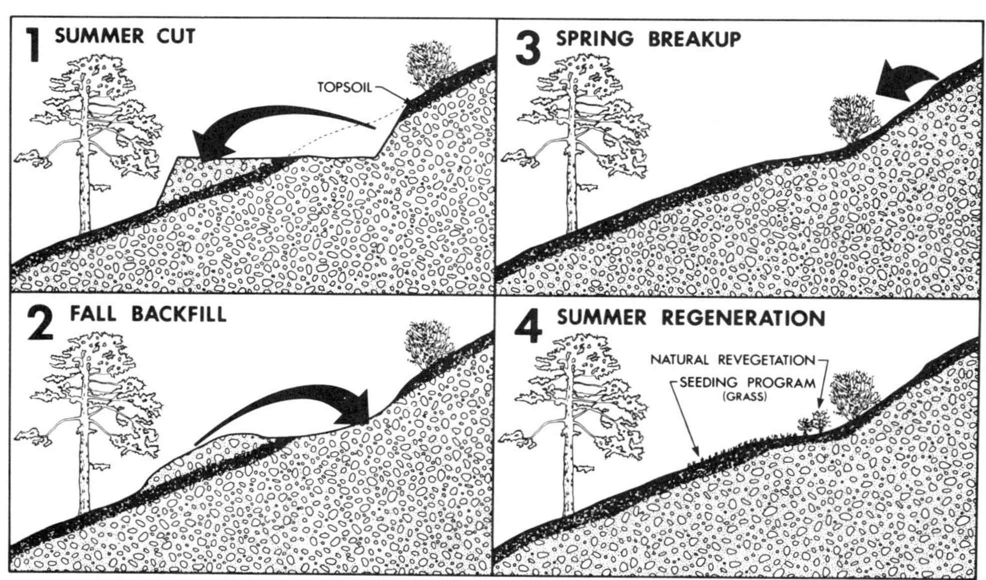

FIGURE NO. 14b Reclaimed road cut on side hill

Trenching and test pits should be covered in as soon as possible after use and seeded and fertilized to prevent erosion.

With respect to adit drivages they should be carefully sited and the surplus coal material should be hauled away from the site or if a convenient depression can be found the material buried and seeded over.

A security deposit or Performance Bond is usually fixed for an exploration program by the Provincial Government with the money usually around $1,000 per acre, refunded on completion of reclamation and to the satisfaction of the Inspector.

SUMMARY

In this paper I have attempted to identify some of the problems experienced in coal exploration in the Canadian Rocky Mountains. Most of the problems are not un-common in many areas of the world but when they are all put together can cause frustrating and often incomplete project data.

Large expenditures are currently being made in Western Canadian coal exploration programs which shows a confidence by companies in the continued growth of coal development for this part of the world.

ACKNOWLEDGEMENTS

1) Halferdahl & Associates Ltd. - Edmonton, Alberta.

2) Roke Oil Enterprises Ltd. - Calgary, Alberta.

3) D. McFarland - Meadowlark Farms, Calgary, Alberta

4) Coleman Collieries Limited - Coleman, Alberta

5) Reclamation Dept., British Columbia Mines & Petroleum Resources.

6) Union Oil Company of Canada Limited - Calgary, Alberta

7) Canadian Longyear Drilling Company - Vancouver, B.C.

DISCUSSION

QUESTION: Is $4,900.00 per mile total road costs or reclamation costs? What is reclamation cost?

ANSWER: The $4,900.00 per mile cost referred to in my paper is for the reclamation of the road only. This figure includes all machine time, involved in, restoring the ground surface back to its original contour, seeding, fertilizer and labor costs. The actual construction cost of the road in the first place would probably be in the order of $2,200 to $2,500 per mile in mountain areas.

QUESTION: Approximately, what is your increase in cost per foot drilled and logged in winter operations over summer operations?

ANSWER: Normally this increase in winter drilling is between $9.00 and $12.00 per foot depending on the severity of the winter conditions. This increase is attributed to the extra cost involved in keeping the roads open from snow and ice, the task of hauling water to the rigs and heating of the enclosed rig unit and mud tanks. The logging cost does not normally increase with a winter operation as in most cases the logger unit is contracted out for that particular program and paid on drill footage, so additional waiting and travelling time in winter causes no extra cost.

QUESTION: Do you plan an all year round or seasonal mining in the mountain areas? Do you plan strip or underground mining?

ANSWER: The mines are operated throughout the year and this includes surface and underground mines at elevations of 4,000 feet to 7,000 feet above sea level. A planned allowance is usually made at surface mines for 3 to 4 days shut-down in winter due mainly to blowing snow causing visibility problems for truck drivers and shovel operators. However, over the past few winters most mines have operated throughout the year with no breaks for bad weather. Various new underground and surface mines are currently being planned in the mountain regions of British Columbia and Alberta, and mainly contingent on obtaining markets for the coal.

QUESTION: Down to what temperatures can you do drilling and how do you avoid freezing and your drilling muds?

ANSWER: Well, we drill in temperatures down to 35 degrees to 40 degrees below zero Fahrenheit, which is cold. We avoid freezing of drilling mud by keeping the drills operating seven days a week, three shifts per day round the clock. Also, the drill rigs are enclosed with heated mud tanks and the mud is kept circulating wherever possible. Diesel fuel and other additions are used with the mud but this can affect the volatiles of the coal and should therefore be carefully watched if this is found to be the case. The water is supplied to the rigs in tanks heated by the vehicle's exhaust air or propane heaters to avoid freezing problems in transit and at the rig.

QUESTION: What are the light-colored lenses and pods shown in one of the slides in the thick and thin coal seam and do you use geophysical logging procedures in drill holes?

ANSWER: As to the light colored lenses and pods, they are shale and clay partings which occur in the thick seams. You can't really correlate them right through the property as they are intermittent and lense out rapidly. Yes, we do use geophysical logging methods in every hole both cored and uncored holes and a full suite of logs would normally be Gamma Ray Neutron, Sidewall Density, Resistivity and Caliper. However, more sophisticated Focus Beam, dip and orientation logs with ash and moisture determinations are being run. Logging techniques originally used in the oil industry in Western Canada were adapted in the late 1960's for coal exploration and used extensively. Due to hole caving problems in mountain areas one run is usually made through the drill pipe before the rig leaves the site.

QUESTION: Do you have to revegetate including planting trees or do you allow it to return naturally?

ANSWER: All we do is to seed the restored trails and drill sites with a mixture of grass seed and fertilizer and after a few years natural revegetation takes place including trees. We are not obliged to plant native species but the grass and legume seed mix must be approved by the Lands and Forests Department for each area. To help natural revegetation and provide a seed source the tree debris and wood slash is brought back onto the restored areas wherever possible.

QUESTION: Why not keep the Alberta rough roads open as fire breaks?

ANSWER: That is a good point except that the Regulations demand full restoration of the exploration roads after completion of a program. This can be amended in the field by a Forestry Officer if the road is useful but as a general rule the roads are fully reclaimed. The security reclamation bond is re-payable on final restoration to the satisfaction of the local Forester, which is usually 3 years after initial completion.

QUESTOIN: What is the quality of the British Columbia coal, and Alberta coal?

ANSWER: The B.C. and Alberta mountain coal is usually of high grade metallurgical quality and when washed has a specification of from 6 to 10 percent Ash, Free Swelling Index between 5 and 9, Moisture from 6 to 8 percent, Volatile Matter from 19 to 26, Sulfur 0.5 percent. Some seams or portions thereof are found to have a lower coke index but are excellent as thermal coals with a heat value of 12,000 to 14,000 Btus per lb.

QUESTION: In reference to the practice of air lifting drills, what problems were encountered in supplying water to wire line core drills and what were your solutions?

ANSWER: Wherever possible a water supply is obtained by pumping from nearby creeks or springs, but in winter when these freeze over and water lines are frozen solid the only way is to airlift water supply tanks slung beneath the helicopter. This has been done on many occasions and does not materially effect costs as the helicopter is based in camp anyway. Even in summer programs the water supply can be miles away from the rig making a water line too impractical to fix up. Also, the majority of holes make water during drilling and this is used to full advantage by the drilling contractor.

QUESTION: Why is coal core recovery so poor? Why can't something be done to improve coal recovery in the thick coal seams?

ANSWER: Well, we've been wrestling with this problem, at least I have, since 1969 and methods of improving core recovery are discussed in my paper. Briefly, some of these techniques are the use of hydraulically powered drills, better instrumentation, triple tube, split inner tube, mud additives and incentive bonus schemes for drillers. Some very good core recoveries in the 95 percent range have been obtained but the program average is usually around 65 to 70 percent. The coal is extremely soft and pliable and easily washed away with the drilling fluid. The raw coal fines below 28 mesh are about 35 percent of the total seam and are difficult to recover. The interbanded harder coal and shaley partings in the seam cause variations in bit penetration when drilling through thick seams often in excess of 40 feet and core is lost in this way. That is why an accurate indication of where the core has been lost from is critical in arriving at overall seam quality. The cores can give an average picture throughout the property, but bulk sampling normally by adits in the seam are a more reliable source of quality determination.

18

Exploring the Metallurgical Coal Deposits of the Canadian Rocky Mountains

By I. P. Dyson
Geological Consultant
Paul Dyson Consultants
and Holdings Limited
Calgary, Alberta, Canada

INTRODUCTION

This paper describes the development of exploration for metallurgical coking coal in Western Canada. It critically reviews what has been achieved and hopefully, thanks to hindsight, may be of assistance to persons engaged in exploration for coal in topographically and tectonically difficult areas.

Some of the methods used, the deficiencies of those methods, and those improvements which have been made, will be discussed. It is realized that the experiences of Western Canadian coal geologists exploring the Rocky Mountains will not necessarily be the same as those of geologists in other areas, but it is felt that the topographic and tectonic environment of Western Canada is very similar to that of certain South American countries which are probably on the brink of an exploration boom.

The paper reviews the general geology of the area as it relates to coking coal and reviews the development of coal exploration and exploitation in the Canadian Rocky Mountains up to the present.

GEOLOGY

So that an appreciation of the location and geology of the area can be realized, the stratigraphic and structural distribution of these metallurgical coals is briefly discussed.

Stratigraphic Distribution of the Coal

The metallurgical coals of Western Canada are primarily found along the eastern slopes of the Rocky Mountains (Figure No. 1).

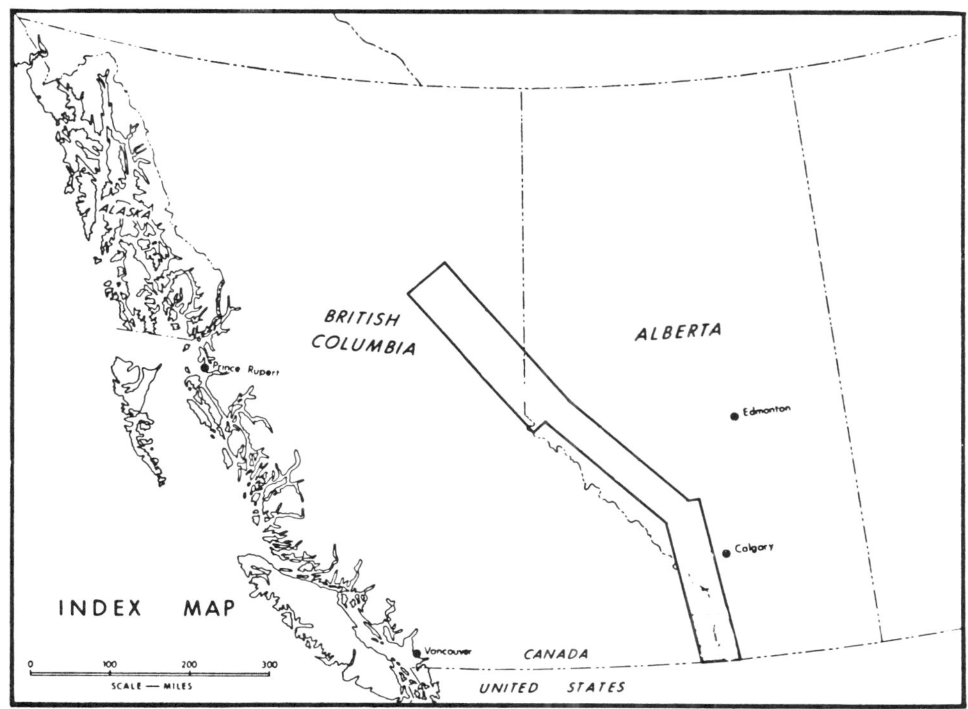

Figure No. 1 - Location Map.

The geographic distribution of the coal bearing formations covers the eastern Rockies coking coal belt from the United States border to northeast British Columbia. The coal bearing formations occur in linear belts paralleling the mountain front. This pattern of formational outcrop is a result of the geological structure of the area which will be briefly reviewed later.

The coking coals occur exclusively within Lower Cretaceous

and possibly Upper Jurassic rocks, but not wholly within the same formation. The rock units which have been recognized as coal bearing are the Kootenay formation, the Blairmore Group, the Luscar formation, the Commotion formation and the Gething formation. Some of these formations are lateral equivalents of one another and their inter-relationship is illustrated in the "Geological Distribution of Coking Coal Reserves" (Figure No. 2). A few general comments can be made with regard to the typical thicknesses and characteristics of the coal found within the different formations.

Kootenay Formation. The Kootenay formation contains potentially economic deposits of coking coal throughout the southwest of Alberta and the southeast of British Columbia. In general, the seams vary from thicknesses exceeding 40 feet in British Columbia to thicknesses of only about 8 feet just north of Calgary. Further north only one potentially mineable deposit of Kootenay coal is reported to exist. Coals from the Kootenay formation vary from anthracites to high volatile bituminous coals, are generally low in sulphur and, generally, contain less than 20% ash in situ. Coal is being mined from the Kootenay formation by Kaiser Resources Ltd., Fording Coal Ltd., Coleman Collieries Ltd., and Canmore Mines Ltd. These mining operations do not all produce coal from the same seam, but certain quality characteristics appear to be a problem which is common to all the operators. The main problem appears to be the difficulty in washing coals of the Kootenay formation to an ash content below 9% and at the same time maintaining an acceptable yield of product coal.

Coals of the Blairmore Group - Luscar Formation. The Blairmore Group and the Luscar formation are treated together as they are, in general, lateral equivalents as can be seen from the "Geological Distribution of Coking Coal Reserves" (Figure No. 2). As shown in this figure the coals of the Commotion formation in northeastern British Columbia are also lateral equivalents of these coals, but an arbitrary cut off has been made at the Alberta-British Columbia border.

The Blairmore Group contains coal seams which have been recognized as far south as a point just south of Calgary. However, coal seams that are potentially mineable are absent i Blairmore Group south of the Red Deer River. From this northwards to the Alberta-British Columbia border, one or more coal seams occur which have varying thicknesses with a maximum of about 40 feet for a coal seam in the Rock Lake area (30 miles south of the Smoky River area). Cardinal River Coal Co. in Luscar area are mining some seams that have coal thicknesses

AREA	FLATHEAD	ISOLATION RIDGE	CLEARWATER	MOUNTAIN PARK	SMOKY-COPTON	KAKWA / TORRENS	QUINTETTE-WOLVERINE	SUKUNKA	PEACE RIVER CANYON	PINK MTN.
OPERATING COAL MINES (1977)	KAISER RESOURCES LTD. SPARWOOD FORDING COAL LTD. ELKFORD	COLEMAN COLLIERIES LTD. COLEMAN CANMORE MINES LTD. CANMORE		CARDINAL RIVER COAL CO. LUSCAR	McINTYRE COAL MINES LTD. GRANDE CACHE					
LOWER CRETACEOUS	BLACKSTONE FM.	BLACKSTONE FM.	BLACKSTONE FM.	BLACKSTONE FM.	SHAFTSBURY FM.	SHAFTSBURY FM.	HASLER FM.	HASLER FM.	BUCKINGHORSE FM.	BUCKINGHORSE FM.
	BLAIRMORE GP.: MILL CRK FM BEAVER MINES FM GLADSTONE FM	BLAIRMORE GP: BEAVER MINES FM GLADSTONE FM	BLAIRMORE GP: BEAVER MINES FM GLADSTONE FM	BEAVER MINES FM GLADSTONE FM	MOUNTAIN PARK FM LUSCAR FM	LUSCAR FM: COMMOTION FM MOOSEBAR FM GETHING FM	COMMOTION FM: BOULDER CRK MB HULCROSS MB GATES MB	COMMOTION FM.: BOULDER CRK MB. HULCROSS MB GATES MB	COMMOTION FM: GATES MB	
							MOOSEBAR FM	MOOSEBAR FM	MOOSEBAR FM.	
							GETHING FM	GETHING FM	GETHING FM	GETHING FM
	CADOMIN FM	CADOMIN FM	CADOMIN FM	CADOMIN FM	CADOMIN FM	CADOMIN FM.	CADOMIN FM	CADOMIN FM	CADOMIN FM.	MINNES GP
?—?—	KOOTENAY FM: MUTZ MB HILLCREST MB ADANAC MB MOOSE MTN MB	KOOTENAY FM: MUTZ MB HILLCREST MB ADANAC MB MOOSE MTN MB	KOOTENAY FM	NIKANASSIN FM	NIKANASSIN FM	MINNES GP.	MINNES GP.	MINNES GP	MINNES GP	
JURASSIC	FERNIE GP.	FERNIE GP.	FERNIE GP.	FERNIE GP.	FERNIE GP.	FERNIE GP.	FERNIE GP.	FERNIE GP	FERNIE GP	FERNIE GP.

MAJOR COAL RESERVES

MINOR COAL RESERVES

Figure No. 2 - Geological Distribution of Coking Coal Reserves much greater than this, but these thicknesses are attributabl to tectonic thickening rather than to true stratigraphic thickness.

Coal of the Blairmore Group - Luscar formation vary from low volatile bituminous to high volatile bituminous, are low in sulphur and appear to have highly variable ash content, both as raw coals and as cleaned coals at acceptable yields. Cardinal River Coal Co. and McIntyre Coal Mines Ltd. are currently producing coals from the Luscar formation.

Coals of the Commotion Formation. The Commotion formation is the lateral equivalent of the upper portion of the Luscar formation. The term Commotion formation has been widely used throughout the northeast of British Columbia, and its relatic ship within this formational terminology is illustrated by the figure "Geological Distribution of Coking Coal Reserves" (Figure No. 2). As can be seen from this figure, it has beer however, subdivided into a series of members and the main coa occurrences are within the Gates Member.

Coal thicknesses found in the Commotion formation vary from maximums in excess of 40 feet in the Saxon Ridge area, bot only about 6 feet in the Sukunka area. It is reported that up to six seams have thicknesses which might be economically recoverable in the northeast of British Columbia.

Coals of the Commotion formation range from medium to high volatile bituminous, are generally low in ash and low in sulphur. While no commercial mining operations are currently recovering coals from the Commotion formation, all data so far obtained from exploration would indicate that these coals are readily washable to ash contents of around 6% and that they have the desirable characteristic of high fluidity.

Coals of the Gething Formation. The Gething formation is the lateral equivalent of the lower part of the Luscar formation It is recognized as far south as the Mountain Park area and extends throughout the northeast of British Columbia coking coal area. Seams within the Gething formation attain maximum thicknesses in the Pine Pass area where seams up to 25 feet thick have been reported. These thick seams generally contain one or more shale partings thicker than one foot. Thin seams are present in the Smoky River area, but in general, the seams of the Gething formation are about 6 to 8 feet thick in the area extending from the Alberta-British Columbia border to the Sukunka River. Coal seams within the Gething formation are thinning rapidly in the Peace River rea and no viable coal seams appear to be present in the ething formation more than 15 or 20 miles north of the Peace iver.

These coals are generally medium to low volatile bitumin-us, are very low in ash and low in sulphur. No commercial ining operations are presently recovering coals from the ething formation, but the Sukunka project is at the stage here an announcement of production schedule is anticipated. he testing of samples from this project indicate that these oals are probably well suited to washing to an extremely ow ash content (less than 5%) at acceptable yields. Coals rom the Gething formation are further reported to have the esirable characteristic of high fluidity.

tructural Distribution of the Coking Coals.

The coking coals of Western Canada occur along the eastern lopes of the Rocky Mountains and in the foothills. This is artly the result of the depositional pattern of the coal earing sediments and partly the result of erosion of the oal bearing beds.

The coal bearing beds were originally laid down along the estern edge of the present "plains" or "prairie" areas and ne depositional basins probably extended westward in the reas where the front ranges of the Rocky Mountains exist oday. In late Cretaceous times, the Laramide orogeny up-ifted the Rocky Mountains and the coal bearing rocks were ubsequently eroded from the present mountain areas. Simul-aneously they were uplifted along the present foothills area nd erosion has today proceeded to the point at which the oal bearing rocks now occur at surface in much of the foot-ills belt.

In general, the geological structure of the foothills elt reflects the crustal shortening that accompanied the aramide orogeny. The area is characterized by folding and aulting parallel to the mountain front of varying intensity. n general, the faults tend to be reverse or thrust faults nd the folds are often very sharp and assymetrical.

A detailed discussion of the structure of the whole area s beyond the scope of this paper, but a few general comments an be made. In general, the intensity of the structural eformation decreases from south to north along the foothills o that broad folding occurs in northeast British Columbia, .g. Sukunka area, but is almost unknown in southern Alberta.

This results in good prospects for conventional underground mining of low dip seams in the northeast British Columbia area but essentially no good prospects for this type of minin in southern Alberta (Figure No. 3).

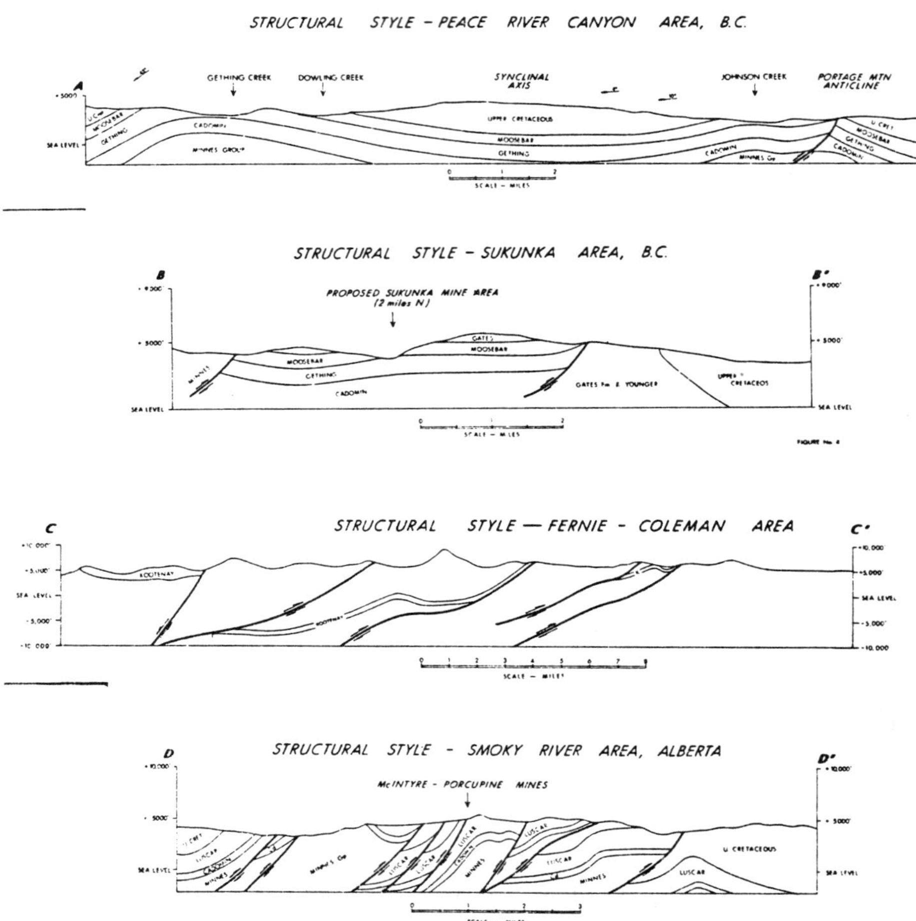

Figure No. 3 - Structural Style, Western Canada Coking Coal Area.

The high dips and associated thrust faulting in southern Alberta, besides making underground mining difficult, result in a very complex geological picture. This complexity, in turn, results in the need for more detailed exploration and resultant higher exploration costs.

A benefit that can result from this intense structural deformation is the "flow" of coal under deformational pressures into large "pods" along the structural axes. The mining of these "pods" has been exploited by Coleman Collieries Ltd. recently at Tent Mountain and an analogous situation exists in the Luscar area.

HISTORY OF PRODUCTION DEVELOPMENT

The increase in world-wide demand for coking coal in the 1950's and 1960's sparked new interest in the coking coals of Western Canada.

Initial exploration and upgrading of existing small mines which had managed to stay in business following the conversion of the railway to diesel commenced in the 1950's. Coleman Collieries Ltd. obtained the first export contract and started to export coal to Japan in 1967.

Early exploration targets were, at this time, in areas which had previously been coal producers or had been subjected to some exploration in previous coal mining periods. The exploration objectives were primarily new open pits and areas of low dipping structurally simple coal seams which could be exploited with conventional underground coal mining equipment available either from the U.S.A. or U.K.

As a result of this surge of exploration activity four additional new mines can into being on the basis of contracts to ship coking coal to Japan. These were the mines of McIntyre Mines Ltd., Cardinal River Coals Ltd., Kaiser Resources Ltd. and Fording Coal Ltd. (Figure No. 4).

The mine of Cardinal River Coals Ltd., owned 50% by Consolidation Coal Co., began production in February 1970. This mine produces approximately 2 mm tons per year from an open pit in an intensely folded area. The area had previously been worked underground, but was abandoned in the early 1950's.

The open pit of Kaiser at Sparwood commenced production in March 1970. This pit adjacent to the old underground mines of Michel and Natal produces about 4 mm tons of coking coal per year. Kaiser also produces about 750,000 tons per year from its underground hydraulic mine.

In May 1970 production commenced at the Smoky River mine of McIntyre from an underground operation scheduled to produce 1 mm tons per year. There were plans to expand production to 3 mm tons per year on the basis of a successful underground operation and new open pits. McIntyre presently has the capacity to produce approximately 2 mm tons per year with about 40% from the underground operation.

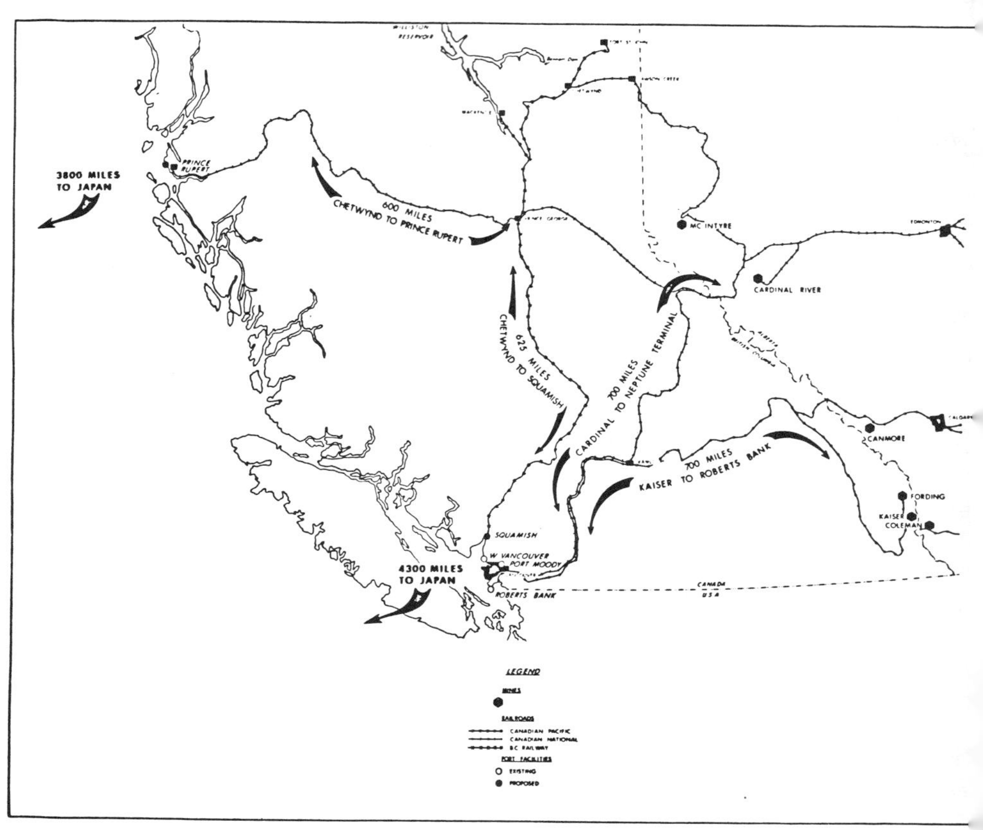

Figure No. 4 - Mines - Transportation

In March 1972 the open pit mine of Fording (60% Canadian Pacific Investments Ltd., 40% Cominco) commenced production. All the production is from an open pit at a rate of 3 mm ton per year but the operation could be expanded to 4 mm tons pe year relatively easily. This was an area previously explore but not exploited as also was the case with McIntyre.

Both McIntyre and Fording had to construct new towns and new railways were required. Fording built 35 miles of new railway spur whereas approximately 100 miles of new railway linked the Smoky River area to the existing rail. In the Smoky River case, the rail serves not only the mine but also provides a new connection to northwestern Alberta from Briti

Columbia. The construction was funded by the Alberta Government and the railway has reportedly been a financial burden since it was built.

It should be mentioned that Canmore Mines Ltd. also produce coal at a rate of less than 500,000 tons per year for the Japanese market. This production is mainly anthracite sold to specialist consumers. Similarly Byron Creek Collieries Ltd. located between Coleman Collieries Ltd. and Kaiser Resources Ltd. are producing this year about 1 mm tons of coal - 75% for thermal use - 25% for a formed coke process in Japan.

The development of these new projects created a rush for coal leases in the 1960's and by about 1970 all the prospective and even non-prospective areas underlain by coal measures were under lease.

For the last ten years coal exploration has been proceeding rapidly. The standard of work carried out has varied greatly. Today several projects are at the stage that if the market were to open up with a corresponding priced increase, then contracts could probably be obtained for the export of the coal and new mines would go into production. (Fording in 1972 was the last new mine producing metallurgical coal to commence production. There are no new mines in the construction stage or scheduled for production.

REVIEW OF EXPLORATION

With the benefit of hindsight it is possible to look back and see the problems encountered in the past and assess what is being done in exploration today.

For the mines that went into production, there were, of course, the numerous start-up problems which plague all new developments. However, rock stability problems both in open pits and underground were greater than anticipated. The raw coal delivered to the preparation plants was not as anticipated either chemically or physically. These type of problems are one which should be minimized by a thorough exploration program.

Early exploration projects were primarily run by geologists unfamiliar with coal exploration. They were either "hard-rock" geologists or petroleum geologists with no background in coal exploration and with little or no experience in the various disciplines associated with the development of a coal mine. Management usually consisted of coal executives with an engineering background who were often excellent at managing relatively small mines which still existed, but who had little or no appreciation of geological exploration. For years these coal operators had had no exploration budgets.

The hard-rock geologist usually performed very detailed surface mapping and sampling of all and every seam down to 20 cm in thickness followed by diamond drilling with no borehole geophysical logs. The samples were analyzed but only on the basis of a simple proximate analysis usually of the raw coal. The petroleum geologist, on the other hand, was usually general in his surface mapping as he had always sought large structure before. He was somewhat scornful of the "primitive" technology of the diamond drill and punched down rotary holes and ran geophysical logs. Once again, the samples were sent to a laboratory for proximate analysis. These laboratories in the early 1960's in Western Canada often knew nothing about coal but were able to read the ASTM test specifications and produce a proximate analysis.

Looking back it is possible to recall numerous exploration programs which did not obtain the data required to really assess the potential of properties from the point of view of a total feasibility. Even the mines that went into production did so on the basis of what today would be an inadequate exploration and testing program. In addition to the lack of coal experience by most of the companies, the customers, in the Canadian case, the Japanese, were not very demanding at that time. Today all this has changed. Hopefully, a better program can be planned to minimize mining and preparation problems after start-up. Additionally, the customer is certainly demanding more data in view of the previous shortfalls in production and the wide choice of new coal sources.

Probably there is no simple answer to achieving a first class exploration program but there is one very important factor. Geology must not be an end in itself. The persons in charge of the program, often geologists, must appreciate surveying, geology, coal mining both open pit and underground coal preparation, geotechnical and environmental problems. They need not and, in fact, will not be experts in all these disciplines, but a basic knowledge is essential so that data required for all these phases of the feasibility is obtained from the start.

Bearing these factors in mind and with the Western Canada experience as a background, the following very basic exploration advice is believed worthy of repetition:

(a) Before attempting any detailed geological mapping make new and accurate topographic maps. Transfer field data from air photos to topographic bases using the stereo-plotter equipment. Many hours of mapping and helicopter expense have been wasted by people making "approximate maps" and then trying to do detailed work.
(b) In structurally complex areas core as much as possible. Rotary holes may be slightly cheaper, but the additional geological, engineering and geotechnical information available from the core more than offsets any additional cost.
(c) Survey drill holes for both degree and direction of deviation. This was not often done in early programs but has been found to be significant.
(d) Run geophysical logs that are useable. Do not select the contractor on price but on quality.
(e) If the core is not all described in detail, preserve it carefully. The geotechnical and engineering staff may want to use it.
(f) Analyze the coal samples in conjunction with the advice of the coal preparation engineer. It is he who has to design the plant based on the data being obtained.
(g) Bring the geotechnical and mining engineers into the program at an early stage. Money and time can be saved by using the exploration program from the start for the benefit of all rather than just for geology.

These seven points are certainly not intended to be an all inclusive guide to exploration but they are items in which many Western Canadian programs were deficient. They are the items which have required unnecessary repetitive work and expense when the deficiencies of the initial exploration on a property have been analyzed. It should not be necessary to remap, redrill, relog and resample which has been the case so often when second rate work was done during the initial program.

PRESENT STATE OF EXPLORATION

Today in Western Canada there are several metallurgical coal projects which have reached the stage at which a production decision could be made. The depressed state of the coking coal market has not allowed the completion of new contracts and all the time the capital costs of projects must be revised upwards.

Several projects are technically ready for development (Figure No. 5). In addition to those shown, other projects such as expansions of the Kaiser and Fording operations could also be put into production. The potential production of

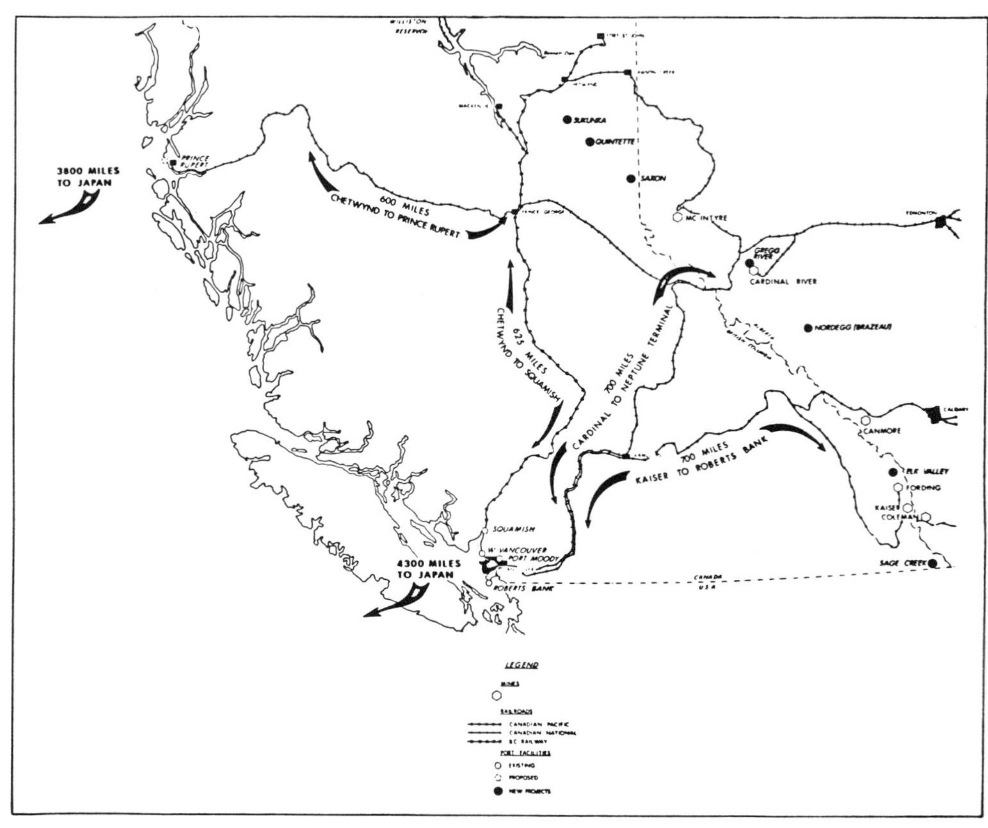

Figure No. 5 - New Projects.

these new operations totals in excess of 20 mm tons per year.

Exploration is continuing today on practically all these properties. In 1978, a minimum of $15 million was spent on metallurgical coal exploration in Western Canada to modify and finalize the engineering designs. In most cases the companies which are completing these feasibilities have on staff mining engineers, coal preparation engineers and other specialists necessary for a comprehensive feasibility study. Excellent studies are being carried out today and when the market develops new mines will hopefully be put into production with a minimum of start-up problems.

ACKNOWLEDGEMENTS

The valuable experience gained from many years of association with numerous colleagues and clients is gratefully acknowledged.

DISCUSSION

QUESTION: What special characteristics of the seam make it desirable for hydraulic mining techniques?

ANSWER: I, myself, am not particularly familiar with the details of the kind of hydraulic operation of Kaiser Resources in the Balmer seam, which is a very successful underground operation. The main things I think they have--and I'm open to correction here as there are some Canadian people sitting in the audience who are more familiar with it than I. They have several advantages. One is a thick seam--40 to 50 feet of coal. They're able to bring this coal to the surface without having to pump coal uphill. Basically the coal is sluiced down to the mine entry with the water which was used to cut the coal, and they also have adequate roof and floor characteristics which enable the caving and the general operation of the mine to be successful. I am sure, for instance, that Eric Beresford with his mining background, if you are interested, would know much more about this than I do. When we started exploring for coal in Western Canada, metallurgical coal--everybody said, "Well, what have we got to find?" The answer was "Well, we've got to find low dip areas, which we can mine with the equipment everybody knows about that is used in the United Kingdom and the United States. We should also look for these structurally thickened pods which might be suitable for open-pit mining."

Later the philosophy was such that everybody jumped on the hydraulic bandwagon for a while. We were looking for reserves above drainage in seams which were thick enough to hydraulically mine. That, I think, is a key phrase. Whoever you talk to has a different opinion as to what is thick enough to hydraulically mine. I think, generally speaking, numbers that I hear talked about by the mining engineer are in the region of 10 to 12 feet. You have the problem of development headings. So one of the things we have in our steeply dipping beds (with the price squeeze that we have with our sort of $15.00 a ton cost to move the coal to our export port) is that we can't afford to be doing any rock development or very little, and so this becomes a very key thing in explorations. We have to have seams which are thick enough to allow us to do our development roadways in coal. This is a somewhat limiting factor for seam thickness.

QUESTION: Are you using angle drilling exploration of steeply dipping coal seams?

ANSWER: The answer to that is "yes." I think if you are at the stage that you know which way the beds dip, presumably almost to the bottom of the hole, I think this is a good technique. You can often shorten the footage to some extent, but I think it can also be a dangerous technique in areas where we don't know what the dips are at the bottom of the hole, because we quite often drill a hole and we'll get a change in dip. A very significant change in dip. If we are not able to pull oriented cores, (I know this is technically possible, but it hasn't been practically possible in coal exploration.) then once you get a change in dip in the hole, you're not really sure which direction the dip is. If you have an angled hole, say at 45°, and you have a dip in the bed, which is at 45° to the drill hole, you don't know whether it's vertical or horizontal. Quite often you can make a good guess, but it is a little bit of a dangerous practice unless you really understand the structure. I think with the advent of the dip meter, which we are talking about here, this could probably be solved. We could get into the discussion, that maybe we want to drill rotary holes instead of diamond holes and then use the dip meter, but I think that you can get a lot of special information from a dip, and I think in a structurally deformed area we are going to have to continue to core the holes nearly all the way, and probably use the dip meter as well. Just as a matter of interest, it is not the complexities that we have in the structure. I've been in places where we've drilled two holes off the same set-up. One is 45° and one vertical. And even at a depth of 300 feet where the drill holes, you know, are probably only a couple of hundred feet apart--my trigonometry isn't instant--we have almost an impossibility to draw a cross section, so any thoughts of wide-spaced holes are very difficult. But the angle hole does have that special application and is very helpful.

USSR: Exploration Drilling

19.

Coal Extraction in Deep Mines: Improving Geological Estimates of Reliability

by V. S. Borisov and E. V. Terentyev

19

Coal Extraction in Deep Mines: Improving Geological Estimates of Reliability

By V. S. Borisov
Deputy Chief of Sojouzuglegeologia
and E. V. Terentyev
Ministry of Coal Mining
Moscow, U.S.S.R.

The production of coal in the USSR increases continuously and will continue to increase in the future. On the background of expanding mining operations in the Eastern parts of this country, mainly at the expense of open pit mining, the existing mines in the developed coal basins, such as Donetzk, Pechora, Karaganda, and Kuznetzk, which produce high BTU and coking coal will be modernized and made more effective.

The increased depths of operation (500 to 800 meters) in these basins brought about specific phenomena which were not common at shallow depths.

These are outbursts of coal, gas, and rock, elevated temperatures and rates of gassing, unexpected buckling of sufficiently hard rocks, and caving of mine workings, etc.

As the frequency of these phenomena with depth increased the problem of the so-called "great depths" has become very important for the geologists and miners.

Let me give you a few examples to illustrate the importance of the problem.

The sudden outburst of 14,000 tons of coal and 250,000 cubic meters of gas occured at the depth of 710 meters during the development of seam "l" in the Gagarin mine in the Donbass Basin.

In one of the mines of the same production association "Ordjonikidzeugol" and in the same basin the sudden outburst of 5,000 tons of coal and gas occured at the depth of 750 meters during the development of a 0.2 meter thick seam "K_6". Prior to the opening of the seam a complex of antioutburst measures had been undertaken which reduced the gas pressure at the place of the opening from 42 to 2 atmospheres.

In the Kalinin mine in 1972 the sudden outburst of 400 tons of coal occured during drilling of the 43 millimeter diameter hole at only 3.5 meters depth.

Mechanization of coal mining and increased loads on each wall brought greater rates of gas emission in mines. Over 70 percent of the Donbass mines are classified as "supercategory" by explosion hazard and some of these mines have the gas release rates of 150 cubic meters per ton of daily production. It is becoming especially important in these mines to have high accuracy methods to predict rates of gas emission and to determine air volumes needed to provide safety for miners. The sampling of coal seams and imbedding rocks had to be done in the recent years to study in more detail gas content of the already assessed deposits.

It will be further shown that some of the above phenomena may occur at significantly shallower depths than 500 to 800 meters which means that the term "deep mines" may define the mines of different depths in different coal fields and even in the same basin making the prediction more complicated.

Thus, the term "deep mines" is supposed to define such depths where under average conditions in the considered strata a regular increase of limit stress states is observed in the elements of the imbedding rock masses manifesting themselves immediately or some time after an opening is made in the form of brittle failure or viscous elastic yield, i.e. extrusion of rock into the opening from bottom, wall, or roof.

Although this definition is too conditional, it has a clear, specific meaning in physical terms which every day gives cause to make our miners worry and to undertake protective measures according to their engineering genius. It means that our mining engineers have gained a certain successful experience in mining at the depths to 1000 meters in the Donbass conditions, though not enough to extrapolate it over "deeper levels".

General characteristics of the parameters which change with depth and affect geological conditions of coal mining will be discussed below. Also some problems of predicting these changes with higher accuracy that may be instructive are discussed.

The criterion which determines the degree of precision required for certain parameters evaluated at the stage of geological surveying is a detailed description of the variability of the parameters in their relation to the geological structural features of any coal field developed at great depths or on the surface. There is no one recommendation about the frequency of sampling that may be spread over all possible situations. In all instances the frequency of sampling must correspond to the variability of the parameters in question.

It is especially important to have a good agreement between the results of subsurface sampling and real values. In this connection, the accuracy of instrument used for sampling and the errors of averaging techniques must be taken into account. The degree of errors introduced by averaging completely depends on the frequency of sampling. Both frequency of sampling and averaging technique are specific for every individual deposit.

It is equally important to establish a quantitative relation between the errors introduced by instrument and averaging. A prospecting geologist must know specific weight of each error. Otherwise, the need for more reliable data will substantially increase the costs of prospecting.

TEMPERATURE

As the mining operations advance to deeper horizons the problem of temperature regime control has emerged, because the natural temperature of rocks at these depths is 45 to 50° C., as in the Donbass coal fields. The registered air temperatures in the deep mines without the means of conditioning is 27 to 34° C. and relative humidity is 85 to 98 percent.

According to MakNII, a comparative analysis of temperature changes in underground air and rocks in deeper mines shows that the air temperature is affected, beside rock temperature, by such factors as the load on the mining fall and rate of development work, increased mechanization of mining, and the use of conveyors to transport coal.

For example, the air temperature increment in the 180 to 220 meter long wall at the depth of 800 to 1000 meters varies from 4 to 10° C. dependant on load.

The increasing rate of development works from 60 to 80 to 350 meters per month at the depth of 900 meters raises the air temperature in the wall by 5° to 7° C. under similar conditions.

In addition to the abovementioned factors, the temperature in the deep mines is significantly affected by the oxidation processes in the mines, compression of ventilation air, mode of ventilation, maintenance of workings, water inflow, etc.

The analysis of a heat budget in the existing mines shows that the specific contribution of the local heat sources is 25 to 50 percent. The other components of the budget come from dry heating or evaporation in the imbedding rocks. However, the specific contribution of local heat sources into the total budget will decrease with the increase of depths and temperatures of rocks despite growing mechanization. This trend should be taken into account at the stage of planning operations at lower horizons and designing new mines for great depths. Therefore, the prospecting geologists remain responsible for obtaining direct temperature data from deep levels in their relation to the structural elements of coalfields, hydrothermal situation, and other geological features. The average geothermal gradient for some coal basins of the Soviet Union is as follows: 3° C. per 100 meters for Donetzk, 3.5° C. for Kuznetzk, 1.7° C. for Karaganda, 1.3° C. for Kizelovsk. However, the geothermal gradients change significantly within the boundaries of individual geotectonic zones of coal basins.

The highest geothermal gradients in the Donetzk basin (3.5 to 3.9° C.) are characteristically found in the area of the Main Anticline of the Central Donbass with complex tectonics. As a rule, such regions are characterized by high gas emissions in the mines. The broadest range of the thermal gradient fluctuations (from 1.7 to 5.4° C.) is observed in the Kuzbass, where it is partially affected by the abnormally high heat flux in the Southeastern part of the basin. At the same time, the geothermal gradient in the Kizelovsk basin is as low as 1.3° C., fluctuating from 1.0 to 1.6° C., which is caused apparently by the cooling effect of ground water contained in the Visean and Turean Karstic limestones directly connected to surface water.

The above information adds to the importance of careful studying of the temperatures of coal strata at the stage of geological surveying coal fields with the aim to help designers to provide optimal environment in the modern highly mechanized mines.

GAS CONCENTRATIONS IN COAL STRATA AND GAS EMISSIONS INTO MINE WORKINGS

The attempts to increase coal production and to move mining operations to greater depths causes considerably higher rates of gas emissions into workings which in turn affects engineering plans to ensure a high rate and safe coal production.

A safety program in a mine is closely related to the composition, concentration, and distribution of gases in coal strata and rocks. Therefore, the study of natural gas emitting conditions and geological factors which control generation, accumulation and distribution of gas in coal strata, as well as the mechanism of gas flow into mines is the first priority task for prospecting new coal fields and developing deeper horizons of the operational mines.

Gas concentrations in coal and rock strata affect, in many ways, the designs of the development methods of deposits and the systems of their development, including loads on wall and coal extraction technology. Therefore, the data on gas distribution with depth becomes especially important.

It is equally important, both for the main developed coal fields such as Donetzk, where 1,200 to 1,300 meter-deep mines are now under construction, and for the Kuznetzk and Karaganda basins where the coal seam gas emitting rates at the present depths have already exceeded those for the same depths in Donbass.

The experience we gained in mining shows that the methane concentration increases down the dip of a coal seam with the increasing degree of metamorphism of the latter. G. D. Lidin in his theoretical analysis of the methane concentrations with depth showed that the methane occlusion by coal became constant at the depth of 800 to 900 meters, while the total methane concentration in coal grew continuously with the growth of gas pressure at the expense of the free methane entrapped in coal pores. Indeed, below 600 meters level the increase in methane concentration per 100 meters gradually stops, because the coal at this depth reaches maximum absorption capacity. It seems that the gas from coal seams at depths around 1,000 meters will decline considerably and further deepening of mines will not cause an increase of the gas inflow rates.

This fact, however, must not be interpreted as the basis for believing that below 700 meters gas inflow will decline. Actual inflow in the deepest gassy mines in the Donetzk-Makeevsk region is considerably greater than in the workings at the depths below 700 meters which is a result of methane release from imbedding rocks. In some instances significant gas flow from imbedding rocks is observed at shallower depths, provided the geology is conducive. For example, the "Tomashevskaja" mine became gassy during initial development. The mine has been sited in a dome structure between two large thrusts. The discharge shafts of the mine have extracted over 24,000,000 cubic meters of methane without exhaust flow equipment during three years of operation. We found in many instances long life gas storages in operating mines developed in brachyanticline and flexure ally young structures or in thrusted areas. The filtration and absorption characteristics of rocks become better under these or similar geological conditions and they act as the main conduits for gas travel into a developed mine.

The presence of gas in imbedding rocks can be detected first at the stage of exploratory drilling. It may present itself either in the form of gas bubbles in drilling mud or explosive release in large volumes. Typically, the prospecting holes for gas are found in the small size flexure or anticlinal structures affected by large tectonic faults.

According to Professor A. I. Kravtzov, G. I. Voitov, and O.V. Tender gas emission from coal bearing rocks will increase with depth as probed by the existing data. Greater concentrations of heavy methane homologues, including oil and bitumen in certain localities, are expected from the rocks below 1,300 meters and even at shallower depths, as was the case in the mines of Donetzk-Makeevsk, Kadievsk, and Krasnoarmeisk areas of Donbass. The clear-cut tendency of this sort will undoubtedly control physico-chemical properties of the underground atmosphere. The existence of such a trend is supported by the discovery of a number of commercial gas deposits at the Northwest and Northern limits of the Donbass region.

It seems that in all coal basins of this country the gas from imbedding rocks will be in equilibrium with the gas from coal seams at the depths of 650 to 700 meters, although deviations from this pattern are possible. For example, the gas flow from coal bearing rocks in the Kuznetzk and Karaganda basins, where coal concentration is high, becomes adequate to the gassing from coal seams already at the depths of 500 to 400 meters and shallower. It means that the estimates of gassing in deep mines must be based on accurate assessment of natural gas distribution both in coal and rocks and potential "explosive" gas from coal and rock outbursts. It is clear that gas in deep mines will rise with depth in direct proportionality to initial gas concentrations in coal and rocks.

ENGINEERING AND GEOLOGICAL ESTIMATES DURING PROSPECTING AND EXTRACTION OF MINERALS

The advancement of coal mining to deeper levels brings about new problems of large size solid and developmental works which need to be supported and maintained under complicated geological conditions without undue overload on capital and production costs.

The engineering solutions found for small and medium depths problems are no longer applicable for deep mines designs.

In the course of the last two decades the engineering geologists have worked out a new field of interest, i.e. engineering geology in mining.

The properties and states of rocks are studied both by geologists and mining engineers. The reports of geologists are of typically descriptive character about petrology, mineralogy, texture, structure, lithology, and facial features of rocks. This information is used to establish a relationship between coal properties and metamorphism of coal and rock at certain depths.

The works of the experts in rock mechanics are usually devoted to strength, strain, and geophysics of rocks that may be easier to quantify and present in the form of mathematical analysis.

Both types of studies are useful for practice and science, although they may have certain demerits such as that "geological reports are difficult to use for engineering computations, which mathematical studies are based on too idealized concepts of medium".

We still do not know which characteristics of rocks must be taken as controlling parameters of rock pressure and how they can be eventually used in mine designs. This is the reason why engineering and geological information may be inadequate in some cases and unused in the others. It has been established that depth affects the behavior of rocks in different ways. On one hand, it decreases porosity and water content, increases density and strength, and reduces the number of open cracks which all should increase the stability of mine workings. On the other hand, the stress in rocks increases and triggers rheological processes, failures and instability, and sometimes outbursts and rockbursts. It is also probable that stresses grow faster than strength with depth.

Various studies showed that the variability of the physico-mechanical properties of rocks in Donbass with depth from 200 to 400 to 1,200 to 1,400 meters was not uniform even in the areas of occurence of the same grades of coal. The absolute value and character of such variability is greatly affected by the lithologic and petrogenical characteristics of rocks, depth of occurence, residual tectonic stresses, geological faults and hydrogeological conditions, etc. Therefore, it is necessary to establish an optimal number of the physico-mechanical properties and their values which need to be determined by geological prospecting.

On the basis of the findings about physico-mechanical properties of rocks obtained by way of a subsurface and core drilling and studying rock pressure in workings, the Soviet researchers (2) have developed a set of parameters to be used for a deep underground working stability evaluation. These parameters are as follows: uniaxial tensile and compression limit strengths, static Young and shear modulus, Poisson coefficient, cohesive strength, and the angle of internal friction. It is also necessary to know the factors which change these parameters positively or negatively. Since the determination of some of these parameters is difficult, correlations between certain individual parameters were established to facilitate obtaining the initial parameters for a geological forecast in an area.

These are the particular data which must characterize a rock structure in a model with the account of their areal variability. The objective of geology is to evaluate this variability with the greatest possible accuracy and establish a correct sampling pattern.

One of the difficult aspects of stability modelling is to determine the effect of water content. It is known that water decreases strength and increases deformability of rocks, as well as encourages rheological processes. At greater depths the stresses in the rock in place become higher and rheological processes more active. However, the rock samples from core drilling after preparation for laboratory testing change their natural water content characteristics. This is the reason why it is difficult to obtain strength and strain characteristics for rocks with natural water content.

Moreover, the strength and strain characteristics of the rocks determined in laboratories are not adequate to those of the real rock in place because of different geometry and structure and distorted stress pattern.

Therefore, the strength properties of rock samples must be tested in three-dimensional compression cells.

The geological studies in the Kizelovsk and Donetzk basins are being made down to 1,500 to 1,900 meters. The engineering and geological conditions of coal mining at these depths can not be evaluated without a comparative geological analysis of different factors, even though there were relationships for medium depths that could be extrapolated to these depths.

One of the approaches to this problem was proposed in the Donetzk basin by local scientists who developed a method to predict geological and engineering characteristics of a potential mining area based on the detailed studying of the composition and sedimentation of coals and imbedding rocks. In the overlying and underlying rocks of a seam, the lithologically and facially related paragenetic complexes of rocks are identified and generalized over the area of potential mining in the form of isofacia zones. The primary material for such a generalization is the observation data on the immediate and main roof and floor of the old faces, drilling data and core test results, as well as samples taken in the mine. The prediction maps which result from such an analysis contain an accumulation of necessary geological information. The validity of the information is verified by the effectiveness of mine operations and actual conditions in the nearest levels.

This method of predicting mining and geological conditions for coal seams finds continuously broader application in the Eastern part of the Donbass area.

The prospecting geologists, however, have not yet resolved the problem of detailed study of the structure of rock in place which directly affect rock behavior in mining. It is important to establish relationships between the state of rock in place, on the one hand, and petrogenic composition of strata, their lithology, incidence linkage, fracturing and other features, on the other. The results must always be verified against "in situ" data.

THE OUTBURSTS OF COAL, GAS, AND IMBEDDING ROCK. METHODS OF PREDICTION.

The first sudden outbursts of coal and gas in the Soviet mines were registered at a depth of 150 meters in the Donetzk, Kuznetzk, Kizelovsk, and Partizan basins, and at a depth of 400 meters in the Karaganda basin.

One hundred twenty five mines in the Soviet Union in 1974 mined 346 coal seams with a potential for outbursts. The number of such seams, increased by 15 percent from 1968 to 1976. The Donetzk basin and its Centralny area are the most dangerous because they host around 80 percent of the total number of outburst potential mines and seams. The outburst incidence in recent years became more or less stable at around 50 to 70 outbursts per year due to special protection measures and despite systematic growth of mining. In the Donbass area, this figure constitutes only 70 to 90 percent. The greatest number of outbursts has been recorded in Donbass, which is the reason for the special attention which is given to this dangerous phenomenon by the local scientists and geologists.

The deposits of the Centralny area of Donbass are composed of Middle Carboniferous and Upper Carboniferous rocks in the shape of a large anticline fold with 40 to 80° steep wings.

There are multiple disjunctions of a thrust character in the area of the Main anticline which cross the rock mass at 35 to 40° from the strike. The 2.0 kilometer thick carboniferous rock is represented by sandstone (40 percent), sandstone shales (21 percent), argillite (34 percent), limestone (3 percent), and black coal (2 percent).

The steep-dipping thin and medium thick seams with multiple tectonic faults are being mined in this area. The greater part of the seams have unpredictable properties. Most of them have a relatively high methane concentration. In some mines in the area gassing is at a rate of 90 cubic meters per ton of production.

With the advancement of mining operations to greater depths, the danger of coal and gas outbursts becomes greater, although shallow depth outbursts are known (in the Number 3 mine of the "Ordjonikidzeugol" association an outburst occured in the seam "K_I" at 24 meter depth).

The character of outbursting and its power changes with depth. For example, in 1963, the average depth of mining in the Centralny part of Donbass was 626 meters, there were 134 outburst potential mine-seams with the average power of one outburst 62 tons. In 1973, at a depth of mining 756 meters, the number of outburst potential mine-seams increased up to 176 with the average power of one burst at 376 tons. An entirely new phenomenon came into being from 700 to 750 meters in depth, i.e., sandstone outbursts which increase in the number and force with depth.

According to the hypothesis which explains the mechanisms of coal and gas outbursts, protection measures must include a foregoing development of protection strata, water pumping into the coal seam, drilling of foregoing holes, a concussion blasting, etc.

The analysis of the cases of sudden coal and gas outbursts shows that the very methods of deep deposit mining must ensure the safety of operations in order to evaluate expensive and sometimes ineffective local protection measures. However, it is always important to have reliable results of prospecting surveys on the gas emission from coal and rocks, gas pressure, structure of seams and other characteristics of the coalfields.

The analysis of outbursts with relation to the grade of coal shows that the most dangerous are the "OC" and "T" grades, according to Russian classifications. The first outbursts from these coals in Donbass were recorded at 210 meters. The first outbursts from grades "G" and "K" (Russian classification) were recorded at a depth of 330 meters. The first outbursts from anthracites and semi-anthracites were recorded at a depth of 176 meters.

There is almost a linear relation between the tendency for coal and gas to outburst at depths down to 700 meters. Further down, the tendency decreases slightly which may be explained by the effect of protection measures or changing mining and geological conditions, although the latter have not yet been studied adequately.

The scientists of the MakNII Institute and mining geologists are now at the stage of introduction of a method to predict deep rock outburst potetial using prospecting drilling data. The specific characteristics of the outburst potential of sandstones have been identified by the scientists with the aim to use them to establish the outburst potentials of deep rocks. The method consists of an evaluation technique to examine core samples taken from the surface and to compare physico-mechanical and petrogenic characteristics of the samples. These characteristics are as follows: porosity, tensile limit strength, rock forming, grain size distribution, and cement composition, which are all outburst symptons, such as core breakage into separate discs or circular cracks.

According to I. I. Kachan and B. N. Nedosekin, the criteria which may characterize the conditions of potential outbursts are as follows: compressive resistance is 1.7 to 2.0 fold lower, effective porosity is 1.5 to 2.0 fold higher, quartz concentration is 1.3 to 1.5 fold greater, cement concentration is 2 to 3 fold lower, rock forming mineral grain size is 2 times greater than that in the sandstones of no such potential. This method was used to predict conditions at a depth of 800 to 1,000 meters in the "Krasnoarmeiskaja-Kapitalnaja: mine and at a depth of 1,200 meters in the "Krasnaja Zvezda" mine. In both cases, the estimates turned out to be correct. Similar analysis has been done for future operations at depths of 1,300 to 1,370 meters in the mines "Shachterskaja Glubokaja" and "Cholodnaja Balke Nidgnaja", etc.

From the mining experience, it is known that many coal and gas outbursts occur in the areas of geological faults. At the same time, there is evidence of no such phenomena in the places where workings cross many of such geological distortions.

The results of many experiments conducted in the mines of the Donetzk basin showed that in certain geological structures, the safe zone of stress release might be even greater in size than under normal conditions of rock in place. There may be certain geologically distorted structures which have smaller stress release zones but not small enough to be of any danger.

However, there are works (P. P. Lutzik) which establish a direct correlation between coal seam outburst potential and coal fracturing, tectonic variability of the thickness of coal seams, such as blow-ups, thinnings, and draw deformations. In general, the outburst potential of the coal seams in the Central Donbass is related to the prevalence of tectonic thrust dislocations accompanied by coal compression and faulting.

These are the complicating circumstances that make the solution of the problem of prediction of the mining and geological conditions under which outbursts will occur on the basis of geological data difficult.

During the last 10 years, the number of mines and mine seams in Kuzbass potential of outbursts almost doubled. All cases of coal and gas outbursts were registered in development workings.

The Karaganda coal basin shows the same trend, i.e., the deeper mines become, the more frequent are the cases of outbursts. All the cases were registered in zones of large tectonic faults or in the accompanying zones of smaller faults. Apparently, this is the main reason behind outbursts recorded in the most tectonically complicated areas of Saransk and Churubay-Nurin, where they started coming from the depths as shallow as 200 and 300 meters from the earth surface. The power of the outbursts grows with coal metamorphism and gas pressure and release.

A phenomenon which is no less dangerous and which deteriorates the safety of mining operations at great depths is rockbursting, which occurs under a certain combination of mining and geological conditions. The rockbursts occur as a result of the concentrated growth of potential elastic compression of rock followed by the sudden release of the accumulated power in the places of sudden failure.

As a result of a number of research projects in some of the coal basins of the U. S. S. R., the scientists of the VNIMI (Prof. I. M. Petuchov, et.al.) have found the solutions to a number of problems of rockbursting under various mining and geological conditions, including the development of recommendations on how to provide safety in deep mining operations.

One of the most important factors which controls rockbursting is coal and rock elasticity which predetermines the ability of the latter to accumulate great potential energy.

A number of the coal basins in the Soviet Union have their own theoretically backed up practical safety programs for the mining operations under conditions of high rockburst potential. The Soviet scientists and engineers have developed the elements of the theory of protection strata which appears as the most effective tool to fight rockbursts and other dynamic developments. A method of predicting rockburst potential areas has been developed and introduced into practice. The method is based on the analysis of rock elastic properties by using prospecting geological data or compression tests in mines. Mining geological data and physical data that may characterize the ability of coal and rock to brittle failure. This method has been used to estimate the rockburst danger of all the main deposits of the country at the stage of design and construction as well as those in operation at greater depths. A rockburst information agency has been established.

The realization of the rockburst protection program has considerably reduced the number of rockbursts in the mines of the country. The number of rockbursts at one mine has been reduced during the last 15 years 13 fold.

However, the problem of rockbursting is far from being completely resolved both on the theoretical level, in areas such as prediction by geological data, and on the practical level, such as establishing specific means to control rockbursts.

We have discussed only some of the characteristics of mining and geological conditions of coal extraction in the deep mines of the U. S. S. R.

The improvement of reliability of estimates of mining and geological conditions can not be attained without the quantitative and qualitative perfection of geological information and the methods of its acquisition. In pursuit of this goal, the geological projects in the U. S. S. R. have increased considerably the density of prospecting and drilling, developed new and perfected old techniques of prospecting, including the development of analytical methods to plan drilling and sampling and verify the results. The operational mines are always used to perfect the working methodologies. The accuracy of geological predictions was verified for all main coal fields of the country. The results have contributed to the improvement of the existing methods of prospecting.

The Geological Service of the Ministry of Coal Mining of the U. S. S. R. has developed a set of requirements from results at initial geological data and published the work in the form of a Regulation under the title "Requirements of the Mining Industry to the Prospecting Geological Works and Initial Geological Material Needed for Designing New and Modification Old Mines and Quarries". The purpose of this document is to improve reliability of geological estimates.

DISCUSSION

QUESTION: Is the strain on the rock in your basin due to structural complexities?

ANSWER: The answer is yes, at least most of our scientists think that tectonics predetermine the strength.

QUESTION: Have you attempted pre-mining methane drainage techniques?

ANSWER: Pre-mining methane drainage. Yes, we use this technique for all mines, either from the surface or from underground workings.

QUESTION: What mining systems predominate? Longwall retreat or room and pillar?

ANSWER: We don't use room and pillar systems; longwall retreat systems predominate. In fact, it is the only one system we use.

QUESTION: What specific alteration or modification in mine lay-out design is being made in very deep mines as a result of your research work, such as natural support in three dimensions, pillar and panel dimensions and artificial support, roof bolting, etc?

ANSWER: We have worked out a special preventive system of mining such seams. We mine first presented seams. The hydrofracturing of seams and drilling degasing holes before mining workings, pre-mining holes. And you see some other technique that Mr. Borisov may discuss with you privately.

QUESTION: What is the rate of the coal mined at 500 or 1000 meters depth?

ANSWER: In this condition in Donbass we mine mainly coking coal, and very high quality coking coal it is. That's why we have to go so deep.

QUESTION: Could you alter these specific geologic factors, which are responsible for the outburst adverse conditions found in Soviet coal basins?

ANSWER: You see in the report that we presented we have described this technique at least partially, so you can look through the report. We are short of time, you see and once again if you are interested in it, Mr. Borisov can explain to you more thoroughly.

QUESTION: What parameters are measured during exploration to determine the reliability of outbursts of coal and gas?

ANSWER: Well, I think the answer will be the same. The parameters are described in the paper, so you may look through it and find out for yourself.

QUESTION: Where do outbursts of sandstone occur? Are the sandstones the gas reserve ore?

ANSWER: Yes. They are.

QUESTION: Can you comment on the actual techniques to predict dangerous mining situations at depth?

ANSWER: The answer is the same. Will you please look through the report.

QUESTION: The question is whether that material that we mentioned in our presentation is available in the West.

ANSWER: Yes, of course it is; if you apply officially to the Ministry for the Coal Industry, I hope you will be able to get it.

QUESTION: How varies the quality of the coal between the first barriers with higher and lower geothermal gradients?

ANSWER: In those barriers the coal, you see, becomes more carbonized (harder, lower volatility). That's how I translate this first. And also with the depth, coal becomes more carbonized.

QUESTION: Do you have in the USSR multiple-seam deep mining operations? If, yes. How does the situation affect the statistics and what are the minimum intervals between seams in order to assure safety and economic mining?

ANSWER: Yes, you see in some of our basins we mine multiple seams. For example, in Kuzbas and Donbass in Karaganda they know those basins that were mentioned in the presentation. The minimum distance between the seams varies--it varies with geological conditions--and also it varies with the variation of dipping of seams. For example, for Kuznetsk Basin where you have steeply inclined seams, the minimum distance is about 20 or 25 meters, and those seams are thick seams. And, of course, they have a special sequence that they should observe--must observe--while mining these multiple seams.

Coal Logging Techniques

20.

Some Improvements and Developments in Coal Wireline Logging Techniques

by D. R. Reeves

21.

Extremely High Resolution Density Coal Logging Techniques

by Ken W. Fishel and Robert Mayer, Jr.

20

Some Improvements and Developments in Coal Wireline Logging Techniques

By D. R. Reeves
Managing Director
BPB Instruments Limited
East Leake, Leicester, United Kingdom

INTRODUCTION

Continuing interest in coal logging has over the last few years resulted in further improvements in equipment and techniques. This paper reviews some of the recent trends and is divided into four sections:

Firstly, improvements to techniques which are now relatively well established are considered. Secondly, the results and experience of some tools which are relatively new in the field of coal logging are discussed. Thirdly, the application of new analytical techniques in handling the large volume of data now provided by coal logs is reviewed and, finally, some of the items currently being researched are described.

IMPROVEMENTS IN EXISTING EQUIPMENT

The first equipment specifically designed for conveniently recording coal logging parameters is now in regular use. Basically, combination tools developed both for speed of operation and quality of measurements are now starting to become standard equipment. Various tools have emerged which combine three or four different measurements into one instrument. The most popular combination is a long spacing density, short spacing density, gamma ray and caliper, although in some cases a resistivity measurement is substituted for one of the densities. The effect of the new tools has had a considerable impact on the business. Apart from the immediate saving in time by making only one run through the borehole for the general logs, combination of tools overcomes problems of lining up logs and when used in conjunction with on board digitisation units can record sufficient information to present detailed logs as well as general logs, all being recorded during the same run.

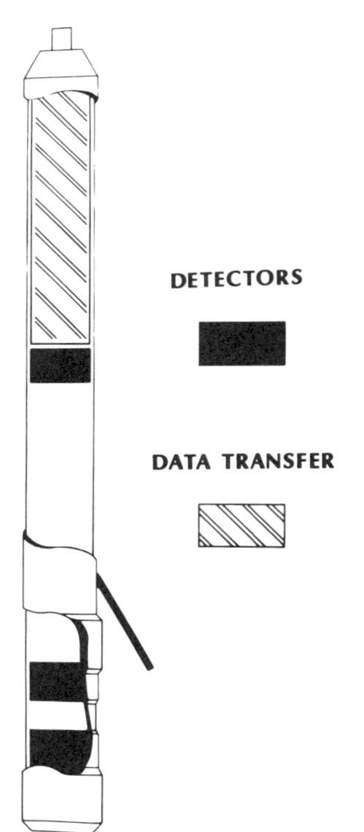

FIG. 1
COAL COMBINATION SONDE

The combination tool shown above, Fig. 1, runs fully calibrated double sidewall density, caliper and gamma ray logs.

The four measurements are usually displayed as three different presentations (or logs).

These are a combination of L.S. density, caliper and gamma ray measurements on a scale of 200 or 100 : 1 and presented as indicated in Fig. 2 as a coal lithology log. This presentation allows rapid identification of coal seams, sandstones and shales.

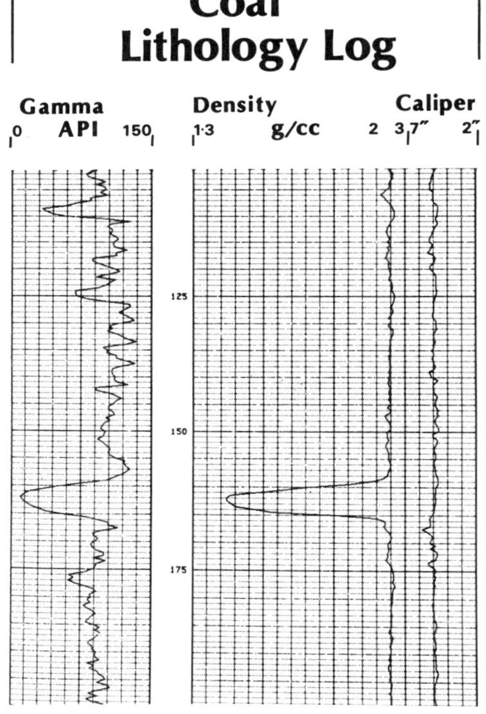

FIG. 2
COAL LITHOLOGY LOG

A second combination of short spacing density and caliper as indicated in Fig. 3, run on an expanded scale over the coal seams, is designed for accurate interpretation of bed thickness and dirt partings and is presented as the seam thickness log.

Seam Thickness Log

Caliper | Bed Resolution Density

Inches 2 | 500 SBRDU 300

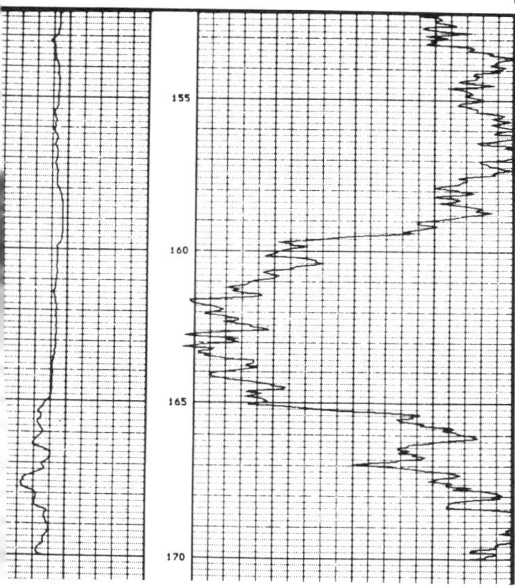

FIG. 3
SEAM THICKNESS LOG

FIG. 4
COAL QUALITY LOG

Coal Quality Log

150 Gamma Ray~API 0

1·3 g/cc~Density 2 3

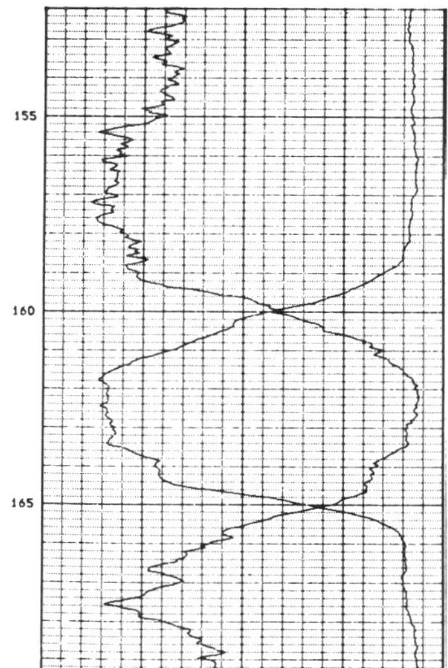

Finally, the combination of he LSD and the gamma ray reversed), presented on detaild scale, is known as the coal uality log and is the basis or measurement of ash content, ig. 4.

All the above information an now be obtained on one run ith logging speeds of around 0 ft. per minute over the eneral sections and 6 ft. per inute over the coal seam etailed areas.

The second significant nprovement to emerge has been ne proper calibration of ensity logs which effectively ermits true rock densities to e displayed on all logs rather than relative density counts.

A calibration accuracy of 0.01 grammes/cc is now the standard which can be guaranteed. Calibrations have been developed using the field base of operations as a primary reference and then, with field jigs, a recalibration at the actual borehole location. As a result of this, log headings can actually prove the calibration accuracy of the tools on each job. Fig.5 shows a standard log heading and how the calibration and jig values are placed on the heading and on the record to prove the logging tool and source combination are properly calibrated.

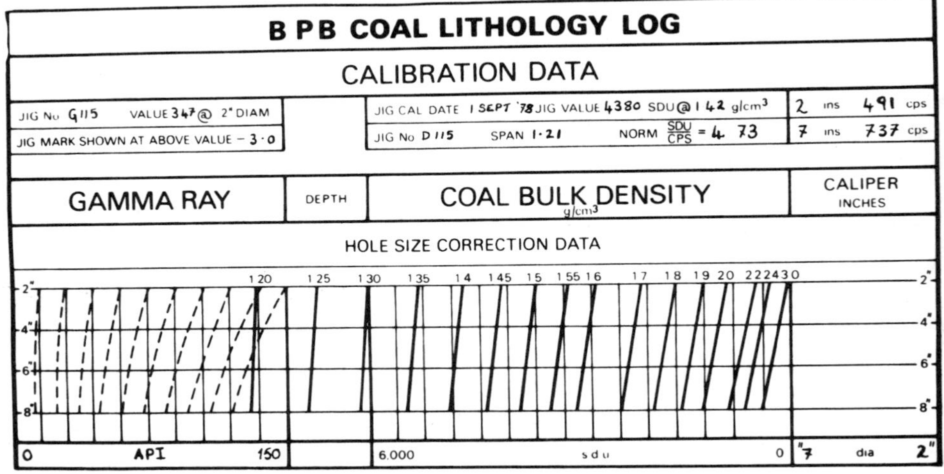

FIG. 5
HEADING DETAILS

Density logs need to be corrected for caliper and mud weight to get accurate values. As can be seen from the heading, Fig. 5, little detail in the rock density range of 2 - 3 grammes/cc can be seen from the log. An improvement has been to linearise the density response so that details in 2 - 3 grammes/cc range can be seen. However, the effect of borehole diameter changes and drilling mud become important in this density range so it is necessary for the measuring system to compensate for these effects. Thus, density logs can be corrected automatically for all the usual drilling variables (but not caves). A comparison between non-linearised and linearised and compensated logs using the same tool data is shown in Fig. 6.

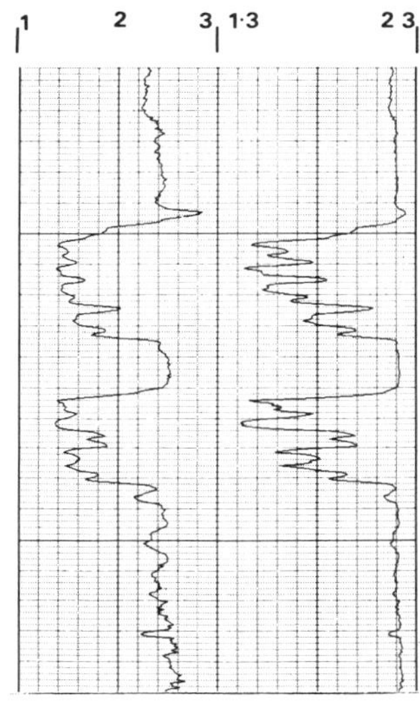

FIG. 6
DENSITY LOG COMPARISON

It is anticipated that use of linear density logs will become dominant in the lithology log display in view of the easier interpretation for shales and sandstones. However, the logarithic display still makes measurements in the low ash range easier to read on an analogue record.

In addition to the full calibration and linearisation of the density measurements, a similar process has been undertaken in neutron logs. In this case the final answer is now presented in terms of rock porosity although it is still necessary to know if the rock

FIG. 7
NEUTRON LOG COMPARISON

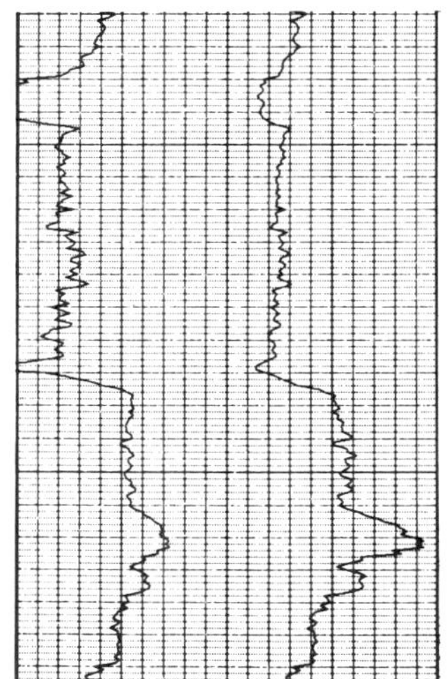

matrix is sandstone or limestone for accurate determinations.

The example, Fig. 7, compares the non-linearised neutron log with the calibrated log which has also been linearised and compensated for caliper changes

NEW TOOLS

Recently two measurements previously only developed for the oil industry have made an appearance in scaled down version for coal logging. These are the sonic log and the dipmeter log.

Sonic Log

A new tool has been developed which is 2" in diameter and approximately 6' in length consisting of one transmitter and two receivers and run in a decentralised situation as shown in Fig. 8.

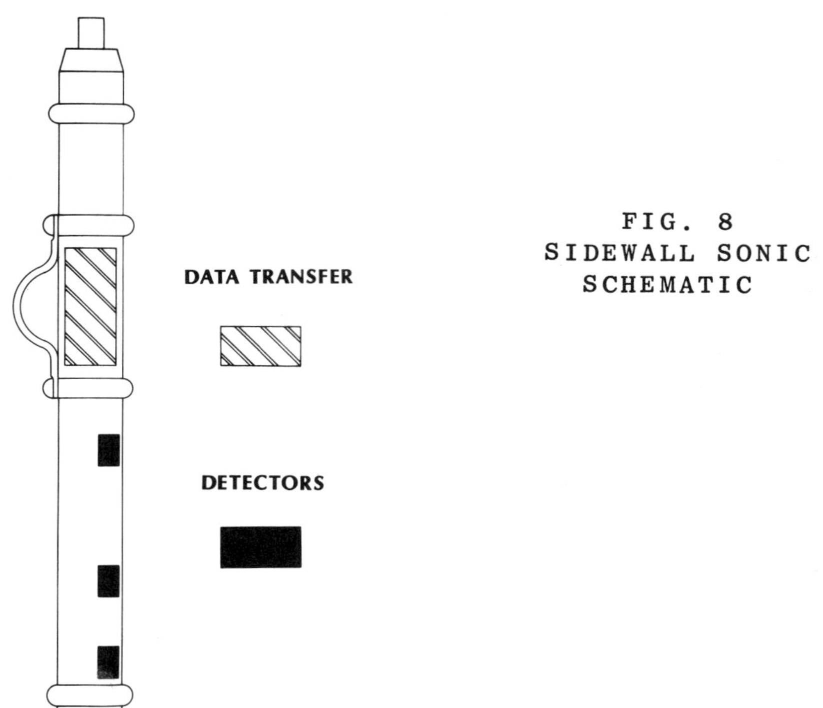

FIG. 8
SIDEWALL SONIC
SCHEMATIC

There are certain technical problems in scaling down the size of a sonic tool but results show that these have been overcome with the introduction of a sidewall version of the oil field sonic measurement. By oil field standards this tool is relatively simple but results to date indicate that it is producing extremely interesting information. Sonic logs are expected to produce three different areas of interest to the coal analyst.

<u>As an indicator of coal rank</u> since sonic measurements respond to changes in compaction and density.

<u>As a method of investigating</u> rock strength.

<u>As a further parameter</u> for use in the evaluation of coal properties as described in a paper by the author in the First Symposium (i.e. Moisture and Carbon).

The examples below are of sonic logs run in the last two years. Fig. 9 shows logs over two

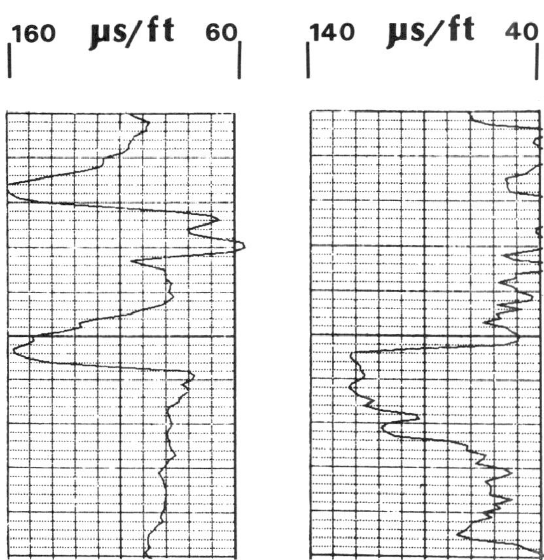

FIG. 9

SONIC LOGS OVER DIFFERENT RANK COALS

seams of different rank. The righthand log, which shows a sonic velocity of around 120 m/secs is a high rank coal in the bituminous category whilst the left hand log shows a sonic velocity of 160 m/secs of a much lower rank sub-bituminous coal. These results can be reproduced in all types of coal fields and show the tool is responding to different ranks.

Fig. 10 shows conventional sonic and density logs which have been combined according to the formula used by the oil industry to produce a strength index log. The characteristics of this log confirm what can be seen in the lithology and preliminary comparison with triaxial strength tests has shown a reasonable correlation.

Sonic logs are far more susceptible to the natur of the formation than radiation logs and in shallow holes and in poorly consolidated material it is not always possible to obtain good quality results.

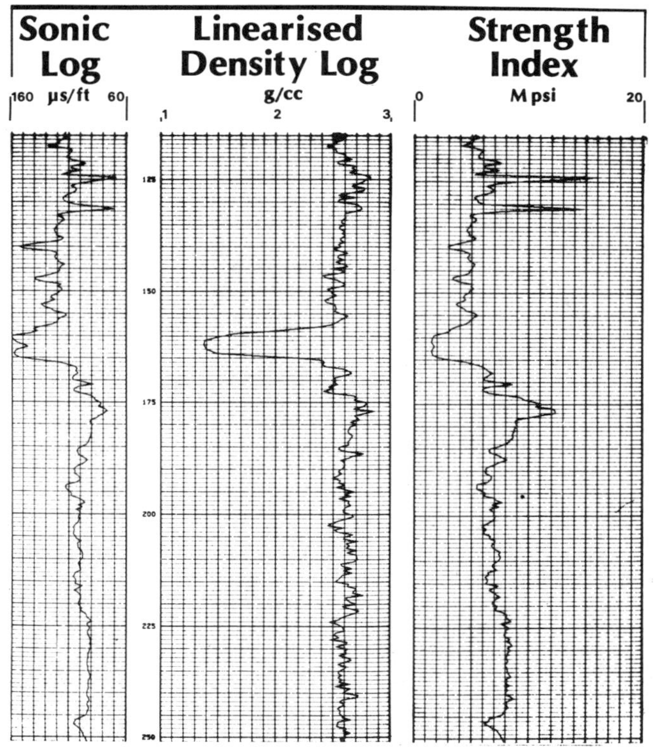

FIG.10
STRENGTH INDEX
DERIVED FROM
SONIC AND
DENSITY LOGS

Dipmeter

There has been a requirement for some time to measure strata dip in the narrow diameter boreholes. Recently a new dipmeter has been under evaluation which has a maximum diameter of 2" and a length of 12'. The tool consists of a three arm resistivity measurement together with a record of caliper and a measurement of data which allows the correct orientation of the tool and borehole deviation to be continuously recorded. A schematic of the tool is shown below. This tool has some important applications for coal exploration.

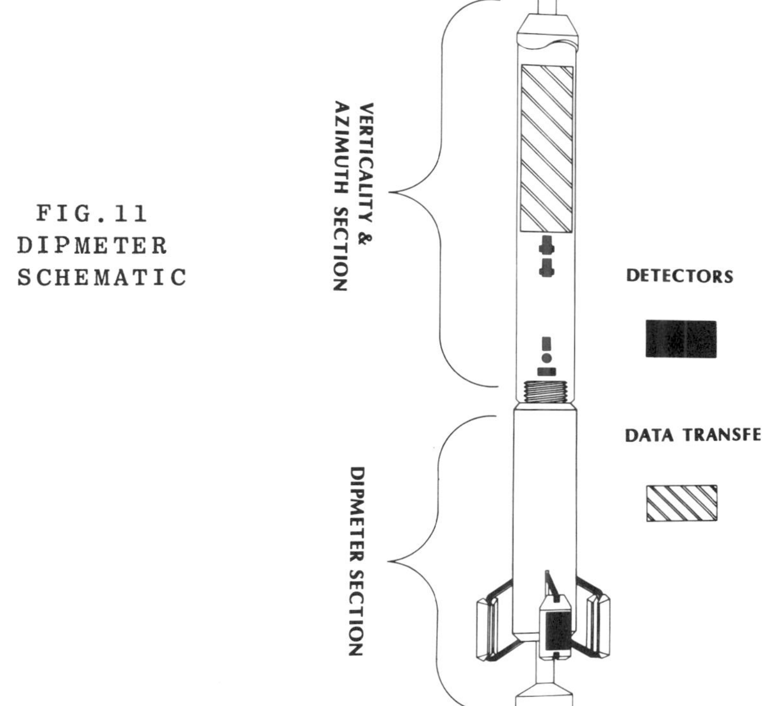

FIG.11
DIPMETER
SCHEMATIC

Accurate measurement of dip is obviously important in assessing seam thickness whilst the additional knowledge on direction of dip will give valuable structural information.

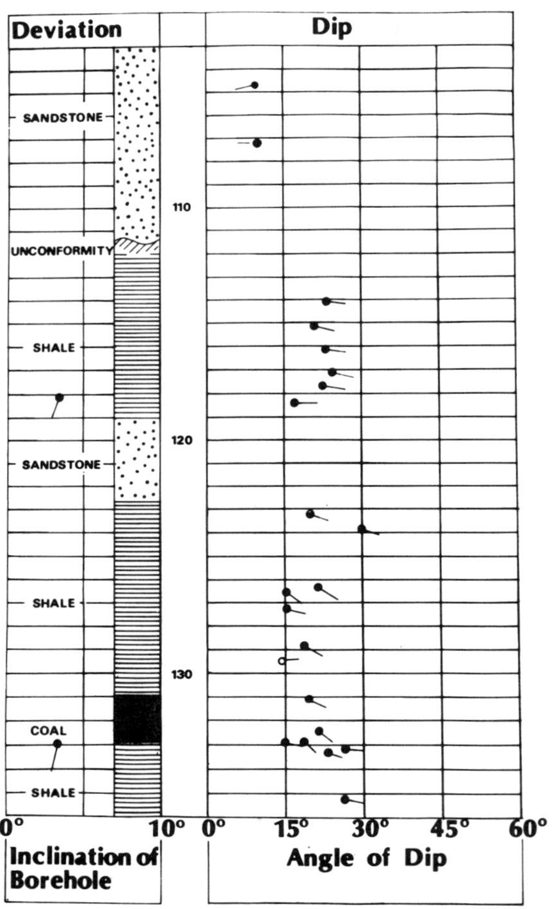

FIG. 12
DIPMETER PRESENTATION

In favourable conditions the continuous dip measurement will reveal bedding characteristics and thus have an application for depositional studies.

Fig. 12 shows the presentation of the dipmeter results as a 'tadpole plot' and illustrates the value of this tool. The position of the 'body' indicates the amount of dip, whilst the tail shows the direction viewed with north pointing to the top of the page. The uppermost reading for example shows a dip of 10° in a direction of 260°.

The log clearly shows the unconformity with the dip of the lower section of about 20° to the east ($90 - 110^o$) compared with the upper section with a dip of 10° to the east (260°).

The log also shows the borehole inclination which is 3° almost due south.

Interpretation of continuous dipmeter reading is - by coal logging standards - relatively expensive, thus the system has been designed to be able to give spot dips as well as continuous records. Spot dips can be obtained on site by the engineers after completing the run.

DATA HANDLING AND ANALYTICAL TECHNIQUES

The increase in intensity of coal drilling plus the additional information now available from logs has produced an awesome volume of data. Whilst operational control and site problems can be handled with conveniently prepared logs, much analytical and geological information of considerable value is not used. To meet this need data handling and computer controlled and assisted analytical techniques have been introduced.

The introduction of field digitisation units - such as the TU1 illustrated below, Fig. 13, allows full recording of data at a relatively low cost. Data is recorded on simple cassettes and converted later into full nine track tapes, headed and arranged to meet a required format.

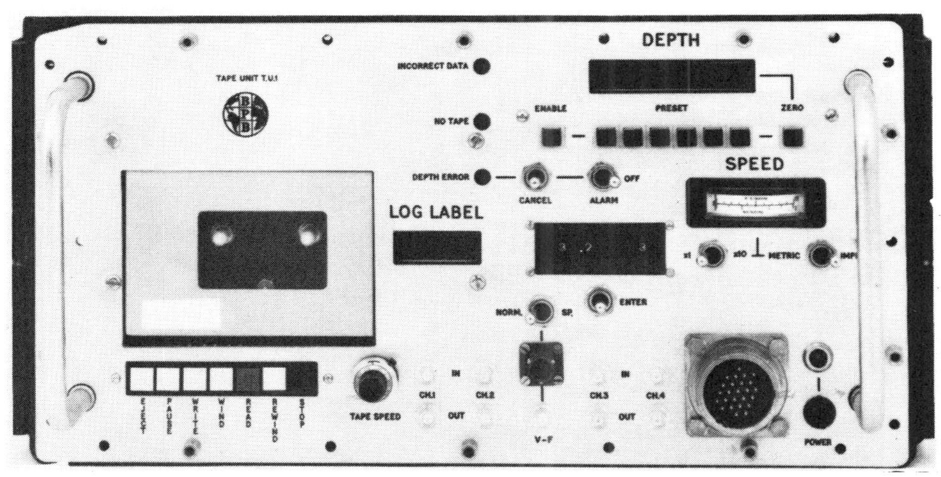

FIG. 13
ON BOARD TAPE UNIT

With the use of very high sampling rates in the field, the data now available for computer processing is substantial. The computing process is ideal for comparing the responses of logging tools, producing recalculated logs and presenting data in a convenient form for interpretation.

Some of the techniques now in use are:

Cross Plots

This technique allows data from different logs to be compared by crossplotting one log against another. The usefulness of this technique can be seen with the density-sonic crossplot shown below, Fig. 14, which shows how different coal types can be recognised. This technique is also valuable in comparing spot analysis for ash with log derived densities.

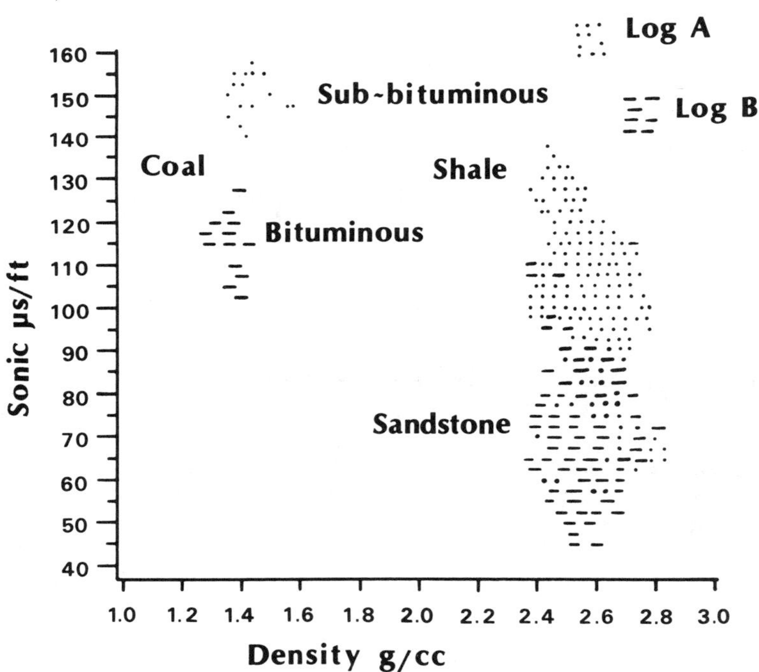

FIG. 14
CROSS PLOT OF LOGS SHOWING COAL TYPE

Lithology Plots

The density - gamma logs can be processed to show the relative abundance of sand, shale and coal with a direct computer interpretation print out.

Fig. 15 is an example of such a print out showing the logs interpreted into coal, shale, rock matrix and rock porosity, all of which can be obtained from a single run with the combination coal tool. Apart from clearly defining all the coal seams, the presentation shows variation of matrix (usually sandstone) with shale in an almost depositional fashion enhancing both correlation and the understanding of the sedimentation.

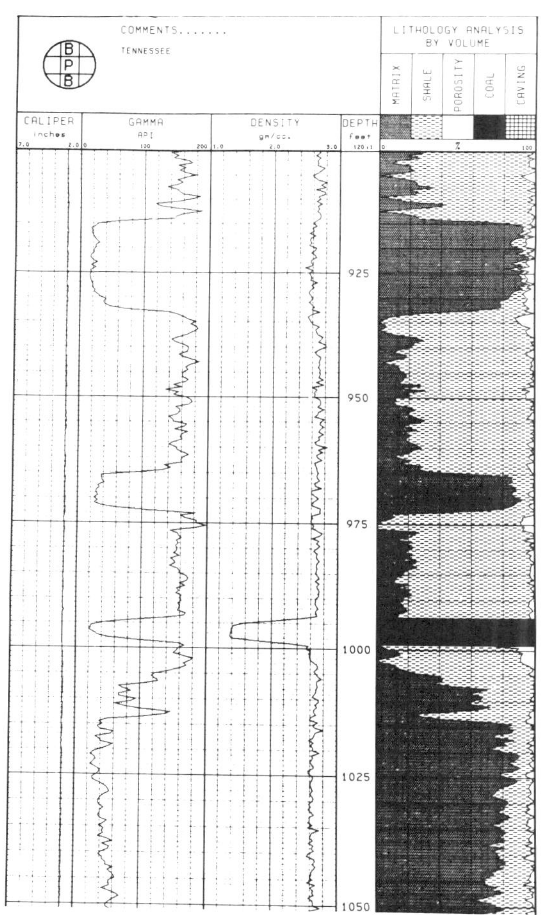

FIG. 15
COMPUTERISED
LITHOLOGY
INTERPRETATION

Strength Index

The density sonic combination required to give this information can be processed and plotted in the form of a continuous log. Fig. 16 shows a most useful display of lithology, strength and rock density plotted from information taken off the field tapes.

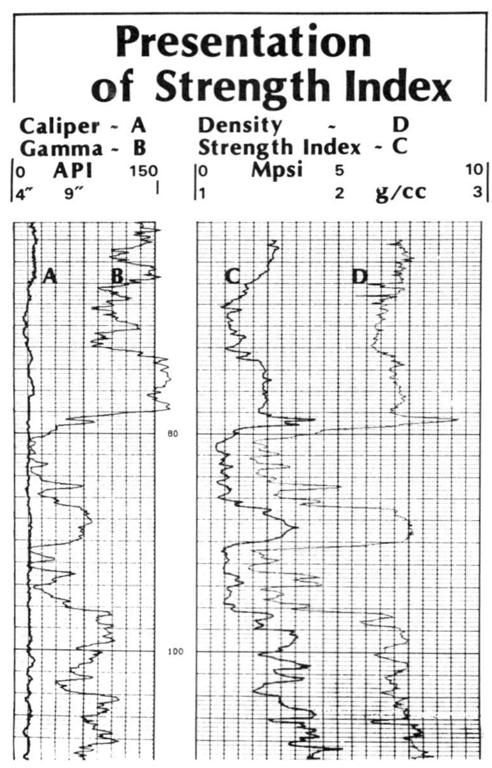

FIG. 16
STRENGTH INDEX PRESENTATION

Data Enhancement

Analogue field recorders cause some distortion of information due to the incorporation of time constants but by digital techniques the distorted information can be reconstructed.

Ash Content Analysis

The development of accurate density values from the compensated density tool allows coal seams to be analysed in detail for density levels and thus in most cases for ash. Completely automatic definition of seam boundaries and natural divisions within the seam are not yet available. However, the manual interpretation procedure can now be greatly speeded up by using computer aided interactive graphics.

FIG. 17
SOME OF THE HARDWARE USED FOR
DIRECT PLOTTING OF INTERPRETED LOGS

In this technique the density log is displayed on a graphic screen and seam boundaries and intra-

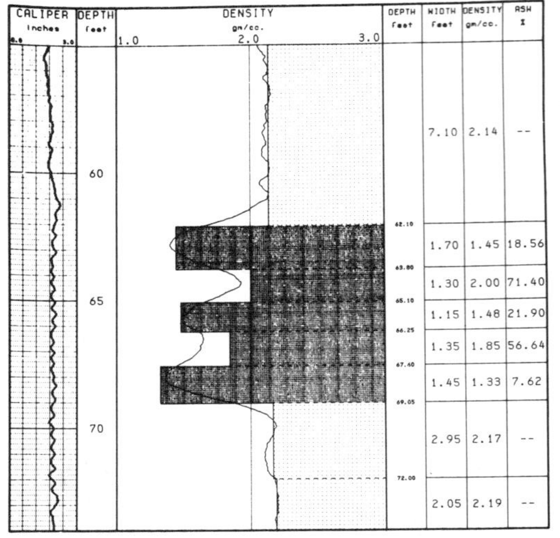

FIG. 18
EXAMPLE OF COMPUTER AIDED
SEAM INTERPRETATION

seam structure are marked on the screen by an electronic 'light pen'.

Thus aided the computer can proceed to analyse ash content of the identified sub-section very quickly and tabulate results and statistical splits.

Fig. 17 shows the computer hardware which is used in the process.

Fig. 18 shows a seam section analysed in this method.

NEW DEVELOPMENTS

Most of the existing tools are capable of further development and over the next few years the followin may be expected to happen.

Density Logs

Higher resolution tools for seam partings and boundaries and possibly some type of compensation for caving by using two density spacings are likely to be produced.

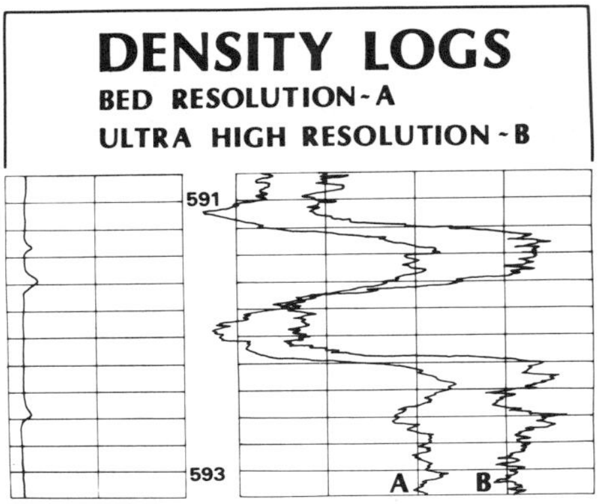

DENSITY LOGS
BED RESOLUTION-A
ULTRA HIGH RESOLUTION-B

591
593
A B

FIG. 19
IMPROVED RESOLUTION
OF DENSITY RESPONSE

Improvements in making calibrated measurement through casing are also likely to be developed. Fig. 19 shows a log from an experimental ultra high resolution tool showing the comparison with an existing BRD log.

Neutron Logs

Much higher resolution will be achieved. The example right, Fig. 20, shows an experimental high resolution neutron log which can already exceed the highest density resolution log (BRD). The improvement of the neutron log response in the high porosity range, and hence for determining moisture content, is another high priority.

FIG. 20
HIGH RESOLUTION
NEUTRON LOG

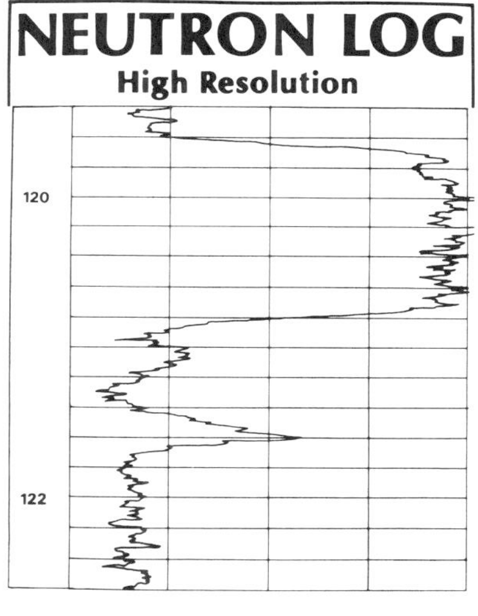

Sonic Logs

Improvements to make this tool more reliable in shallow holes and in less consolidated ground are under investigation. Measurements of the difficult sheer wave as opposed to the larger and easier transverse wave will be investigated which, if successful, will provide more rock mechanics information.

Sulphur

The industry still awaits the first practical down-hole sulphur measurement. Whilst no new 'physics' technique can be expected, improvements in electronics and data collection techniques are rapidly improving the chances of a viable method being developed.

Computer Analyses

With the high volume of field data now being digitally processed, considerable efforts in software developments can be expected. The main effort will be in developing a system for detailed analyses of seam structure, coal types and evaluation of a complete coal 'model' as described by the author in the First Symposium.

ACKNOWLEDGEMENT

I am grateful to various members of BPB Instruments Limited who have helped in the preparation of this paper and to various clients who have allowed information to be published.

REFERENCES

D.R. Reeves Application of Wireline Logging Techniques in Coal Exploration

Proceedings of First International Coal Exploration Symposium 1976 pages 112 - 128

DISCUSSION

QUESTION: Will a high resolution neutron log in the near future be adequate to show moisture variations in non-caving silt stones and muddy silt stones and as an indication of the fractured condition? Perhaps in confirmation with gamma ray logs?

ANSWER: High resolution neutron will probably give a better response in high moisture rocks such as shale and silt stones provided shale content is high. However, it will have much less penetration.

I know the NCB have been able to correlate the neutron log with rock fractures. I must admit that I am not entirely sure why this should be as I would have expected the sonic log to be more affected. However, the increased resolution of the neutron log may help this technique.

QUESTION: How do you handle drop out and noise on the sonic log and what effect does this have on the strength index?

ANSWER: In the field we record the raw sonic log which means that drop outs and noise affect the log and give it a somewhat spiky appearance. From the tapes we can, in the laboratory afterwards, re-run the log and suppress the noise, effectively de-spiking the log. Normally with the major oilfield logging companies this process is carried out in the truck and indeed if required we will shortly be able to do this ourselves in truck. However, it is probably important not to miss the effect of drop out noise, particularly in near surface areas where the sonic log is often affected by unconsolidated material.

I think we would recommend that the strength index is always computed from a de-spiked log since this will tend to give a smoother and more averaged result for the strength index.

QUESTION: What is the effect of drill hole irregularities on the dipmeter log?

ANSWER: Borehole irregularities can have a serious effect on the dipmeter log because in order to get a high resolution boundary determination by resistivity measurement it is necessary to have electrodes placed very closely together which means that penetration is small. In badly caved drill holes therefore it is likely that one of the pads will not be in contact with the borehole wall and results will be difficult to interpret. In some cases it may be possible to use the caving for the boundaries but this should only be used when the geologist has an excellent knowledge of the formation.

QUESTION: Can you correlate any of the log information for the petrographic features of the coal seam?

ANSWER: In many cases geologists will know the usual structure of a coal seam and when this is the case it is often possible to correlate the log features with the structures. For example, a harder durain coal often shows an increase in density and thus also has a relatively lower gamma ray count than might be expected and may also have a very good caliper. It is still very difficult to recognize the nature of these changes unless you have a clue. However, provided enough good logs can be run it should ultimately be possible to obtain ash content by two different methods, say density and sonic, and once this is possible an indication of the coal type can be made.

QUESTIONS: Would you elaborate on the dipmeter logs for determining hole deviation?

ANSWER: The deviation section of the dipmeter consists of five transducers, all of which are carefully mounted with respect to each other. Two transducers are level cells which are mounted at right angles to each other and three transducers are magnetic coils which are also mounted in a fixed relation to the level cells. The magnetic transducers define the magnetic orientation of the tool and the level cells define the tilt of the tool in respect to the magnetic orientation. From this information the hole deviation can be calculated. All the transducers produce continuous information which is recorded on tape. The third magnetometer is required so that the tool may be used in a variety of areas with different magnetic inclinations without re-calibrating. The tool will not of course work in a cased hole where a gyroscope will be required.

QUESTION: To what degree of accuracy can the rank of coal be determined by the sonic log? Or rather does that give an indication of rank?

ANSWER: I would agree that at this stage it is an indication of rank rather than an accurate determination. Rank of course in itself is a somewhat complex function but if it can be accepted that it is related to the metamorphism of coal, which in itself increases density and reduces moisture content, then the sonic log will show an increase in speed as the same quality coal is metamorphosed. I have seen many examples around the world where we expect coals of certain ages to give typical sonic velocities. For example carboniferous coals are often about 125 microsecs/ft while quarternary lignites are often as slow as 160 microsecs/ft

QUESTION: How close is the ash content determined by the quality log to actual analytical results?

ANSWER: In seams in which quality bands are thicker than say 3 ft. and borehole caliper is good, the correlation is excellent and once the nature of the coal is calibrated and understood by the equipment, comparisons to within 1%-2% are easily possible. As seams get thinner more interpretive adjustments have to be made and also in bad borehole conditions more corrections are required. In very thin beds of 6 inches or less, significant errors can occur. However, if beds are alternating in quality the tool tends to give an average of the bands and thus, quite often over thick seams composed of alternating thin beds of coal and inferior coal, the overall comparison is quite good.

QUESTION: How much of the equipment described requires specially trained geophysicists? In other words what are the logs an exploration geologist for coal could use with a reasonably short training period?

ANSWER: The equipment described in this paper does require special training in order to be used efficiently and properly. I personally do not believe that anyone other than a trained operator can provide logs of good quality and maintain the equipment in serviceable condition. Simple resistivity logs require very little training but we are now talking of sophisticated calibrated tools which are complicated laboratory instruments, thus training is necessary.

Further, I believe that geologists will have to have special training to understand the full value of the logs and to extract the information contained in them.

QUESTION: When will the dipmeter be available in Canada? Could you indicate some idea of the costs? How many runs are required and could you elaborate on processing the logs and the sensitivity of hole condition for this system?

ANSWER: The first tools will be available in Canada in the Spring of 1979 and the price will be in line with the Combination Coal Sonde currently used for coal logging exploration. The cost therefore is in line with the normal aspirations of the coal exploration geologist. However, this tool may carry an additional cost if complete interpretation on a continuous basis is required. To use the tool for measuring significant dips in coal seams and adjacent rocks should only require on site interpretation which should not really add a great deal to the cost. I anticipate that the coal geologist will find the sedimentation information which can be unraveled by the dipmeter of considerable interest and therefore in some cases I anticipate a full continuous interpretation will be required. All the information required for both strata dip and borehole deviation can be obtained from one run

21
Extremely High Resolution Density Coal Logging Techniques

By Ken W. Fishel, Chief Geologist
Island Creek Coal Company
Lexington, Kentucky, United States
and Robert Mayer, Jr., Director
Well Reconnaissance Division,
Gearhart-Owen Industries, Inc.
Dallas, Texas, United States

INTRODUCTION

A new technique, based on the EHR, or extremely high resolution logging probe, providing significantly greater resolution in the delineation of coal seams and partings, has been developed. While this probe has been proven effective by several years of down-hole logging in the eastern United States, its existence outside the eastern United States is little known.

The purpose of this paper is to present to the foreign (and U. S.) coal explorationist data on this probe. It is felt that this probe would be a worthwhile addition to the array of down-hole logs available to the coal explorationist.

This paper will not present data on the general subject of the value of wireline logging for delineating coal measures, as this has been well documented by a number of authors; it is our intent to show how this new technique, used in conjunction with previously described logging methods, will enhance coal exploration.

EVOLUTION OF THE DENSITY SONDE

The density sonde has evolved in a few years from a tool capable of determining the thickness of a coal seam to within ± 30 CM and delineating a seam at least 48 CM thick to a tool capable of determining the thickness of seams to within ± 5 CM and delineating the presence of seams a minimum of 15 CM thick. This evolution of increased resolution has been the result of decreasing the spacing between the source and the detector. The Long Spacing Density (LSD*) probe typically has a spacing of 48 CM and a boundary resolution down to 22 CM.

Figure 1 illustrates a typical seam signature for a LSD probe. In this example the thickness of the seam (which was cored) would have been understated by over 10% utilizing normal thickness determination procedures (Reeves, 1971), and the thin (2.5 CM) shale band easily misinterpreted as being an 18 CM thick band of shaley coal.

The High Resolution Density (HRD) typically has a spacing of 24 CM and a boundary resolution down to 9 CM. The Bed Resolution Density (BRD) typically has a spacing of 15 CM and a boundary resolution down to 5 CM (Lavers and Smits, 1976). The need for the construction of a tool capable of delineating coals and impure bands as thin as 2 CM has been stated. It is felt that Well Reconnaissance has developed such a tool, referred to as an Extremely High Resolution (EHR) density tool. It is wondered whether the lack of available adjectives will preclude the development of increased resolution tools.

* References to LSD, HRD and BRD probes are intended to provide continuity to previous literature. Illustrations used do not necessarily characterize logs obtained by other companies.

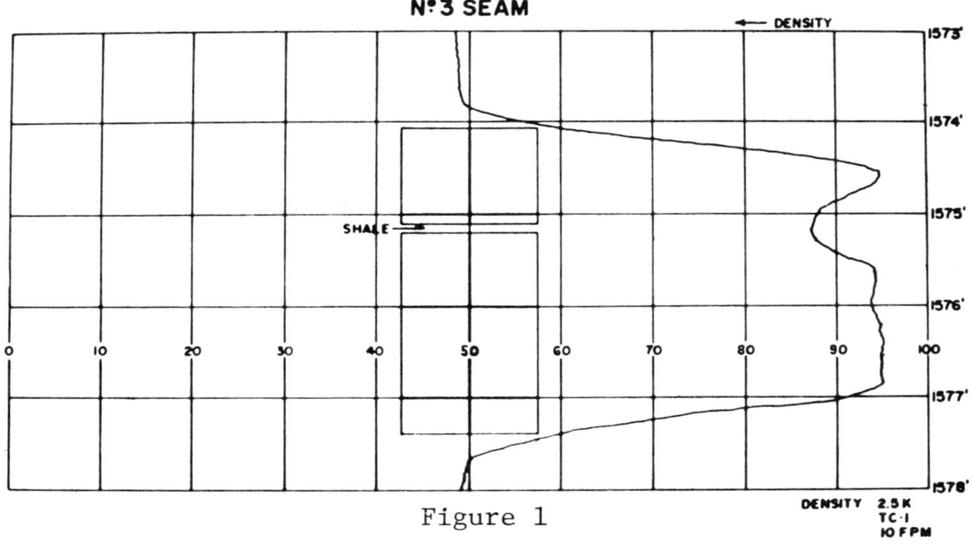

Figure 1

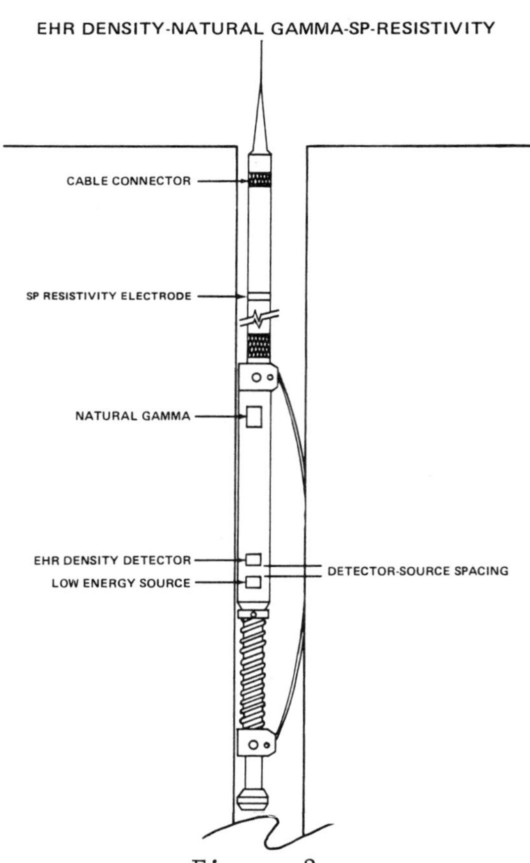

Figure 2

THE EHR PROBE

This probe was developed by Well Reconnaissance in response to the request of a major coal company to face up to specific problems in Appalachian coal exploration. Much of the remaining reserves are in seams that are thin and laterally inconsistent. This represents a problem both in evaluating reserves and in extending existing mines or in planning new deep mines. Existing wireline logging methods, when used to supplement core information, did not provide sufficient refinement of data necessary for reserve calculations or economic mine planning. One indicated direction would be the development of greater resolution density logging.

An obvious approach in accomplishing greater resolution is a decrease in detector-source spacing as shown in Figure 2. A plethora of design decisions follows from this approach. For effective collimation, the source must be of an energy much lower (less than 100 KeV) than those ordinarily used. For great statistical stability, essential to the repeatability of the minute detail to be encountered, the source must have a high radiation output. This low energy, high radiation source must be so encapsulated to meet the rigid specification for well log use set by nuclear regulatory authorities. The detector must be very sensitive to the low energy radiation from the formation and shielded against back radiation. The tool must provide not only the EHR density log but natural gamma, self potential and resistivity, and sometimes caliper logs, all run at the same time. The tool must be small, in the 1¼ inch - 1½ inch (3.17 to 3.81 CM) range.

Notwithstanding the design problems, a tool was produced which met the original requirement for much better seam and parting resolution. In addition, the far sharper lithological signature achieved by the EHR makes for better seam correlation.

APPLICATIONS OF EHR

Figure 3 is an example of the increase in resolution that results from the use of the EHR probe. This hole was logged with the normal density probe (i.e. LSD) and the EHR probe. Both probes were run at a speed of 10 feet (3 M) per minute and at a scale of one inch (2.5 CM) to

LONG SPACING DENSITY
EHR
0
10'
20'
30'

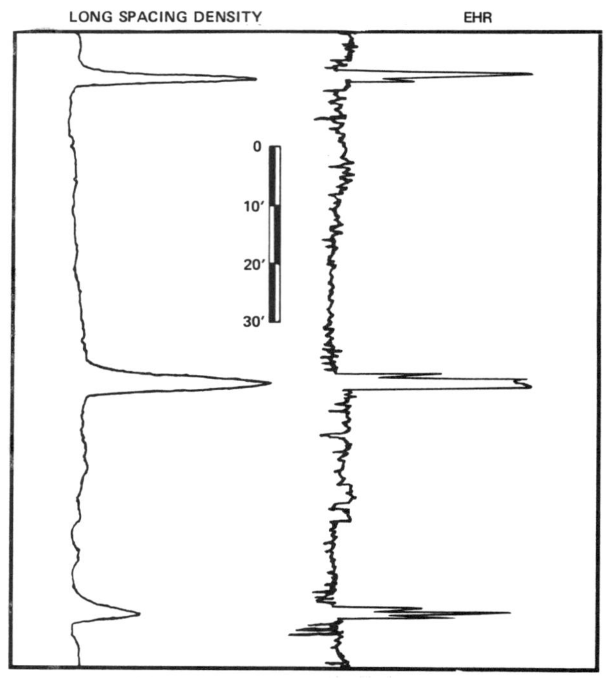

COMPARISON LSD WITH EHR
DENSITY PROBES
BUCHANAN CO., VIRGINIA

Figure 3

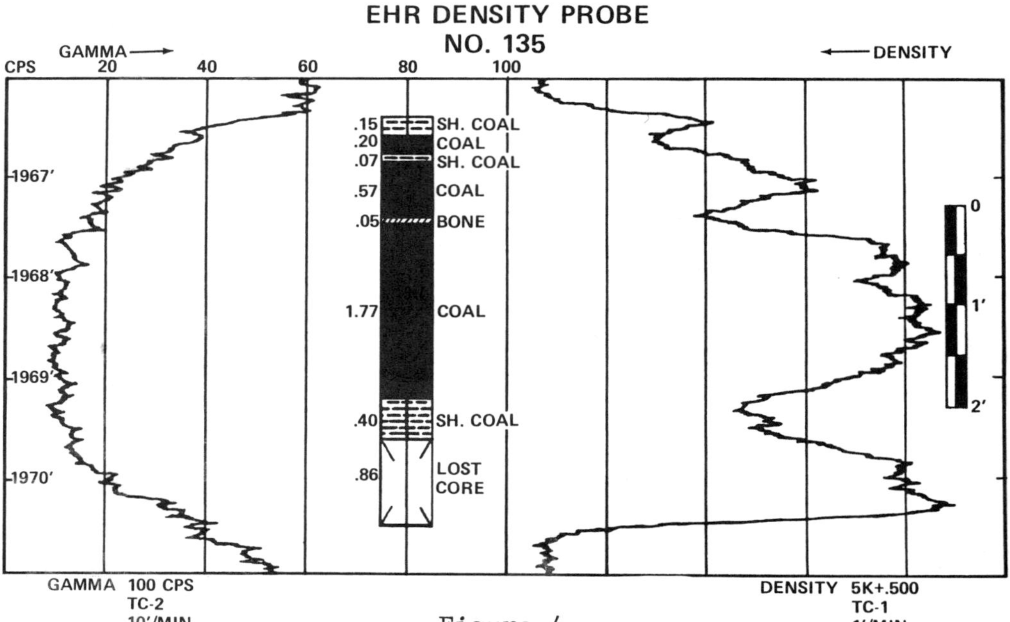

EHR DENSITY PROBE
NO. 135
GAMMA →
← DENSITY
CPS
20
40
60
80
100
.15 SH. COAL
.20 COAL
.07 SH. COAL
.57 COAL
.05 BONE
1.77 COAL
.40 SH. COAL
.86 LOST CORE
1967'
1968'
1969'
1970'
0
1'
2'
GAMMA 100 CPS
TC-2
10'/MIN.
DENSITY 5K+.500
TC-1
1'/MIN.
6-27-75

Figure 4

10 feet (3 M) of hole. There is an obvious slope on the density curves with the LSD probe, while the EHR curve is nearly horizontal. On the LSD probe the three seams appear to be composed of a single bench with the lowermost seam being very shaley. In actuality, the EHR log indicates each seam to be comprised of several benches with the lowermost seam being comprised of three coal benches with two shale splits which the LSD probe united into a very shaley coal. The EHR probe improves the accuracy of thickness determinations, improves the definition of impure benches, may improve ash content determinations, and can improve seam correlations.

SEAM THICKNESS

For thickness determinations it is desirable to relog the significant seams at an expanded scale. In the U. S. it is common to use a scale of one foot (30 CM) of hole to one inch (2.5 CM) of chart paper and run at a speed of one foot (30 CM) per minute. Figure 4 is an example of an expanded EHR density log that was run on a core hole. The seam boundaries are easily determined due to the lack of a significant slope to the density curve. In coal basins where the coal is friable, breaks easily and difficult to core, it has been found that the seam thicknesses determined from the EHR log are often more reliable than those determined by measuring core in the field. It is felt that the seam thickness from the log is accurate to within 0.1 feet (3 CM).

IMPURE COAL BENCHES

In addition to accurately determining the seam thickness, the expanded EHR log differentiates between impure coal benches in a seam. This is illustrated by Figures 5 and 6. Each case shows a portion of a two inch (5 CM) coal core and its corresponding signature on the EHR density log. Figure 5 shows band of shale less than one inch (2 CM) thick on top of a bench of clean coal. The EHR log readily distinguishes the two benches. Figure 6 shows the log discerning between a bench of relatively clean coal and a band of shaley coal less than one inch (2 CM) thick.

The expanded scale EHR log, it might be said, puts the core into written form, helping the explorationist determine the extent of core sampling.

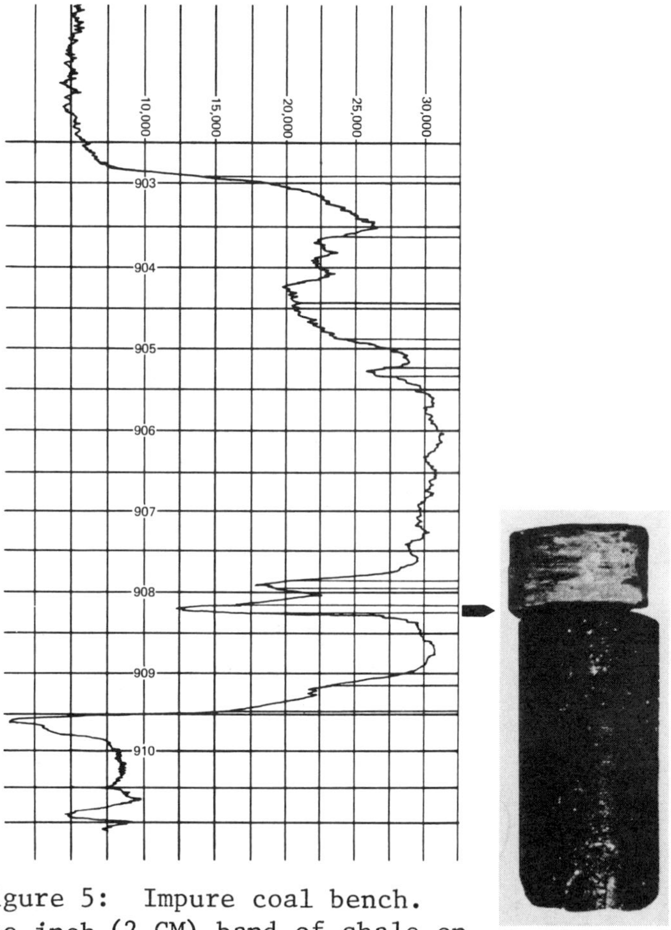

Figure 5: Impure coal bench. One inch (2 CM) band of shale on top of a bench of clean coal.

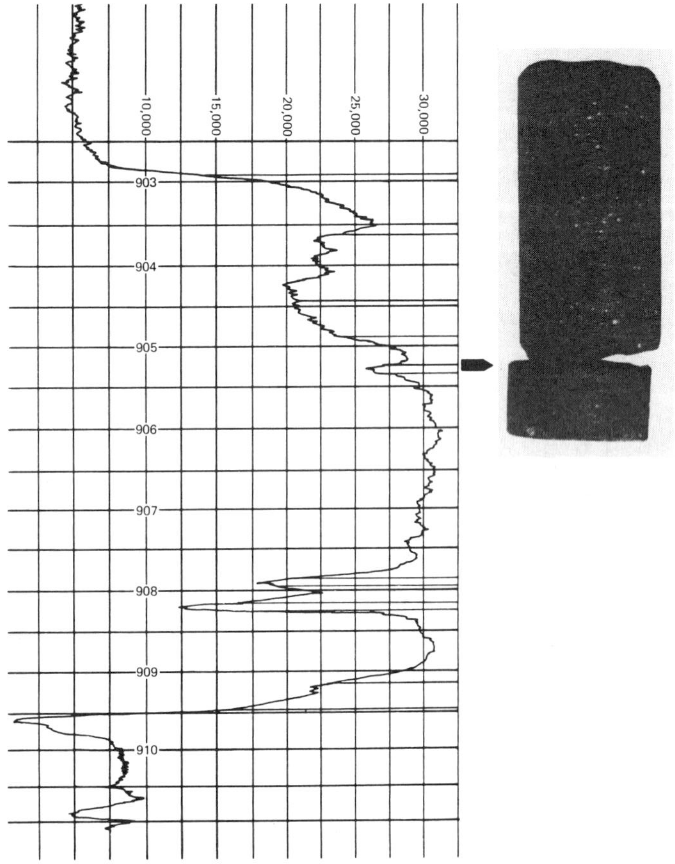

Figure 6: Impure coal bench. Shaley coal less than one inch (2 CM) thick and bench of relatively clean coal.

ASH DETERMINATIONS

An attempt was made to correlate the EHR density curves with ash and spg. Three eastern U. S. cores were chosen at random and sampled by bench with a raw ash and spg analyzed for each bench. The EHR logs and analyses are shown on Figures 7, 8 and 9. Figure 7 is a low volatile coal, Figure 8 is a medium vol coal, and Figure 9 is a high vol coal. A perusal of each illustration suggest that there is a good relationship between ash content and count rate, a relationship which has been previously determined by a number of authors (Lavers and Smits, 1976; Kowalski and Holter, 1975; Peeters and Kempton, 1977). For the three cores presented herein the graphical relationship between ash and count rate is shown in Figure 10. The correlation coefficient is 0.92, which is good considering the differences in the ranks of the coals that were tested. It appears that additional work with the EHR log could prove fruitful for determining raw ash and spg contents.

SEAM CORRELATION

One of the major responsibilities of coal explorationists is to properly correlate seams. Miscorrelating seams for an underground operation would be disastrous and embarrassing, and with the EHR logs the chances for a miscorrelation are significantly reduced. The cross-section shown on Figure 11 illustrates some of the correlation difficulties in the eastern U. S. The lower seams and upper seams are laterally persistent and easily recognizable on the logs. However, the No. 11 seam is replaced by a sandstone unit in the middle log, while the No. 11C seam stratigraphically rises over the channel. Note that the characteristic density signature of the 11C seam in holes 4 and 5 almost precludes a miscorrelation of the 11C seam in hole 4 with the No. 11 seam in hole 5. Also, note that the gamma signatures of the seams are not very distinctive. A miscorrelation of the No. 11 and No. 11C seams would result in a difficult mining operation.

SUMMARY

The EHR density log has been found to be a valuable asset for the evaluation of coal seams in the eastern U. S. Its usage has resulted in more accurate determinations of seam thickness, definition of impurities, and more accurate correlations. In addition, it has the potential to enable the coal explorationist to quickly calculate reasonable raw ash or spg values for non-cored zones or as a check on lab analyses.

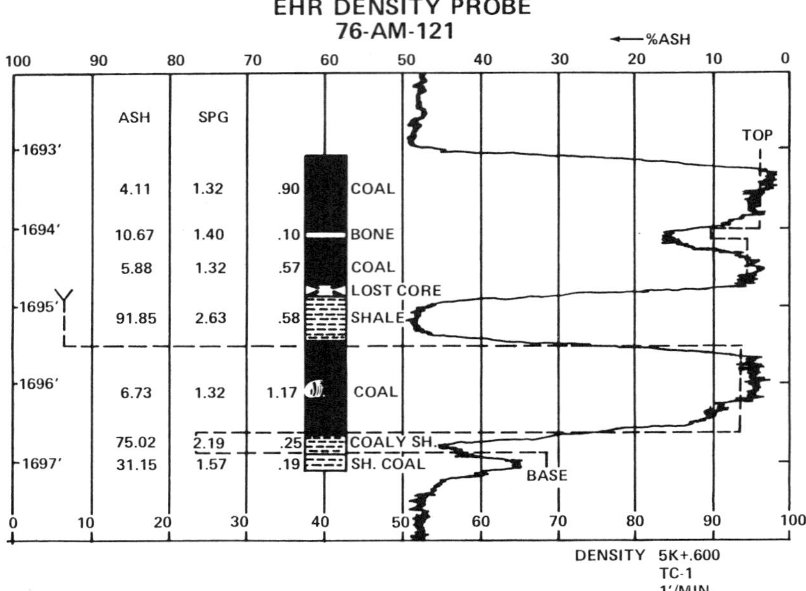

Figure 7

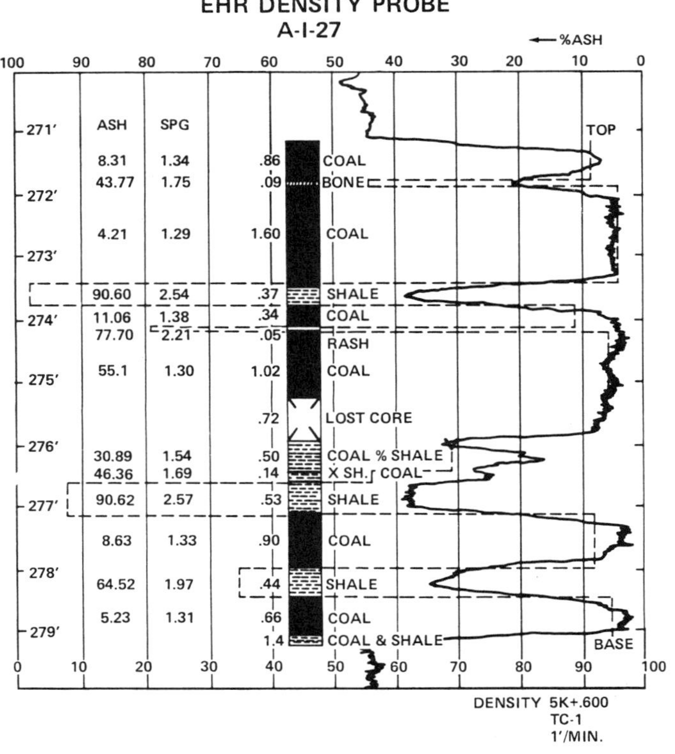

Figure 8

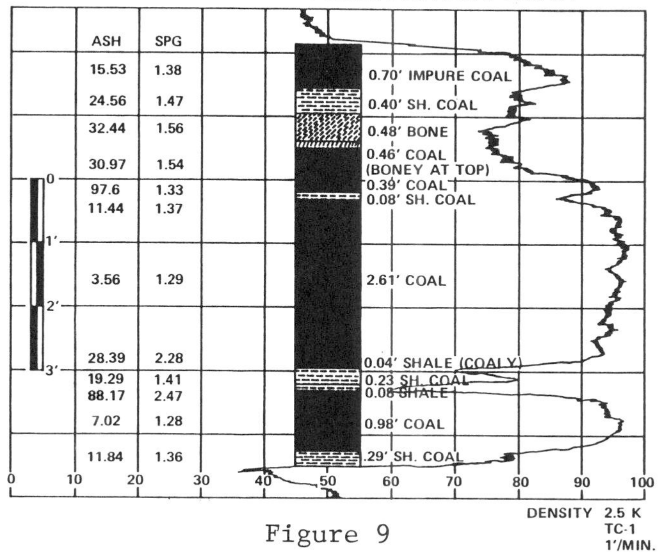

Figure 9

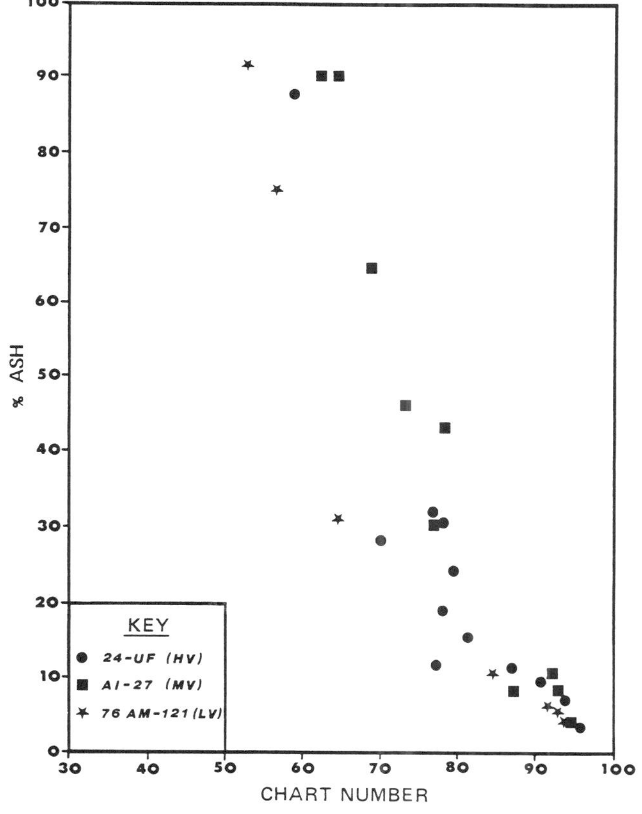

Figure 10

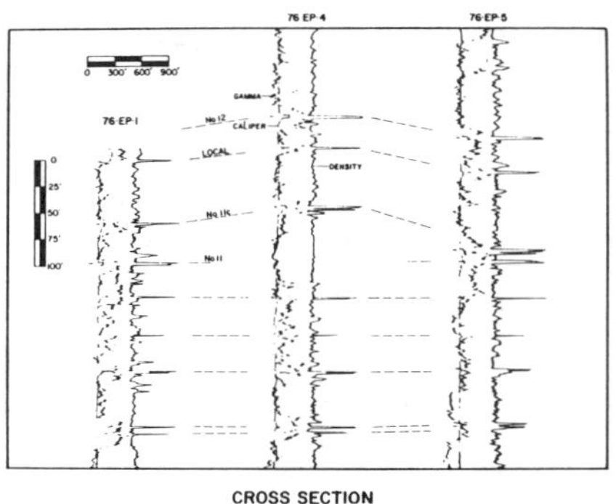

CROSS SECTION

Figure 11

Use of this technique in other areas, along with other wireline logging methods, should provide a refinement of data that would be advantageous in many instances.

ACKNOWLEDGEMENTS

The authors are indebted to Mr. Joe A. Andrews, Geologist, Island Creek Coal Company, Lexington, Kentucky, for his evaluation of data and extensive assistance in the writing of this paper.

Grateful acknowledgement is also made to Consolidation Coal Company, Eastern Exploration, Library, Pennsylvania, for generous aid in the preparation of this paper, including the use of illustrative logs and cores; to Consolidation Coal Company, Bluefield, West Virginia, and to many logging service companies in the area, for cooperation in the development of the equipment; and to Continental Oil Company, Ponca City, Oklahoma, for defining the need for the new technique.

REFERENCES

Kowalski, J. J., and Holter, M. E., 1975, Coal analysis from well logs: SPE AIME paper no. SPE 5503, 50th Annual Fall Meeting of SPE of AIME, Dallas, Texas, Sept. 26 - Oct. 1, 1975.

Lavers, B. A., and Smits, L. J. M., 1976, Recent developments in coal petrophysics, in Muir, William L. G., ed., Coal exploration, proceedings of the first International Coal Exploration Symposium, London, England, May 18-21, 1976; San Francisco, Miller Freeman Publications, Inc., p. 129-152.

Peeters, M., and Kempton, N. H., 1977, Wireline logging for coal exploration in Australia: The Log Analyst, May-June, p. 24-29.

Reeves, D. R., 1971, Coal logging in the U. K.: Canadian Mining and Metallurgical Bulletin, Feb.

Coal Exploration, 1976, Application of wireline logging techniques to coal exploration, in Muir, William L. G., ed., Proceedings of the First International Coal Exploration Symposium, London, England, May 18-21, 1976: San Francisco, Miller Freeman Publications, Inc., p. 112-128.

DISCUSSION

COMMENT: Before I read the first question, I want to explain something that will be applicable to this question and to some of the others. My connection with the EHR and other logging systems has been as the designer, or the developer, or the manufacturer of the equipment and systems as opposed to being involved with providing a logging service. It happens that there are a number of our customers here. The customers would include, not only logging services, but coal companies themselves. There are a number of customers here who have run the equipment and their practices may vary slightly. They may know more about the answers to some of these questions, and if they have anything to add I hope they will do so.

QUESTION: Do you log your entire hole in one foot per minute or do you log the entire hole at a faster speed, and then relog the coal intervals at a slower speed?

ANSWER: Yes, that's exactly what is done. Usually the entire hole is logged at a pretty good clip, ten or 20 feet per minute, or sometimes an even higher speed; then, you go back and relog the coal intervals. I might add that when equipment is provided for digitizing the logs on site, it will then be possible to run the coal intervals from the tape. It will be possible to run the entire log at a comparatively fast speed. It will not be necessary to make more than one run, or to re-log the coal interval.

QUESTION: What source strength is used for the EHR?

ANSWER: At present we are using a strength of 500 millicuries of Americium^{241}, although there are other sources that are being considered.

QUESTION: Have there been any abnormal drill hole effects on the EHR, using such close spaced instrumentation?

ANSWER: The EHR would definitely be affected by a thin washout. Fortunately, experience shows that most coal seams do not have such washouts. If they do appear in a hole, they're usually connected with shales or some part of the hole other than the coal seams. There are several things I'd like to add to this. If it is a smooth hole, the EHR is less affected by the changes in hole diameter, gradual change in hole diameter, than most other density logging methods I am familiar with. It is affected not at all by the changes caused by going in and out of fluid. This derives from the fact that we are using such a low energy source that we can shield it so well on the back side. Generally speaking, there is little effect from washout or cored holes. In large rotary drill holes that have been drilled rather carelessly, there may be the rugosity effect that might affect the EHR log. But, usually, if there is such an effect of course it is obviously on the caliper log, you would expect it, then you would generally be able to see it on the density log itself. There's nothing you can do about it other than to ignore it, or to not use that particular piece of data. But to sum it up, generally it has not been a problem in coal seams.

QUESTION: Can you comment on the use of EHR logs on open holes with particular reference to the effects of hole irregularity?

ANSWER: Well, it has to be used in open holes. It cannot be used in cased holes or inside drill stems.

QUESTION: You appear to be using BPB terminology for logs -- LSD, BRD. Are you using BPB in your equipment?

ANSWER: No. The use of those terms was simply to be in terms of frame of reference of several papers that Mr. Reeves and others with BPB have given, where they have been used, just to make a comparison. There were several illustrations in the paper where the term LSD was used once in an illustration of a log by itself, and another in comparison with the EHR, and I do wish to say that this was not intended to imply in any way that it was run by BPB. Where the log is termed LSD, it may not have been exactly the same spacing as what BPB would use, and the resolution obtained might not be exactly the same.

QUESTION: What are the accuracies of an ash analysis sing the HRD logs?

ANSWER: Well, the figure that was assigned by the erson who did the analysis that was used in the illustration as a correlation coefficient of 0.92. As I mentioned, there s a lot to be done here before assigning quantitative values. hat will be facilitated by using digitized logs. One thing hat complicates the digitizing of the EHR log is the ecessity for sampling at a much higher number of times per oot than would be with a less detailed log. In other words, is necessary to sample for digitizing on the order of 00 times per foot if you are going to retain the extreme etail, the fidelity of the detail, that you should have, hen the log is reconstructed from the tape.

QUESTION: When applying EHR, what is your normal logng speed and what is the time constant of the receiver? assume the questioner meant detector.

ANSWER: Well, the logging speed of course has been overed and when using the scale, is one foot per minute and e entire hole is 20 feet per minute. As far as the time nstant, we use no time constant at all in the sense that u would, say for a natural gamma log where your statistics e some percentage of the overall count. As I mentioned, e of the things achieved here was to have a high enough unting rate so that the statistics are just a negligible rcentage of the overall counts. So. there is no time nstant at all used, except for the mechanical time constant the recorder. And that is really more than anything else ich causes the extremely slow speed to be run in the coal ctions. As the logs are digitized, I think that it'll be ssible, as I indicated, to run the entire hole at a speed at least 20 feet a minute.

QUESTION: What is the cost of the EHR density probe?

ANSWER: It is reasonable and a rather small percentage the cost of the overall system. In terms of the overall uipment, the logging system and the logging truck, the ditional cost for the EHR density probe as compared with me of the other tools is just a small increment.

QUESTION: What is the diameter of the probe?

ANSWER: The basic diameter of it is 1¼ inches, although

we have worked on a slightly different version that is 1½ in Even at 1¼ inch the effective diameter is 1½ inches because the bow-spring system used.

QUESTION: Can the EHR be calibrated?

ANSWER: Yes. It is a standard practice. We provide calibration. It is standard practice to calibrate it at the beginning of each set-up. The actual calibration in terms o density or ash content will follow from comparison such as shown.

Computer Technology

22
Microcomputer Technology for Coal Exploration

William G. Miller, Technical Director,
and William H. Smith, President,
GeoGraphics Corporation
Champaign, Illinois, United States

Coal exploration is a complex, time consuming and often expensive process. Once the basic field survey data has been collected, many weeks of effort can be involved in evaluating this information to provide a reasoned estimate of the parameters of the site. The timeliness of reports is crucial to the further acquisition or development of a property, and the ability to assimilate and rapidly digest large volumes of geologic and engineering data is at the heart of producing timely reports. This ability can be strengthened by adding more labor, which drastically increases costs, or by supporting existing personnel with available technology.

The application of technology to increase productivity is not new. Computers, for example, have greatly increased the productivity of banks and other financial institutions. Computers have not been widely used to increase the productivity of the exploration phases of the coal industry because of a lack of personnel familiar with their application and the high cost formerly associated with them.

Large scale integration, or the ability to produce electronic circuits of very small size and low cost has, as a result of the U.S. space program, produced a new generation of small, inexpensive computers. This technology has matured to the point that it is being applied in the most unexpected places, from automotive fuel injection to planimetering coal reserves maps.

In the context of this paper, a microcomputer is defined as a physically small, inexpensive computer system composed of very large scale, integrated circuits. Micro in this context refers to the physical size of the device and not its computational power.

Micro, or small computer, technology became important for nonmilitary applications in the period 1972-74 with the introduction by INTEL corporation of the first of its series of single chip computers. At that time they were simple devices with little applications support. Since then, however, memory technology as well as processor technology has progressed rapidly.

When microcomputers were first introduced, they were considered to be nothing more than dedicated device controllers which could perform such mundane functions as stop light sequencing and automatic temperature regulation.

Since their introduction, the costs of small computer systems have dropped to such a point that individuals now can afford personal systems. The price reductions have been on the order of one hundredfold over the last four years. The proliferation of small systems has brought about a flurry of activity in the development of programs to make use of the newly available hardware. In fact, most common programming languages available for large scale computers--FORTRAN, BASIC, COBOL, PASCAL, APL and others--are available for certain of the microcomputers. Almost every semiconductor manufacturer produces a microcomputer, each with different capabilities.

At present, there are two major types of microcomputer. The first type uses an 8-bit word size and the second uses a 16-bit word size. One may consider the 16-bit version as a compact reproduction of current minicomputers, while the 8-bit version is somewhat slower than present minicomputers and fills a gap in performance between the programmable calculator and the high speed mini. In general, the same functions can be performed by the 8-bit micro, 16-bit micro, traditional minicomputer and large main frame. The only difference between them for a specific application is the cost/benefit ratio.

The main benefits of this new technology arise from the ability it affords to use several dedicated computers to provide direct interaction with the data at low cost. Interaction with data presented in the form of maps is central to the ability of the coal resource analyst to provide reliable estimates of coal occurrence and quality.

One of the most common types of interaction encountered in resource evaluation is the calculation of areas and volumes to estimate reserves. This particular task is very labor intensive and can be quite time consuming when traditional methods of planimetry are involved. Verification of results for computational as well as technical accuracy often requires more effort than the original planimetry.

This type of procedure is a prime example of how tradition methods of investigation can be simplified and advanced by microcomputer technology while allowing the geologist or engineer to maintain control over the procedure. Historical resistance to the use of computers has arisen because of a lack of programming expertise on the part of most coal resource analysts and a lack of understanding of the basic engineering principles required on the part of the programming staff of the "DP" departments of most corporations. The personnel conflict this situation spawns, together with the high price of program development, formerly large running costs and inconvenience of batch mode (i.e. punched card) processing, has prevented widespread computer applications in this area.

Microcomputer technology can reduce the resistance to new applications, because the programs required to perform a function like planimetry need not be rewritten for each application. Programs, once developed, can be retained by the microcomputer until they are needed. The main functions of scaling, area calculation and coordinate transformation, in the case of microcomputer planimetry, can be used by the operator as "building blocks". Formerly these functions were fixed in a device such as a planimeter or digitizer, and the operator required additional expertise and a calculator to compute volumes from areas and thickness displayed in the form of maps, or to convert from one scale to another, etc. Microcomputers can, because they are in fact computers, request the desired parameters and automatically provide converted final answers to the operator without additional assistance.

The cost of applications development like map planimetry can be much reduced with microcomputer "building blocks". For example, one can automate the data collection phase with an intelligent planimeter, the data recording phase with an intelligent memory system and the final computation phase through a local computer or "time share" system. Each of these modules could be purchased as a stand-alone "black box" and plugged together (programs and all) to provide a total solution to the problem. Little development time would be required compared with developing each module separately and then integrating them.

The real impact of the technology is in removing the inconvenience to the operator. The key is conversational interaction. If the operator cannot converse with the data, much of the power provided by the computer is lost, diluted by the many manual processing steps required. Formerly the equipment required to provide interactive capability cost on the order of $100,000 per unit. Microcomputers have brought this figure down to a much more realistic $10,000, and thereby opened vast new areas for the profitable application of computer technology.

Conversational interaction is the procedure by which the operator can ask questions about the map being planimetered, receive an immediate response and then alter his methods or approach based on previously acquired data. This ability is extremely important in being able to check and verify results.

A very attractive advantage of the microcomputer is that it need not be connected to an external phone line or other peripheral processor in order to perform its function. Although it has the potential to communicate with a time-share computer, such connections are not necessary. Since the great advantage of microcomputers is their ability to provide continuous, cheap interaction with the operator, they can in many ways parallel the activities of larger machines without the current costs of about $10.00 per connect hour for the timeshared machines. A connection to a timeshare vendor is attractive in order to provide auxiliary programs and services either beyond the capability of the micro (i.e. large data bases) or to access proprietary software not otherwise available.

An example of a system to perform coal resource evaluation by either microcomputer or timeshare is the CREAS (Coal Resource Evaluation and Analysis System). CREAS was developed to allow more accurate and rapid resource evaluations using microcomputer technology to automate the traditional methods of map planimetry.

The measurements are taken by a digital measuring device which is not dependent upon physical contact with the map medium. The ultimate accuracy of a digitizer is established at the time of manufacture and is not influenced by slippage, as is the accuracy of a conventional planimeter. A digitizer provides the X and Y coordinates of points directly, with no dials to set or read. In order to be useful, a digitizer must be connected to a computer or calculator. CREAS is the link between computer, digitizer and the person doing the resource evaluation. CREAS not only computes areas on a coal resource map, but also factors the areas into tons of coal in-place and stores that information on disk for future reference. After all of the areas on a map (or maps) have been measured, reports can be generated quickly and accurately from the stored data. These reports are free of the transposition and arithmetic errors normally associated with manual generation of such reports. With the operator freed from the task of recording and manipulating the data, he or she is then free to concentrate on more accurate collection of the areas (automated planimetering). Because of less fatigue, the overall performance of the operator is enhanced. The net result is a more exact job performed in less time than by traditional polar planimeter methods.

CREAS is but one example of the application of a new technology to the business of finding and mining coal. As new applications are developed, other tasks will be automated and made less expensive, allowing more direct interaction between the geologist or engineer and the data. This, in turn, will result in more rapid and accurate resource evaluations.

REFERENCES

1. Noyce, Robert N., "Microelectronics," Scientific American, September 1977, Vol. 237, No. 3, p. 62.

2. Noyce, Robert N., "Large-Scale Integration: What is Yet to Come?," Science, March 1977, Vol. 195, No. 4283. p. 1102.

3. Bowers, Dan M., "Systems-On-A-Chip," Mini-Micro Systems, May 1976, Vol. 9, No. 5, p. 74.

4. Kaplan, Alan R., "Personal Computers: The Outlook For Home and Hobby Computers," Computer Decisions, May 1977, p. 36.

5. Kaplan, Alan R., "The Computerized Automobile," Mini-Micro Systems, June 1976, Vol. 9, No. 6, p. 34.

6. Simpson, Henry K., "Getting Small: Microcomputers," Digital Design, November 1977, p. 43.

7. Staff Report, "The Minicomputer in Map-Making," Mini-Micro Systems, February 1978, Vol. 11, No. 2, p. 72.

8. CREAS User's Guide, 1978, GeoGraphics Corp., Champaign, Illinois.

DISCUSSION

QUESTION: How common and useful will these computer systems become in the near future, and what are their anticipated costs?

ANSWER: That really is a three-fold question. The first is how common. The only way I can answer that is to look at what we've been talking about in the last three days. How many times have you heard of a computer being used in some phase of the explorations? Especially with geophysical people, the chemical people, and what I'm trying to allude to is also in the analysis of the geological data. You've seen a lot of computers. What has the development of the small calculator--hand held programmable calculator done for science in general? It has become very common, and I anticipate that the small micro-computer will find a niche between the calculator, which will not go away--everybody is going to end up using them--and the large mini. I think the small and the mini computer will end up being taken over by the microcomputer and that we will be seeing them at our individual desks. I think that the costs will be on the low end for the simplest systems, $5,000.00 or thereabouts. The more complicated ones will be on the order of $20,000.00, which is where they are now.

But what we will see is the advent of graphic capability, the ability to store and manipulate maps in a graphic form for much the same cost. Whereas these uses now cost a quarter of a million, they are going to come down in price and be added to the existing systems. So they will be at our desks. They are going to be able to help us in a clerical manner to store, retrieve and display data for us to look at and interpret ourselves. This leads into the second question and the time frame on this, I think, is on the order of two to three years, that we will see them at our desks. General usage, like we see programmable and hand calculators now.

QUESITON: In programs I've used, the user inputs data points into the computer, which in turn contours the information prior to integration. Most of the time this contouring is too mechanical, because it does not contain the human interpretation element. Does your system allow the geologist to draw his own isopach map, which can then be digitized and stored in the computer to be integrated?

ANSWER: I think the thought behind this question shows an awareness of what's going on with computers these days. It's the fact that this person asks about interaction and the human element, I think it's the one point that I would hope to get across about small computers in general. It is a fact that they're going to provide us the ability and vehicle through which we can add our own interpretation and interact directly with the data. The answer directly is, "Yes, we do allow the person to integrate the hand-drawn map," but beyond that the thought behind this question is, I think, a very important point. We will be using them. Back to the first question of why are they useful. We are going to be using them for clerical support, drawing support, drafting-type support, and some interpretation support, but we must keep in mind that computers can't think, although sometimes I think they have a personality. That's somewhat different. So, the thinking is left to the geologist, and those people who tend to think that computers can do it all, I think, are in error, and are going to see over the next couple of years a rapid increase in the interaction between geologists doing the thinking and computers doing the leg work and clerical work, together coming out with more accurate and more rapid evaluations.

23
Computerized Coal Laboratory Management

By Troy F. Stallard, President,
and Alan White, Vice President,
Standard Instrumentation, Inc.
division of Standard Laboratories, Inc.,
Charleston, West Virginia, United States

MANAGEMENT PROBLEMS

Management has often had difficult problems in obtaining satisfactory analytical information on exploration projects. Nowhere is this more evident than in the coal industry. Most of the time the coal to be explored is located in a remote area considerable distance from the laboratory. This situation can pose transportation or logistic problems and thereby lead to slower turnaround on results. For economic reasons, the utmost speed is often required when the decision as to whether to proceed with drilling is at stake.

Establishment of a laboratory near the work site is often problematic, since qualified employees are difficult to find or relocate. Reduction of the data usually requires extensive calculations and sometimes transformations to prepare for computerization. Naturally, as always with any laboratory, the data must be reliable since the result of exploratory work is frequently the basis for multimillion dollar decisions.

GLOBAL SOLUTION

A total lab management system is needed to address these problems for laboratories in support of exploration projects. This system must be capable of providing quick, efficient processing of samples through setting of priorities, scheduling of work flow, maintaining proper quality control, close control of equipment and competency assurance for technicians. Within this system, the work can then be performed by strict adherence to fixed, detailed methodologies, i.e, American Society for Testing and Materials (ASTM) standards. The elements of this system are keys to success for any lab, including those involved in exploration. However, as the subject of this paper is directed toward support of coal exploration, the emphasis will be placed there.

COMPUTERIZATION IN OTHER INDUSTRIES

An often-used method of addressing the problems of control of equipment, work scheduling, calculations, and generation of reports is through computerization. The literature relating to laboratories other than coal testing labs is full of references to computer applications. At the 1978 Pittsburgh Conference on Analytical Chemistry and Applied Spectroscopy, there were at least 75 papers on some aspect of laboratory computerization representing over 10% of the total. An amazing number of articles are appearing in periodicals such as *American Laboratory*, *Instrumentation Technology*, *Science and Analytical Chemistry*. The interested reader should begin with *American Laboratory* (Vol. 7, no. 2, Feb. 1975 and Vol. 8, no. 2, Feb. 1976) and *Instrumentation Technology* (Vol. 24, no. 2, Feb. 1977), since these issues are devoted entirely to lab automation/computerization.

The science of computerization is not a stranger to the testing laboratory but, to this point, has been little used in those laboratories that are involved primarily in coal testing.

MORE SOPHISTICATED USES OF LAB COMPUTERS

With some programming effort, computers can be utilized to train and monitor personnel actions by prompting technicians to follow very detailed procedures by insisting on the performance of specified actions at prescribed times. Computers working at electronic speeds can then foster the fastest possible performance of the work by always prompting a next step. Training is easily accomplished, since the technician is tutored through a step-by-step procedure and can actually run samples on a "cookbook" level. Computer reports to management on technician productivity or analysis on long-time trends in sample results are also possible.

Finally, diagnostic programs can interrogate operating equipment to identify problem areas.

COMPUTERS AND COAL TESTING

Although the coal industry has been somewhat in the backwaters of progress in the push toward computerization, a number of developments have taken place. Programs are currently being used by coal companies for payroll, financing, reporting, customer lists, report generation and, even on occasion, the operation of an isolated piece of equipment such as a balance. Many labs are also using processors or programmable calculators to do calculations and write reports.

The Bureau of Mines has gone further and computerized its analytical section at Forbes Avenue in Pittsburgh, PA. This system provides for the operation of an entire laboratory, including its major equipment, its scheduling and quality control techniques. One large independent laboratory firm has adapted a similar system. Moreover, a software house is now advertising nationwide its services in building tailored systems specifically for coal labs.

Up to now, however, such systems are for large, complex labs with many technicians and various operations predicating a sophisticated computer system. Practically no programs have been developed commercially for the small coal laboratory supporting coal exploration.

COMPUTERIZED LAB MANAGEMENT

Standard Instrumentation, Inc., following almost three years of intensive developmental work, has assembled a microcomputer system consisting of a special set of equipment, software, and test methods which provides for the over-all operation of a small coal lab. "Small lab" is defined here as a laboratory having 1 to 3 persons involved with actual hands-on lab work.

This system enables an operator(s) to schedule and perform the major test parameters required in coal testing as efficiently and accurately as possible and to do these tests using the current ASTM standards. The computer hardware consists of a microprocessor with peripheral printer and cathode ray tube terminal. The processor is tied directly to an electronic balance, an oxygen bomb calorimeter and a sulfur analyzer, and operates them in an interactive mode. The processor also controls and monitors the muffle furnace, ash fusion furnace and the moisture oven.

A management system provides for the logging of samples and handling of data, including a final typed report for any of eight parameters: moisture, ash, calorific value, sulfur, volatile matter, fixed carbon, free swelling index, and ash fusion temperatures. (Plasticity, dilation and grindability can be added as options.) The operator is not required to write down or maintain a single number during lab testing or reporting and is automatically alerted to specified equipment malfunctions.

FEATURES

Hardware

The processor is a Data General microNova (TM) minocomputer and includes CPU, up to 64K of Random Access Memory, system buffering and power supply. It operates a Real-time Disc Operating System (RDOS) and Digital I/O interfaces. The multi-tasking operating system combines data from more than one subsystem producing speed and precision not available from a single system. The microNova microprocessor has demonstrated ability for high performance regarding accuracy and reliability and real-time applications through many existing instrumentation systems.

A CRT terminal provides for laboratory technician prompting through test procedures and an Interactive Communication Mode. The terminal is a 1,920-character display with detached keyboard and two-plane swivel flexibility.

A terminal printer capable of 30 characters per second with keyboard is included. The printer has a full 132-column line.

A dual drive diskette system provides for mass memory (752,000 bytes) and rapid storage access and retrieval. A hard disk drive is available as an option for larger systems or extensive historical files.

The use of the system also eliminates the need for expensive accessories on the calorimeter (printer, digital thermometer and programmer) and furnace controllers wherever required. An electronic balance with BCD output, replacing a conventional balance, is the only additional lab equipment requirement.

Software

A set of programs based on ASTM methodologies has been developed. The display console prompts the operator on every action to insure the proper procedure is followed exactly. Two levels of verbosity are available for use by the beginner or the advanced technician respectively. An example of the console commands is shown in Appendix 1, Chart 1. Upon each command, the operator will perform the desired action and press an "execute" key to permit the computer to store information and proceed immediately to the next required action.

As previously stated, all information is stored by the computer for processing; the operator does not need to write down a single number throughout all testing procedures. Calculations are, therefore, virtually error-free.

The system can be interrupted at any time to reset priority of samples or take personal time off. Naturally, upon any interruption, the operator will be notified of the next critical event and how long before it takes place. A volatile sample coming out of the furnace is one example of an action which must be performed exactly on time. The computer warning system will notify the operator in time to insure compliance with the precise time requirement of the ASTM standard for this test.

Continuous monitoring of the laboratory equipment will automatically provide early warnings on equipment problems. For example, a problem report will be issued if a furnace does not turn on properly or if it does not hold the specified temperature with an established tolerance. In the event that the system or any part of it is inoperative, provision has been made to permit the manual operation of all equipment until the system is restored.

The software has been designed in a modular fashion so that additional testing parameters, equipment, or even personnel can be added with minimal effort. One needs only to specify the new equipment configuration and/or staffing level, and new programs will be generated ready for implementation.

Optional business packages are available for payroll,

accounts receivable, accounts payable, general ledger, and inventory control.

Management

The system, being computer-based, is much faster than any human and can thereby prompt an operator as fast as actions can be executed. The next step in the program is always immediately presented. Therefore, the system can prompt the operator to work as fast as humanly possible, increasing the capacity for higher productivity. Moreover, the length of time taken to perform tasks is measured and reported to management.

This system permits the training of non-professionals through verbosity levels to perform coal analysis in minimum time. This feature should allow companies to use less experienced -- and therefore less expensive -- personnel in their labs with greater confidence.

A standard sample program is also available, whereby samples with known parameters can be submitted to the system regularly (daily, every 20 samples, etc.) and reports with calculated variances generated for management review and control.

Tracking of historical data can be obtained as an option, wherein sample results from different seams, mines or contractors can be automatically analyzed for trends or significant differences.

SUBSYSTEMS

Reports

The report module of the package presents a "menu" that lists the daily reports generated by the laboratory. This listing consists of displaying a trial summary, printing that summary, printing all the test results of the day, displaying the test results for a specific sample, and printing the test results for a specific sample.

The trial summary allows the operator to take a quick look at all test results for the day prior to printing out final reports. This report can be displayed on the CRT, or the operator may choose to have a hard copy generated by the printer.

When all testing for the day is finished, the system will print the test results for each completed sample. The operator will also have the option of displaying on the CRT the results of any completed sample at any time during the day.

Additional reports will be automatically generated. These daily reports include a log-in summary, a comparison of results of known standards, and a summary of the work completed for a particular day.

The log-in summary report will be available to the operator immediately after he has completed the daily sample log-in. This summary will give the operator a hard copy of the work for that day and the parameters to be tested on each individual sample, as well as a priority listing of the work.

At the end of the testing day, the operator will receive a report on the standards run during the day. This report will compare the results for that day with the known/expected results. If results are not within acceptable limits, the operator will be informed immediately of the need for corrective action. A check on the ability of the lab to reproduce itself on results can be run on demand or randomly. These comparisons will indicate to both the operator and the lab management the lab's precision and reliability.

The final report of the day will be a summary of the work completed. This report will include the number of tests run that day, the amount of time it took to complete those tests, and the number of samples completed that day.

Balance

The operator will have available several measured "scoops" to help in the weighing-out process. The computer will tell the operator which sample is to be weighed, which crucible to use, and which scoop to use, and it will automatically record the tare weight of the crucible. For the weighing back procedure, the computer will call for a specific sample in a specific crucible to be re-weighed. The operator will place the sample on the electronic balance and press a switch whenever ready for the computer to read the weight.

This switch sends a strobe pulse to the computer which tells it there is a new weight on the balance. The balance will send a BCD signal to digital output interface boards in the computer. The computer will read the BCD signal from these interface boards and convert this value into its decimal equivalent.

The laboratory weighing device is any digital balance, such as the Mettler Model HL32 Electronic Digital Balance. The digital output signal must be 32 lines of BCD parallel, TTL compatible data. This digital output makes it possible to transfer the weighing results to the computer, thereby eliminating both weighing and recording errors.

Calorimeter

The system directly interfaces the Parr Adiabatic Calorimeter. It replaces the programmer which takes temperature readings at successive intervals, fires the bomb and signals the completion of the test. The digital thermometer is also unnecessary, as well as the printer. The Preiser/Mineco Model 1 Calorimeter Controller can be interfaced directly if desired.

The calorimeter subsystem programs direct the weighing-out of samples, indicate which bomb is to be loaded and when to insert that bomb, and perform the test sequence. A real-time routine takes temperature readings from thermistor probes every 30 seconds and compares them until equilibrium (zero change for 90 seconds) is achieved. The bomb is then automatically fired, and temperature readings are sampled again until equilibrium is re-established. Temperature readings can be displayed on-demand on the CRT. If zero is not reached in an appropriate time frame or the bomb does not fire, error messages will be generated to alert the operator of calorimeter malfunction, and the test is rescheduled.

At the completion of the run, the bomb is unloaded on computer cue. The information on wire length and acid titration is entered directly on computer demand.

Sulfur Analyzer

Any sulfur analyzer having BCD output can be interfaced. For example, Fisher Analyzer or Leco infrared systems are existing apparatus capable of direct readings. The BCD signal is converted into a numerical display on demand. The furnace is computer controlled by comparing thermocouple signals with a programmed signal and subsequent commands to external SCR to power or de-power as necessary. Temperature readout would be on-demand on the CRT.

Furnaces and ovens

Ash muffle furnaces, volatile furnaces, moisture ovens, and ash fusion furnaces send thermocouple signals to the computer which are compared with a programmed signal and controlled as in the sulfur furnace. Naturally, real-time commands are given to the operator in order to execute the proper procedure for these determinations.

Additional Tests

Other test parameters are expedited by this system. The volatile procedure is a weigh in/weigh out method like ash determination. FSI determinations will be scheduled and prompted and the lab numbers input directly to the computer. Grindability, fluidity and dilation can be run in the same manner as the FSI as an additional option.

SPECIAL PROGRAMS

Since the data in the system is already in computerized format, little effort is required to do whatever calculations or analyses are required (e.g., washability studies). This data could also be directly and almost instantaneously input to a company's central data banks for use by other segments of the company.

CONCLUSION

The use of this system offers many advantages over conventional methods. Enhanced control and accuracy of results are evident. Employee productivity and turnaround time on testing are also affected positively. Moreover, the system enables the utilization of less experienced personnel and speeds their training. The computer-aided coal testing laboratory is expected to have an increasingly important role in the analytical support of coal exploration programs.

APPENDIX I

LABORATORY MOISTURES

10 gram - weigh out

1. Line up these samples in ascending order

   101
   102
   103
   104
   etc.

2. Line up these crucibles in ascending order

   1
   2
   3
   4
   etc.

3. Examine each crucible to be sure it is dry and clean.

4. Clean, level and zero the balance.

5. Press "execute" switch when the balance displays all zeros.

Press any key to continue

DISCUSSION

QUESTION: How would you like to work for a computer? Or do you see potential employee morale problems? This question ties very nicely with the next question, which is: "As an industrial psychologist, are there negative impacts on the lab technician's motivation and job satisfaction? He would have minimal decision-making responsibility and latitude." Are there any problems in associating and directing human beings by a computer rather than other human beings?

ANSWER: Those are good questions. We have seen them a great deal. However, I think that what we might talk about first is: What is an analytical lab trying to do? We have found in 30 years experience with several labs that the most often occurring error coming out in the data is a human error. Once you get your equipment in good shape, run it well, and get your methodology down pat, then what you want is very little human discretion regarding what the operator's next action is. In fact, the ASTM methodologies are set up in order to make sure that one does one thing and then a second thing, a third thing, in sequence so that there's some very fine, highly detailed methodology. When we at Standard Laboratories hire people, the first thing we do is sit the applicant down and say, "O. K., if you want to be a chemist working on the frontiers of knowledge doing fundamental research, you're in the wrong place." What we do is perform a test; a determination that is very specifically spelled-out, and then we do it again, and again; then after a thousand times we do it a thousand and one and a thousand and two, because we want the same answer every time. The only way you can do that is by not building in a lot of discretionary kind of decision making."

I think it's really a question of scale. The person who operates this computer lab will be more advanced in terms of decision-making, in a decision-making kind of dimension, compared with the person on an automated automobile assembly line in Detroit. That person is spot-welding one place every 15 seconds. But you are somewhat on down this scale from where the analytical chemist is working on larger problems. What we're really trying to do here is enable people with less formal education and less in the way of logical decision-making capabilities to be able to generate good, accurate numbers. I think the analogy for this job should be a computer operator. There are lots of jobs in the world today that call for computer operators, and those people put on discs, turn on the machine, and feed the machine. Whatever the machine wants, they oblige it. Our person will be much like that computer operator. There is no question that we're taking out a lot of the kinds of things that we think tend to make the errors, but I think there'll be enough for that person to do in such a job.

QUESTION: How many samples, and these are approximates, per day can be handled by the computer system you propose?

ANSWER: I really can't give you a very definite answer, because that depends upon your equipment configuration and, also, on your sample requirements. For example, if you were to run moisture determinations or ash determinations you could run a great many such determinations with this system. If you throw Btus in there, and you have one calorimeter, then you'd better count on 30 moisture, ash, sulfur and Btus a day. You throw another calormieter in there then you can have double that capability. It really does depend on equipment limitations and your sample loads. If you have a whole lot of ash fusions to run you're not going to get many total samples out.

I would say that the very basic system that we see, which would have one calorimeter, one balance, one sulfur analyzer, and again that is dependent on what options one selects, will cost approximately $35,000.00. That's a lot of money. But that price does include the hardware and the software. There is also about an $8,000.00 possible savings on conventional laboratory equipment; therefore, the net cost is something less than $27,500.00. Again, different options might add to that figure depending on how much additional programming would have to be done.

QUESTION: How do you computerize ash fusion temperature analysis?

ANSWER: What we've really done in this basic model, is have the ash fusion furnace completely controlled by the computer and have the computer, of course, run up the temperature to a certain level, and then require the operator to read the cones at certain time intervals. When the operator reads the temperature one of a matrix of switches in front of the ash fusing device can be selected. If one wants a softening temperature, after reading it as softening he merely punches the button and that value is carried out to the computer. So you don't have to know the value of the temperature, you just have to indicate when you get to that decision point.

Ash-fusion is somewhat an archaic kind of approach to the problem in the boiler systems that many power utilities, and other users, have. I think there's going to be a lot of developments in this area, and pretty soon you will see automatic sensors read those temperatures directly. These will be, again, completely computer integrated. It'll make the whole system a lot easier to handle.

QUESTION: How would we support the hardware and the software?

ANSWER: The computer manufacturer will be doing all of the hardware support, and a person having one of these labs, even if you had one in Australia, would have support through their Data General offices there. They will provide a service agreement based on what you want, 24-hour turn around, 48 or whatever. But the main point is that the computer manufacturer maintains the hardware. On the other hand, Standard Instrumentation, Inc. would maintain the software.

One final thing that was called to my attention is that I had seemingly emphasized the need for a hard copy final report, and it was quite rightly brought to my attention that with the computer systems you really don't need hard copy. Your final product can be other than a hard copy document. For example, you might want to take that data straight into a central computer or some place else without generating any kind of hard copy documentation. In fact, some people might want to just know average values over a period of time and not be concerned with specific analysis. Fortunately once you have it in the computer in its relocatable binary format, then it's no problem to send it wherever you want.

24

On-site Computer Analysis of Coal Logs

By J. K. Hallenburg
Manager, Applications Engineering
Century Geophysical Corporation
Tulsa, Oklahoma, United States

ABSTRACT

On-site computer analysis of coal logs is reviewed. The advantages of capturing raw data on tape and of on-site analysis are examined. It is shown that the on-site computer results in a significant saving in turn-around time for necessary data and processing. In addition, aquisition time is shortened and mistakes are fewer.

INTRODUCTION

We are at the beginning of a new era in logging for coal exploration. For the first time we have available in one place, the technology, the calibrated computer based equipment, and the proper economics for coal exploration. It is interesting to see how we have gotten here.

BACKGROUND

Historically, the use of logging has not been very extensive in coal exploration. Most of the emphasis has been on traditional methods; surface geology, coring, core analysis, and sample analysis. The logs which have been used have been largely uncalibrated and qualilative.

This situation began to change in the early 1970's with the use of the sidewall density tool. This tool gave a curve which was definitive for coal, and, in combination with the single point curve allowed more accurate measurements of depths and coal bed boundaries.

The use of calibrated logs and rigorous interpretation techniques has been demonstrated for a number of years. Schlumberger has published a number of papers relating to this subject. In 1964 they described the use of an on-site computer; in 1967 they suggested the use of simultaneous equations for solving lithology problems. In 1969 they suggested interpretation methods relating specifically to coal. In 1975 Dresser-Atlas published a detailed paper for coal log interpretation involving both the coal and the overburden lithology. Reeves of BPB suggested single curve methods in 1971 and again at this meeting in London in 1976, as did Lavers and Smits of Shell.

The present state of the art of coal log interpretation is surprisingly advanced. The general scheme is to measure several parameters and then use single curve analysis, simultaneous equations, cross plotting and other methods. The equations shown in figure 1, illustrate the method of deriving density from the counting rate of the density tool and then porosity from the resulting density values. This is a big subject. Notice, however, that the relationship between counting rate and the useable parameters is quadratic and not simple. There are other similar relationships for neutron derived porosity, resistivity derived porosity, formation water solids content, shale content, and many others. There are similar relationships for multiple curve analysis.

Quality logs and rigorous analysis have been available from the major petroleum logging contractors for a number of years. However, their methods have naturally been oriented toward petroleum exploration and not coal. This means their systems are designed for big holes, for deep holes, for high temperatures and not for the small, shallow holes we have in mineral exploration. They operate large tools with big crews. Thus, their prices are commensurated with the needs of the petroleum industry. Also, logistics is often a problem since coal and oil fields do not usually occur near each other.

Uranium exploration, on the other hand, has been handled by smaller contractors. Fast, low cost techniques have been evolved to suit minerals exploration projects. The uranium equipment is small and light and usually logging is done by one man crews. Thus, uranium logging prices are significantly lower than petroleum logging prices on a per foot basis. However, except for the gamma ray curve qualitative logs have been sufficient. The gamma ray curve is used to calculate the amount of uranium near the borehole.

The use of logging in coal exploration has been limited. Although the crews, equipment, techniques, and prices have been the same as for uranium exploration, the qualitative nature of the logs has been unsuited to the needs of the coal industry. And, there has been a general lack of knowledge in the coal industry of the potential uses of borehole geophysical logs. On the part of the contractors there has been little incentive to develop tools and techniques suited to coal exploration because of the limited market existing before 1972.

Now however, the market is expanding and new and better techniques are being demanded.

OBJECTIVES OF COAL LOGGING

Coal logging must reflect the economics of the coal industry. Quantitative techniques are needed but the logging costs must be significantly lower than those of the petroleum industry. The logged information must be

DENSITY

SINGLE CURVE EQUATIONS

$$d_{bulk} = a_1 c^2 + b_1 c + c_1$$

$$\emptyset d = \frac{d_{rock} - d_{bulk}}{d_{rock} - d_{fluid}}$$

WHERE C IS THE DENSITY COUNTING RATE AND d IS THE DENSITY VALUE.

Figure 1

Figure 2

Figure 3

gathered quickly and processed rapidly. The data must be in easily useable forms. And, the log must contain at least as much information as the traditional methods, such as core or sample analysis.

There is some specific information which is needed about the coal. The coal depths and thicknesses must be determined accurately. Today's exploration leaves little room for error.

The log must furnish data to evaluate the characteristics of the coal in situ since coring can change the sample physically and chemically.

As a minimum, a log must have a short quantitative analysis. This must contain the heat content, the moisture content, and the ash content. If it is possible to include other components it should do so. Certainly, more detailed methods will follow later.

Logging must allow the evaluation of the total reserves of a project. This is best done by recording the logged data on tape and using computer methods for daily updating.

The ancillary information from a log is equally important. The overburden characteristics can be determined from the log. This includes water content for anticipating problems, and pump system design. Roof and floor competance can be determined for underground mines. The thickness and rippability of the overburden can be determined for open pit mines. Shaft sinking problems can be anticipated. And of course, logs must aid in cost evaluations.

MODERN COMPUTER BASED MINERAL LOGGING EQUIPMENT

These requirements indicate that large amounts of information must be gathered and processed rapidly.

The downhole tools must use digital data handling techniques designed for direct communication with a computer. As a result, these probes have more channels available and wider effective signal ranges.

Figure 2 shows such a multisensored, calibrated probe. This probe measures and transmits to the surface signals from a temperature, neutron-porosity, single point resistance, S. P. natural gamma radiation, three flux gate and two accelerometer sensors simultaneously.

Figure 3 shows a similar tool designed for coal exploration. This one measures density, hole diameter, natural gamma radiation, and a 3 electrode guard log resistivity.

The truck and truck mounted equipment needed for this type system is shown in figures 4 and 5. The capabilities of the computer make it possible to gather this amount of data at high speed with high reliability and good quality control. The end result is good field calibrations and permanent data records which can be reprocessed.

After the required data have been recorded, the computer becomes a data processing system. Logs are plotted out with different formats and different scale options. The raw counts and millivolts are converted to easily useable units, such grams per cc and percent porosity, on the log.

Analytical techniques, such as the cross plot (figure 6), are used in the field with a turn-around time of minutes. Thus, the usefullness of the information is greatly increased.

The prices of these types of systems are roughly comparable with those of good analog systems. But, because of the higher speeds, more data, and greater reliability, the costs are lower.

CONCLUSION

Thus, a new era is beginning in coal logging. For the first time we have available the technology, the calibrated, truck mounted, computer based equipment and the appropriate economics for the coal industry.

REFERENCES

1. In-situ Analysis of Coal by Borehole Logging Techniques, D. R. Reeves. Canadian Mining and Metallurgical Bulletin, February 1971.

2. Computer Evaluation and Classification of Coal Reserves, Wm. H. Smith, International Coal Exploration Symposium, London, England, May 1976.

Figure 4

Figure 5

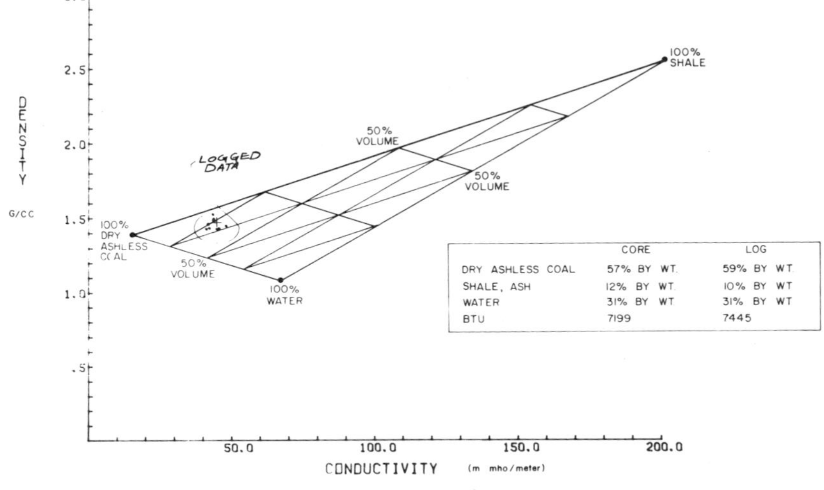

	CORE	LOG
DRY ASHLESS COAL	57% BY WT.	59% BY WT.
SHALE, ASH	12% BY WT.	10% BY WT.
WATER	31% BY WT.	31% BY WT.
BTU	7199	7445

Figure 6

3. Recent Developments in coal Petrophysics, B. A. Lavers and L. J. M. Smits, International Coal Exploration Symposium, London England, May, 1976.

4. Well Log Applications in Coal Mining and Rock Mechanics, L. O. Bond, R. P. Alger, A. W. Schmidt, American Society of Mining Engineers of A.I.M.E. Preprint 69-F-13 (1969).

5. Coal Analysis from Well Logs. John J. Kowalski, Milton E. Holter, American Institute of Mining, Metallurgical, and Petroleum Engineers, Inc. Paper No. SPE5503, (1975).

6. Automatic Log Computation at Well-Site: Formation Analysis Logs, M. P. Tixier, F. M. Eaton, D. R. Tanguy, W. P. Biggs, Society of Petroleum Engineers of A.I.M.E. Paper No. SPE 987 (1964).

7. Log Evaluation of Non-Metallic Mineral Deposits, M. P. Tixier and R. P. Alger, Schlumberger Well Services, Houston, TX.

8. Correlation of Elastic Moduli Dynamically Measured by In-Situ and Laboratory Methods. D. P. Helander, D. J. Myung, Birdwell Division, Seismography Service Corp., Tulsa, Oklahoma.

9. Application of Wireline Logging Techniques to Coal Exploration, D. R. Reeves, International Coal Exploration Symposium, London, England, 1976.

DISCUSSION

COMMENT: I would like to make one comment, if I may. Mr. Stallard spoke about the ability of computers to allow non-technically-trained persons to do technical work. This works the other way, too. The use of computers allows technically-trained people to be relieved of routine duties, so that they can use their technical abilities to better advantage.

QUESTION: What is the relationship between your computer-generated field analyses and approximate or ultimate analyses of a given coal?

ANSWER: I am sorry that that wasn't clear on the diagram. In general, logging tools measure parameters volumetrically. The core analyses from a lab--the sample analyses in general are weight related. The relationship between the computer-generated analyses and the lab analyses are both aiming toward the same point. We are attempting to define what are the component amounts of the formations. That is coal or sand--whatever they are. We are trying to define what these are. And, of course, since we are measuring in a different set of terms than the laboratory reports, one has to be converted to the other. And in the case of my diagram, I converted our volumetric measurements to weight measurements for comparison.

QUESTION: What general technique will be used to figure percent moisture?

ANSWER: I can't give you an exact answer to this. The programs that we're using allow the entry of the fixed information, like the density of the water or the density of the shale, the various components, the resistivity, the gamma ray reading, whatever they are. The programs are set up so that the operator can enter these to suit a particular field. At this point it looks to us like this type of trimming of the diagram will have to be done for each individual field. Perhaps, after more experience has been gained, we'll find a way around it. One way of determining the percent moisture in the coal, independent of the type of cross-plot that I've presented would be to use a neutron-porosity tool. Certainly, this can be run in a coal equally as well as it can be in a sand. Like most logging tools, the density tool, or the neutron tool, any of them that you'd care to mention, more than one parameter is being measured at once. The neutron tool not only measures the moisture content in the coal, but since it measures the hydrogen content, it measures the other non-aqueous hydrogen content of the coal. So at this point if the--if the hydrocarbon content of the coal is low, we can use the neutron tool directly. If the hydrocarbon content is high we'll have to find some way of separating that from the moisture content. At this point, except for cross-plots, we don't have a good way of doing it.

Exploration in India

25.

Strategy of Exploration in the Concealed Coalfield of Kamptee in Central India

by T. N. Basu, T. K. Chandra,
and B. B. P. Shrivastava

25
Strategy of Exploration in the Concealed Coalfield of Kamptee in Central India

By T. N. Basu, Chief of Geology and Drilling,
T. K. Chandra, Deputy Chief of Geology,
and B. B. P. Shrivastava, Superintending Geologist,
Central Mine Planning and Design Institute Limited
Ranchi, India

INTRODUCTION:

The chance discovery of workable Permian(Lower Gondwana) coal deposits in a tubewell drilled in 1940 in the neighbourhood of Kamptee town ship (North Latitude 21°13' and East Longitude 79°12'),north-east of Nagpur(North Latitude 21°8' and East Longitude 79°5'), (Figure No.1-A) brought this area on the coalfield map of the country and the area came to be known as the Kamptee Coalfield (Figure No.1-B). Since the first discovery of coal in this area,several mines have been opened but the knowledge about the extent of the coal-field till recently had been very much limited. This was mainly because of the fact that the entire area is blanketted by a thick(30 to 40 metres)alluvial cover excepting the detached exposures of the Kamthi Formation (younger than the coal-bearing formation) and the rare exposures of the Talchir Formation (older than the coal-bearing formation) of the Lower Gondwanas. Thus, the Kamptee Coalfield can truly be regarded as a 'Concealed Coalfield'.

* Chief of Geology and Drilling,** Dy.Chief of Geology, *** Superintending Geologist.

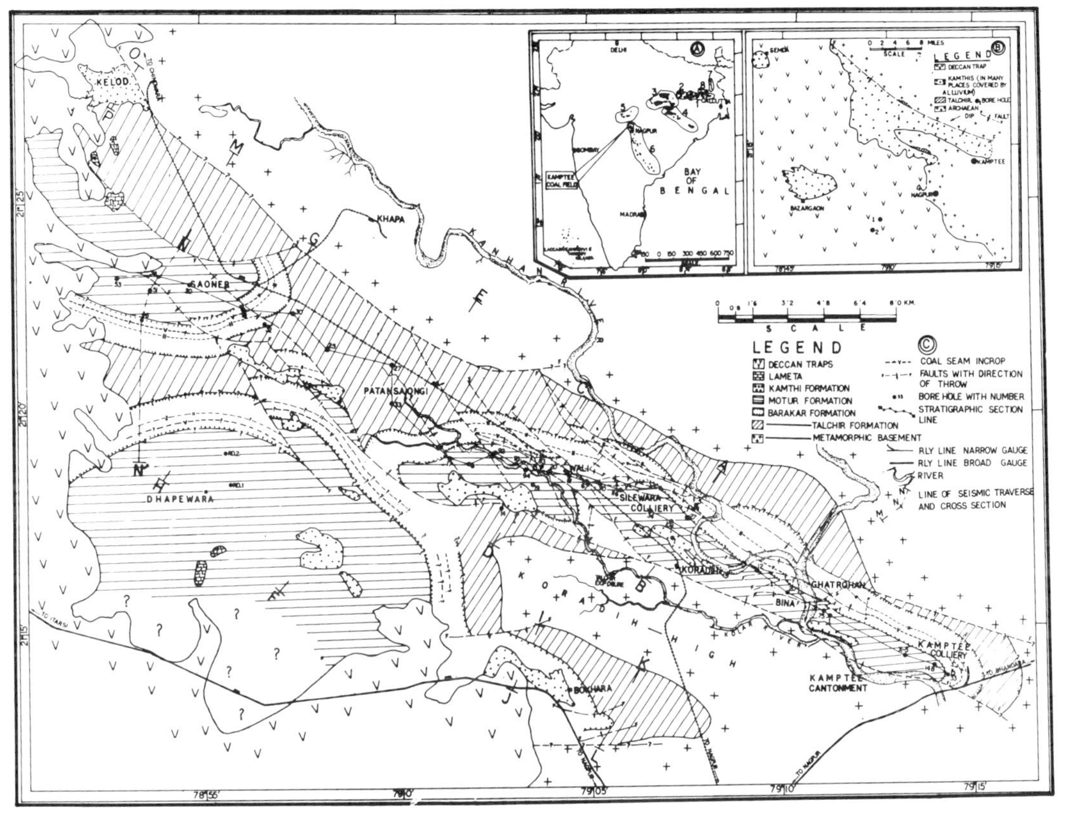

Figure No.1-(A)Gondwana Coalfields of India- 1)Damodar Valley,2)Koel Valley,3)Sone Valley, 4)Mahanadi Valley,5)Satpura Area,6)Pranhita-Godavari Valley,7)Rajmahal Hills,8)Deogarh Group (B)Geological Map of Kamptee-Nagpur Area after E.R. Gee(1952), (C)Geological map of Kamptee coalfield showing sub-crop pattern of various formations,faults,coal seams,etc.below alluvial cover.

In the context of the industrialisation initiated at the beginning of the Second Plan(1955-56),greater emphasis was laid on additional coal production from new and virgin areas in the Public Sector. The Kamptee Coalfield(practically virgin)was taken up for exploration during the Third Plan Period (1961-62 to 1965-66) for this purpose.

Gee(1) had shown the tentative outline of the coalfield in view of its concealed nature. Nothing was known about the geometry of the basin and,therefore,the exploration by conventional methods had necessarily to be taken up in the eastern part only,adjacent to the known areas.

In order to adopt a proper exploration strategy,it became imperative to ascertain the extent and geometry of the basin and prepare a geological map. It was in this context that a seismic refraction survey was carried out by the Geological Survey of India, at the instance of the Division of Geology(now Exploration Division of CMPDI) of the then National Coal Development Corporation Limited, to delineate the basement profile. The seismic data integrated with the available drill hole data helped considerably in evaluating the basin geometry,the total thickness of the sedimentaries and the potential coal-bearing areas of the Kamptee Coalfield.This paper deals with the geology of the coalfield and the surrounding areas and presents for the first time a composite geological map of the Kamptee Coalfield. The map,however, depicts the broad outlines of the formational boundaries and the structural set up and is subject to revision when more detailed information become available. This map formed the basis for evolving the exploration strategy in the north-western part of the coalfield and it is hoped that it would also considerably help in formulating the future exploration strategy in the southern part.

GEOPHYSICAL SURVEYS:

There were attempts in the past to delineate the Gondwana-Archaean boundary so that the outline of the coalfield could be deciphered more precisely. However, the past surveys,electrical resistivity and magnetic, were confined to the probable margins of the coalfield (2,3,4 and 5). Thus, its geometry could not be established till the recent seismic refraction, supplemented to a limited extent with seismic reflection, survey was carried out to delineate the basement profile and assess the variation in the thickness of the sediments along the profile lines. However, the formational interfaces of the sedimentaries could not be deciphered because of the "insignificant variation in the seismic velocities"(6).

In all eight seismic traverses across the basin and approximately along the dip of the sedimentaries were taken (Figure No.1-C).

GEOLOGICAL MAP-
BASIS AND PROCEDURE:

The available seismic and the drill hole data(7 to 15) have been integrated in evaluating the formational boundaries, the geometry of the basin and the faults for preparing a composite geological map of the coalfield. In the areas where control points are lacking, the formational boundaries have been projected on the basis of the observed behaviour in other areas where a large number of control points are available. The thickness of the individual formations has been evaluated from cross sections along the seismic profile lines coupled with drill hole data. The faults have been delineated on the basis of borehole data but a few of them have been postulated on the strength of the seismic data. The structure contours on top of the basement, based on the seismic data, have revealed the basin geometry. The geological map thus prepared depicts the subcrop pattern of the various formations, faults, coal seams, etc., below the alluvial cover(Figure No.1-C).

GEOLOGIC SETTING:

The Lower Gondwana Coalfields occur in linear belts aligned along prominent river valleys in the Peninsular India(Figure No.1-A). The Kamptee coalfield forms the northernmost extension of the Pranhita-Godavari Valley. The Lower Gondwana Group in this coalfield comprises the Talchir, Barakar, Motur and Kamthi Formations in the ascending order and as elsewhere in the subcontinent, these rest unconformably on the metamorphic complex of Archaean age. Besides these, other younger formations are also present within and outside the coalfield. The stratigraphic succession and the thickness range of the individual formations are given below :

Age	Formation	Thickness Range in Metres
Recent/Sub-Recent	Alluvium (Detrital Mantle)	30-40
	— Unconformity —	
Eocene to Upper Cretaceous	Deccan Traps	A few metres to tens of metres
	Lameta Formation	?
	— Unconformity —	
Upper Permian (?)	Kamthi Formation	0-100
	— Unconformity —	
Middle Permian	Motur Formation	200-300
Lower Permian	Barakar Formation	250-300(?)
Upper Carboniferous	Talchir Formation	150-600(?)
	— Unconformity —	
Archaean	Metamorphic complex	

Extent of the Coalfield:

The Kamptee Coalfield(North Latitude 21°10'-21°30' and East Longitude 78°50'-79°15')extends from near the Kamptee cantonment in the east to as far as Kelod in the west and covers an area of 505 square kilometres(Figure No. 1-C). The metamorphic complex forms the northern and the eastern boundaries. The western and the southern boundaries of the coalfield are not yet known as these are covered by the basaltic lava flows (Deccan Traps). Within the coalfield area,a few scattered exposures of the Talchirs,Kamthis and Lameta Formations are exposed. The coal-bearing Barakar Formation is not exposed anywhere in the coalfield.

Outside the presently known limits of the Kamptee coalfield, two inliers of the Kamthi Formation within the Deccan Trap country are exposed in the west around Semda and Bazargaon(about 22 and 16 Kilometres respectively,Figure No.1-B). Since the western and the southern boundaries of the coalfield are obscured by the volcanics,it is not known whether the Kamthi Formation around Semda and Bazargaon continues below the trap cover and forms an integral part of a very large basin or indicates detached sedimentary basins independent of the Kamptee Coalfield.

In this context it is worth mentioning that the existence of the workable coal seams directly under a shallow cover of Trap rocks has recently been reported from two drill holes viz. 1 and 2,about 15 km.south-east of the Bazargaon exposure (Figure No.1-B). In borehole No.1, the Talchir Formation has been encountered at a shallow depth suggesting that the basin margin in the north and east may not be far away from this hole. The other borehole was suspended in the Barakar Formation itself after encountering a few thin coal seams at comparatively greater depths thus indicating that this hole is possibly located on the dip side. A separate sub-basin in this region extending perhaps upto Bazargaon in the west cannot,therefore,be ruled out.

Thickness of the Sedimentaries:

The eight seismic traverses across the basin broadly reveal the variation in the thickness of the sedimentaries (including the alluvial cover). From the isochore map of the total sedimentaries (Figure No.2), it is observed that besides a general southerly trend of increase in the thickness, there are three distinct sectors containing thick sediments viz.(i)Patansaongi-Silewara, (ii)Saoner and (iii)Dhapewara-Bokhara. The Patansaongi-Silewara and the Saoner Sectors occupy the northern part of the basin while the Dhapewara-Bokhara Sector occur in the southern part.

In the Patansaongi-Silewara Sector, the maximum thickness is a little over 700 metres (Traverses-AB and CD). In the intervening patch between the Patansaongi-Silewara and the Saoner Sectors the thickness decreases, indicating that the basin is shallow. Further west in the Saoner Sector, the thickness of the sedimentaries in the northern part along traverse MN is considerably less, being 300 metres. The thickness is inferred to increase in the southern part. Along traverse G-H,in Saoner Sector, a maximum thickness of 750 metres has been recorded and it increases westwards of this location.

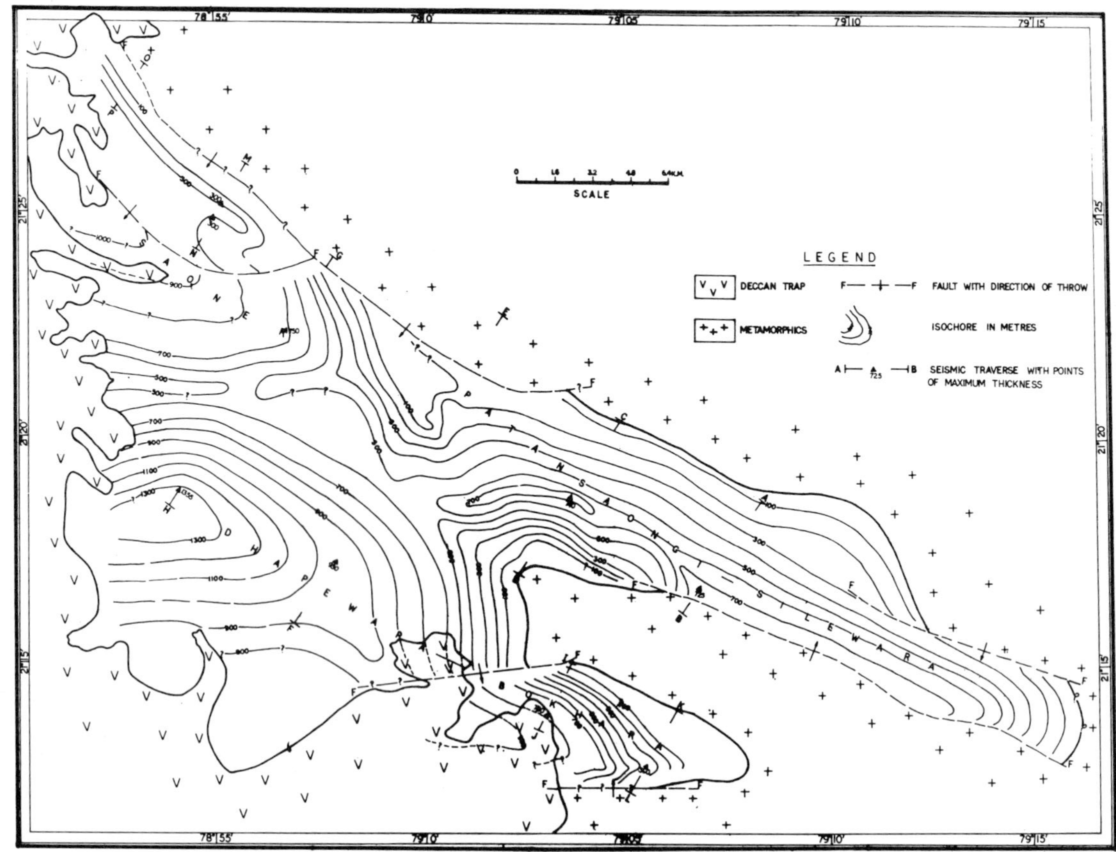

Figure No.2- Isochore of the total sedimentaries based on seismic and borehole data.

In the Dhapewara-Bokhara Sector, the thickness gradually increases both from north to south and from east to west. The thickness increases from 360 metres in the east to as much as 1355 metres in the west(Traverse G-H). Thus, the deepest portion lies in the south western part of the Kamptee coalfield.

Since no single borehole has intersected the entire sequence, the thickness of the individual formations has been extrapolated by integrating the borehole and the seismic data. It is inferred that the thickness of the Talchir Formation is likely to increase considerably(600 metres)in the western part of the Dhapewara-Bokhara and the Saoner Sectors. Such a large thickness probably has not been recorded for this formation in the other Lower Gondwana Coalfields. The thickness of this formation along traverse G-H has been inferred based on the data of borehole No.32(Figure No.1-C). This borehole was suspended in this formation only. If there has been any misidentification of the Talchir in this borehole, it is possible that the Talchir Formation may not be as thick as inferred and a part of it may be attributed to the overlying Barakar Formation.

While the existence of the coal-bearing Barakar Formation has been proved in the Patansaongi-Silewara and the Saoner Sectors and partly in the Bokhara area, the existence of the upper part of the Barakar Formation in the Dhapewara area has been proved in one of the two unsuccessful boreholes drilled recently. Although, the same set of coal seams as available in other areas has not been encountered in the borehole RD-2(Figure No.1C), the presence of a coal band overlying the topmost workable coal seam elsewhere,lends support to the view that the chances of striking workable coal seams in this area are indeed bright.

Lithological Character:

There are two notable exposures of the Talchirs- one in the Kolar river,south of Silewara colliery and another about three Kilometres NNE of Patansaongi along the northern faulted contact. The base of the formation is marked by a 'boulder bed' which is overlain by greenish needle shales and fine to medium sandstones. The 'boulder bed' which occurs as a heterogeneous unit is composed of cobbles and boulders of gneisses,quartzite-sandstones and granites. In the Kolar river exposure,a boulder of nearly 3 metres(longer axis) was observed to lie in a mass of greenish muddy sandstone. The exposure NNE of Patansaongi is composed of hard greenish shales and contains pebbles and cobbles of metamorphic rocks. The rocks are highly jointed and are intruded by quartz veins.

The coal-bearing Barakar Formation,conformably overlying the Talchir-Formation, is composed dominantly of sandstones with subordinate shales,sandy shales,carbonaceous shales,laminates and the coal seams. The lower part of the Barakar is generally composed of fine-grained sandstones and the coarse grained sandstones constitute the upper part. This formation contains five correlatable and commercially exploitable coal seams,numbered I to V in the ascending order,generally varying in thickness from over one metre to about ten metres. All the five seams are confined to the middle part of the formation (about 100 to 120 metres) and this has facilitated the classification of the Barakars into Upper,Middle and Lower (Figure No.3). The Upper Barakars (60 to 80 metres) do not contain any coal seams. The Lower Barakars (70 to 100 metres),though devoid of workable coal seams,contain a number of thin coal bands.

The coals are sub-bituminous to bituminous in rank and are non-caking in character. The ash content in the Upper two seams(Seams IV and V) is higher than that of the lower three seams. The coals from these seams have low sulphur content. The Seams IV and V are relatively thick horizons but are highly interbanded with carbonaceous shale/shale. The coal seams generally reduce in thickness towards west with increasing interseam partings. Some of the seams either deteriorate to carbonaceous shales or split into uneconomical sections or completely pinch out towards west. On the other hand, consequent on the reduction in the interseam partings towards east(Kamptee colliery), the seams attain higher thickness and tend to coalesce.

The Motur Formation overlies the Barakar Formation conformably. The formation consists of an interbedded sequence of clay,mudstones and fine to medium grained

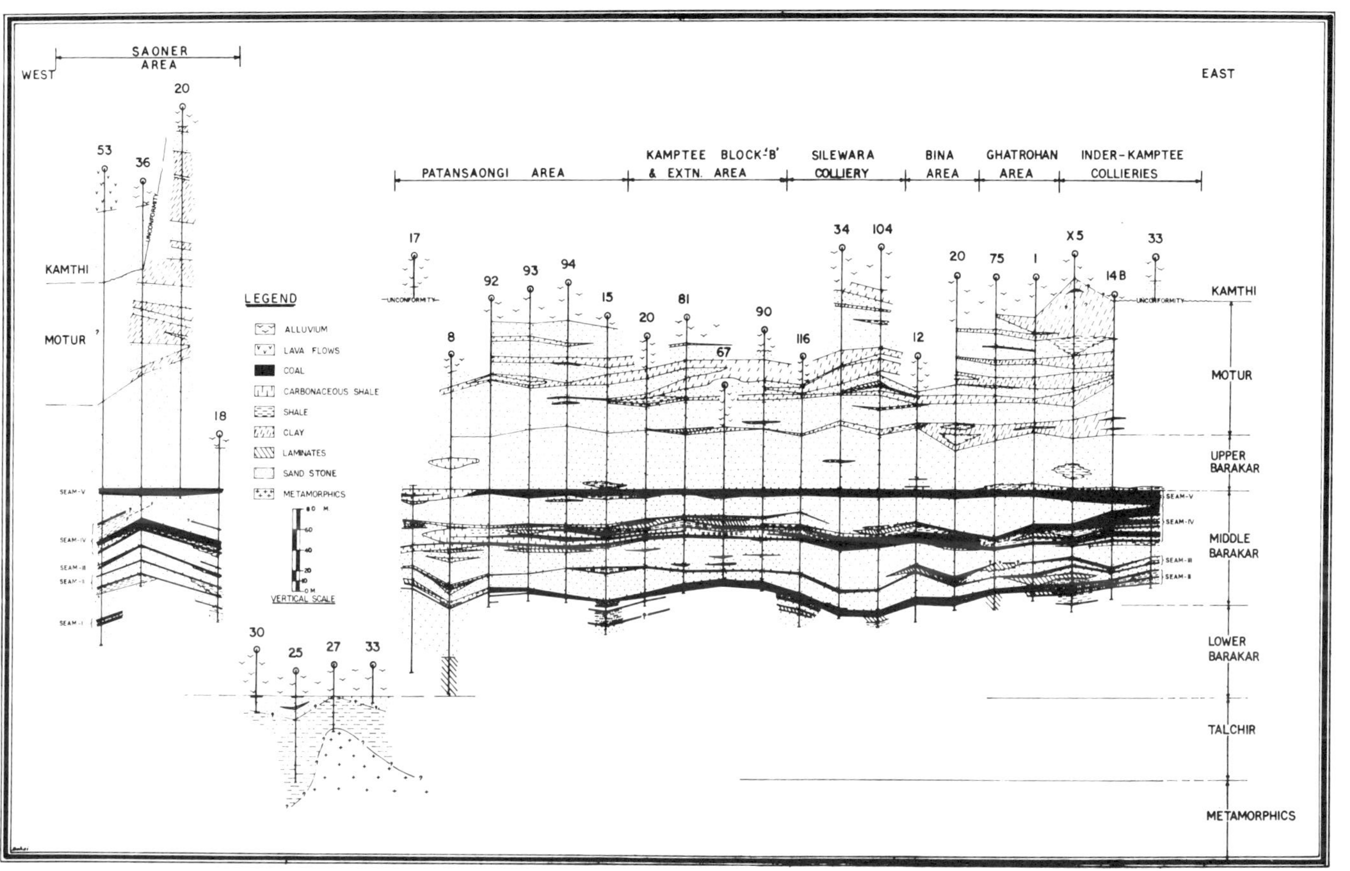

Figure No.3- Stratigraphic section showing the disposition of various formations and coal seams from Saoner in the west to Patansaongi-Silewara sub-basin in the east.

sandstones. The clays display a variety of colours ranging from bright red to green and even grey. The clay beds are more predominant in the upper portion of the formation compared to those of the lower portion which is more arenaceous. The Moturs also contain rare grey shale and thin coal bands but otherwise are completely devoid of any workable coal deposits.

The Kamthi Formation unconformably overlies the underlying formations and overlaps all of these at places. A few scattered exposures are found forming small hillocks. Its spread below the alluvial cover appears to be discontinuous as it has been encountered in drill holes in isolated patches. These perhaps represent the erosional remnants and, therefore, no attempt has been made to show the spread of this formation below the alluvial cover (Figure No.1-C). The Kamthis consist of conglomerates, medium to coarse grained feldspathic sandstones which are hard and siliceous and are often interbanded with hard ferrugineous layers of dark brown colour. The siliceous or quartzitic type sandstones are confined to the upper part and the middle part, though dominantly composed of medium to coarse sandstones, contain a few hard red shale beds.

The Kamthi Formation regarded as equivalent to the Raniganj Formation(Upper Permian)of the Damodar Valley(16) is distinctly different in lithological character from the latter. It is interesting to note that the rock types similar to the Kamthi Formation have been reported around Mangli, about 80 Kilometres south of Nagpur town(17). The floral and the faunal assemblages of this formation indicate Lower Triassic age. The Kamptee coalfield occurs within the same belt (Pranhita-Godavari Valley)and since the Kamthi exhibits similar lithological character, they should in all probability be regarded as equivalent to the Mangli beds(equivalent to the Panchet of the Damodar Valley).

Structure:

The Kamptee coalfield is elongated in a NW-SE direction. In the south-east, the Patansaongi-Silewara and Dhapewara-Bokhara Sectors are separated by a metamorphic wedge. This wedge of metamorphics termed as the 'Koradih-High'(Figure No.1-C) plunges in a north-westerly direction. Its subsurface continuation along the direction of the plunge is postulated on the basis of the seismic data(Figure Nos. 4 and 5). Compared to the axial region, on either side of this 'High', the thickness of the sedimentaries increases, notwithstanding the local variations.

Structurally, the Kamptee coalfield is composed of three sub-basins (coinciding with the three sectors containing thick sediments described earlier)within the overall structural frame work of the field.

Patansaongi-Silewara Sub-basin: The axis of the Patansaongi-Silewara sub-basin coincides with the general direction of elongation of the Kamptee Coalfield. The southern boundary of the basin is truncated against the northerly hading fault (700 m.throw) along the northern flank of the Koradih-High (Figure No.1-C). The deepest part of the basin lies close to this fault in the Silewara and the Kamptee colliery areas and occupies almost the central part towards the Patansaongi end. The northern boundary is faulted in the western and eastern parts but in the central portion it is marked by natural depositional contact. However, the fault which marks the boundary in the eastern part cuts through the Talchirs and for all practical purposes the Patansaongi-Silewara sub-basin lies in a structural trough. The sedimentaries strike in a W N W-ESE direction and dip at angles of 10 to 15 degrees towards south, excepting the closure points in the east and west. This sub-basin has been faulted on a large scale. The faults in the northern part have a W N W-ESE alignment and hade against the dip of the strata, barring the boundary fault in the north. In the western part the faults have a NN W-SSE alignment and hade in ENE direction. These are at angles of 30 to 40 degrees to the Koradih-High.

Saoner-Sub-Basin: The Patansaongi-Silewara and the Saoner sub-basins are separated from each other by an intervening patch of the Talchirs. The total extent of this sub-basin is not known as the western boundary is obscured by the volcanics. The northern margin of this sub-basin in marked by a southerly hading fault. Nearly parallel to this fault, there is another major southerly hading fault against which the northern limb of the basin is faulted bringing the younger Motur Formation against the Talchir Formation, a throw of nearly 800 to 900 metres (?) (Figure No.5). The axis of this sub-basin is on the same elongation as that of the Patansaongi-Silewara and has a north-westerly plunge.

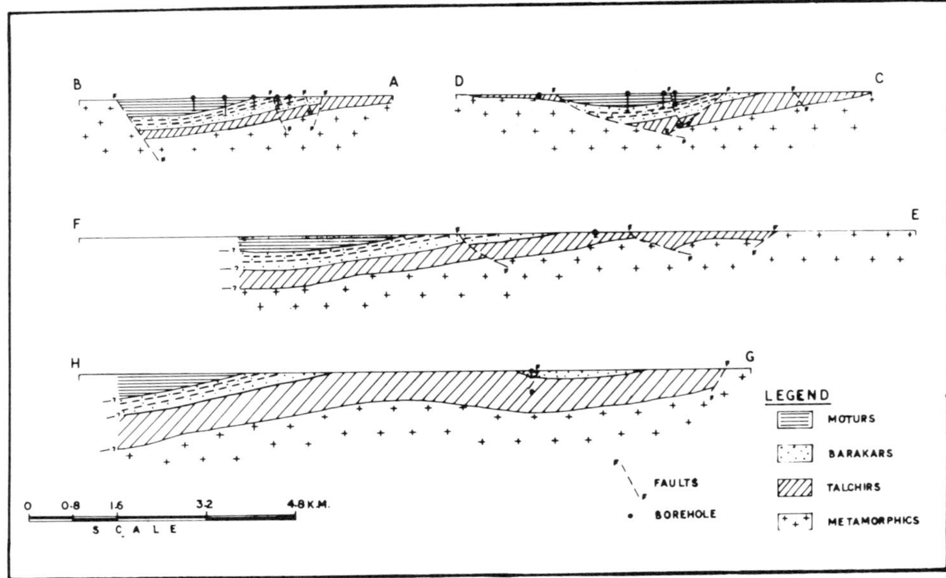

Figure No. 4- Cross Sections along seismic traverses: A-B, C-D, E-F and G-H.

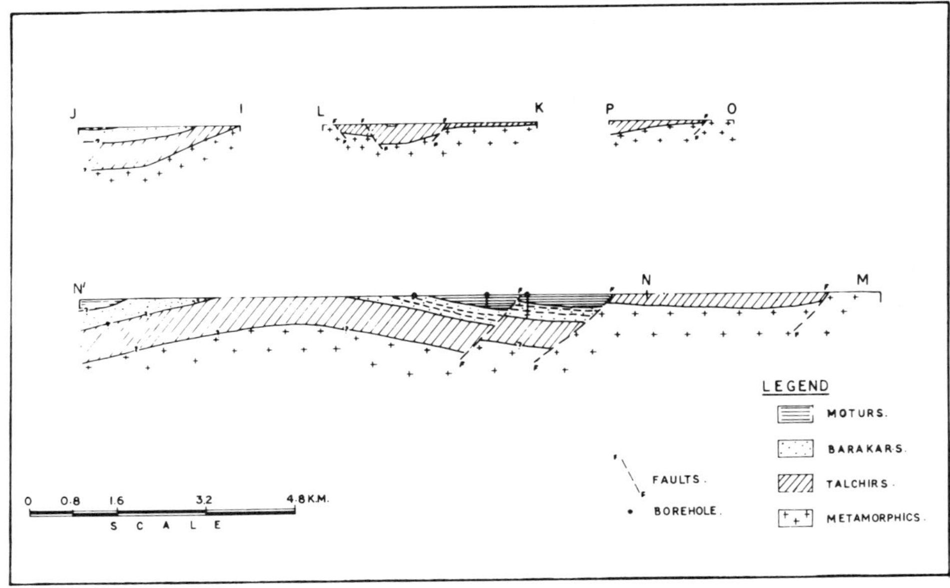

Figure No. 5- Cross Sections along seismic traverses: I-J, K-L, M-N-N' (N-N' built up from borehole data) and O-P.

Dhapewara-Bokhara Sub-basin: The westerly sub-surface continuation of the Koradih-High separates the larger Dhapewara-Bokhara sub-basin from the other two. Its western extension,like the Saoner Sub-basin is obscured by the volcanics. The eastern end of the basin lies near Bokhara,where the southern flank of the Koradih-High marks the northern boundary and the sedimentaries rest unconformably over the metamorphics with a normal depositional contact. It is only at the south-western notch of the Koradih-High that the contact is faulted. The axis of the Dhapewara-Bokhara sub-basin trends in a WNW-ESE direction and has a westerly plunge. The northern limb of this sub-basin is affected by faults north of Dhapewara. The deepest part of this sub-basin lies slightly west of Dhapewara. On the present evidences, this sub-basin has been conceived as one large sub-basin but the possibility of the occurrence of two sub-basins,one around Dhapewara and the other at the Bokhara end,in this tract can not be ruled out. This can be established only when more data become available.

The evidences that have come to light after large scale drilling in Patansaongi-Silewara sub-basin suggest that the faults in the Kamptee coalfield are normal faults and are post-Barakar and pre-Kamthi in age as these faults do not affect the Kamthi Formation. No igneous intrusives have been reported so far from this coalfield,although the volcanics border the western and the southern fringes.

While the structural set up of the Patansaongi-Silewara sub-basin is fully known as a result of large scale exploration,the same in respect of the Saoner sub-basin is partly known because of limited exploration. Since practically no exploration has been carried out in the Dhapewara-Bokhara sub-basin, its detailed structural set up is yet to be deciphered.

EXPLORATION STRATEGY:

The spread of the coal-bearing formations within the Kamptee coalfield was not known,concealed as they are under a thick alluvial cover. This posed problems in the selection of potential exploration blocks and placed severe limitation on adopting a proper exploration strategy for structural, quantitative and qualitative evaluation of the coal deposits in the coalfield.

The existence of the coal-bearing formation in the eastern part of the coalfield was proved in a tube well and based on this find, Kamptee colliery was opened in the immediate vicinity. At that point of time, there was no other alternative but to select exploration blocks for locating coal reserves west of the Kamptee colliery and/or in the neighbouring areas where some scouting boreholes had proved the existence of the coal-bearing formation as a part of an overall exploration strategy. Inspite of the selection of the exploration blocks adjacent to the known areas, the potentiality of the blocks could not be forecast. While certain exploration blocks turned out to be potential for the purpose of mine development, in a few cases either a part or the whole of the exploration block did not prove to be an attractive prospect. Two case studies in this context may be cited. The Ghatrohan Block, west of the Kamptee colliery, taken up for detailed exploration turned out to be highly faulted and the project had to shelved. Similarly, the Silewara Block(now Silewara colliery) was taken up for detailed drilling and it was detected that the north-eastern part was highly faulted and development plans for this part had to be abandoned. It would, therefore, be obvious that no choice could be exercised in the selection of the blocks.

The prime necessity in evolving a meaningful exploration strategy, therefore, was the preparation of a geological map which would lay bare the sub-surface geology of the coalfield. This could be achieved, only if the structure of the basin was known. For this purpose, there were two options: either to cover the entire coalfield with a large number of regional boreholes drilled down to the basement or to introduce seismic methods on a trial basis, which could yield information about the basement-sedimentary interface and reveal the basin configuration. Of the two options, the obvious choice was for the latter because of the fact that it would be less expensive and larger areas could be covered in a shorter time. The seismic survey was, therefore, carried out in the known areas in the east and extended subsequently to the unknown areas in the west. It may be mentioned that seismic surveys are an established part in oil exploration and as their application in the field of coal exploration was a recent development,this technique was introduced in the later phases of exploration in the Kamptee Coalfield.

Subsequent to the preparation of the geological map, it became known that the Patansaongi and the Saoner areas are separated by an intervening patch of the Talchir.In the light of this,the strategy of exploratory drilling in the Saoner area was reoriented and considerable amount of infructuous drilling could also be avoided which otherwise would have been inescapable if the exploration had proceeded on the earlier assumption that the coal measures were continuous between Patansaongi and Saoner.

Further,in the light of the knowledge gained from the seismic surveys,an appropriate exploration strategy has also been devised for the Dhapewara-Bokhara sub-basin of which the Dhapewara area appears to be more potential. The exploration would be initiated from the Dhapewara end in the area enclosed by the 800-metre isochore line of the sedimentaries. In the Bokhara area,the existence of coal seams has been proved but it is observed that the seams have considerably reduced in thickness. In view of this,the Bokhara area appears to be less potential.

To delineate the western and the southern boundaries of the Saoner and the Dhapewara-Bokhara sub-basins,suitable geophysical methods would be employed, as the area to the south and west is covered by the volcanics. This may go a long way in proving the existence or otherwise of the coal-bearing rocks around Semda and Bazargaon areas also.

Further,for deciphering the detailed structure of the Dhapewara-Bokhara and the Saoner sub-basins,it is intended to introduce the new technique of "High Resolution Seismics". The high resolution seismic surveys are reported (18) to be very useful in deciphering the stratigraphic disposition of the formations,the coal seams and their thickness, "wants" (i.e.buried stream channels) and their pattern and the faults of smaller magnitude. This technique is proposed to be used in the Patansaongi-Silewara sub-basin also, where mining is going on in certain areas, to decipher faults of smaller magnitude which might have escaped detection during the course of normal exploration by conventional methods. With the introduction of this technique it would be possible to predetermine the structural setup in the Dhapewara area which would facilitate in proving large resources of coal deposits by orienting the exploration strategy accordingly. This would also result in considerable savings in exploration in terms of both money and time.

ACKNOWLEDGEMENT:

The authors take this opportunity to thank Mr.Naeem Ahmad,Geologist,CMPDI, for his all round assistance in the preparation of this paper.

The authors wish to record their thanks to the Geological Survey of India, Indian Bureau of Mines and Directorate of Geology and Mining, Govt. of Maharashtra, for freely drawing upon from their departmental reports/ records, made available to CMPDI.

The authors are grateful to Mr. R.G.Mahendru,Chairman-Cum-Managing Director,C.M.P.D.I.Ltd.,for his kind permission to present the paper at the Second International Coal Exploration Symposium. The views expressed in this paper are the authors' own and are not necessarily those of the C.M.P.D.I.Ltd.

R E F E R E N C E S

1. GEE,E.R. KHEDKAR,V.R. 1952 The Kamptee Coalfield,Nagpur District,Central Province, Records of the Geological Survey of India,Vol.78,Part-4.

2. KAILASAM,L.N. 1948 Report of Electrical Resistivity Survey in the Kamptee Coalfield Area,Nagpur Dist., Geological Survey of India (Unpublished).

3. ARAVAMUDHAN,R. 1960 Report on further Geophysical investigations of the Kamptee Coalfield,Maharashtra, Geological Survey of India (Unpublished)

4. MITRA, S.K. 1961 Report on further Electrical and Magnetic Survey of Kamptee Coalfield, Maharashtra, Geological Survey of India (Unpublished)

5. ROY, A.K. 1961-62 Report on Geophysical Investigations in Kamptee Coalfield Area, Nagpur Dist., Maharashtra State, Geological Survey of India (Unpublished)

6. PANT, P.R., LAHIRI, S.M., KHARE, A.C., PATHAK, R.C. 1974 Report on Seismic Investigations in the Kamptee Coalfield Area, Maharashtra, Geological Survey of India (Unpublished)

7. HUNDAY, A., RAO, V.V.S.R.H., DESHPANDE, M.L., PATIL, R.D. 1964 Report on the Detailed Exploration for Coal in Silewara Sector, Kamptee Coalfield, Maharashtra, Indian Bureau of Mines (Unpublished)

8. HUNDAY, A., RAO, V.V.S.R.H., SURYANARAYAN, M. 1964 Report on the Detailed Exploration for coal in the Ghatrohan Sector, Kamptee Coalfield, Indian Bureau of Mines (Unpublished)

9. KULKARNI, S.G., 1963 Report of the Exploration for Coal in Bokhara Area, Nagpur Dist., Dept. of Geology and Mining, Govt. of Maharashtra (Unpublished)

10. MAGGIRWAR, C.N., CHOWDHURY, K.K., POHANE, G.R., TAMHANE, V.S., MUDHOKAR, A.M. 1969-71 A Progress Report on Investigations for Coal in Pipla-Patansaongi and Saoner Blocks of Kamptee Coalfield, Directorate of Geology & Mining, Govt. of Maharashtra (Unpublished)

11. CHANDRA, T.K., VERMA, B.K., GEORGE, B. 1970 A consolidated Geological Report on Silewara Area, Kamptee Coalfield, Maharashtra, National Coal Development Corporation Ltd. (Unpublished)

12. CHANDRA,T.K., KHAJURIA,N.N., SHRIVASTAVA,B.B.P., SHINGTE,S.K. 1973 A Preliminary Geological Report on Patansaongi Block and a part of Zone-3 of Kamptee Block 'B' & Extension,Kamptee Coalfield, Coal Mines Authority Ltd. (Unpublished)

13. CHANDRA,T.K., SHRIVASTAVA,B.B.P., SHINGTE,S.K., KRISHNAN,S., WAHI,A.K. 1974 Preliminary Geological Report on Bina Block,Kamptee Coalfield, Maharashtra,Central Mine Planning and Design Institute Ltd., (Unpublished)

14. SHINGTE,S.K., KRISHNAN,S., CHANDRA,T.K. 1976 A Consolidated Geological Report on the area covered by Kanhan (Walni)-Pipla Inclines,Kamptee Coalfield,Central Mine Planning & Design Institute Ltd. (Unpublished)

15. SINGH,B.B., PALKAR,S.G., KRISHNAMACHARALU,T. 1974 Preliminary Geological Report on Inder-Gondegaon Sector of Kamptee Colliery,Kamptee Coalfield, Maharashtra,Coal Mines Authority Ltd.(Unpublished)

16. FOX,C.S. 1934 The Lower Gondwana Coalfields of India,Geological Survey of India, Memoir LIX

17. PASCOE,E.M. 1959 A Manual of the Geology of India and Burma,Vol.2, Geological Survey of India.

18. FARR,J.B., PEACE,D.G. 1977 Surface Seismic Profiling for Coal Exploration and Mine Planning,Third IIASA Conference on Energy Resources,Moscow.

INDEX